FUNCTIONAL ANALYSIS AND ITS APPLICATIONS

PROCEEDINGS OF THE INTERNATIONAL CONFERENCE ON FUNCTIONAL ANALYSIS AND ITS APPLICATIONS DEDICATED TO THE 110TH ANNIVERSARY OF STEFAN BANACH, MAY 28-31, 2002, LVIV, UKRAINE

NORTH-HOLLAND MATHEMATICS STUDIES 197

(Continuation of the Notas de Matemática)

Editor: Saul LUBKIN
University of Rochester
New York, U.S.A.

ELSEVIER

2004

Amsterdam – Boston – Heidelberg – London – New York – Oxford
Paris – San Diego – San Francisco – Singapore – Sydney – Tokyo

FUNCTIONAL ANALYSIS AND ITS APPLICATIONS

PROCEEDINGS OF THE INTERNATIONAL CONFERENCE ON FUNCTIONAL ANALYSIS AND ITS APPLICATIONS DEDICATED TO THE 110TH ANNIVERSARY OF STEFAN BANACH, MAY 28-31, 2002, LVIV, UKRAINE

Vladimir KADETS
Kharkiv National University
Kharkiv, Ukraine

Wiesław ŻELAZKO
Polish Academy of Sciences
Warsaw, Poland

ELSEVIER

2004

Amsterdam – Boston – Heidelberg – London – New York – Oxford
Paris – San Diego – San Francisco – Singapore – Sydney – Tokyo

ELSEVIER B.V.
Sara Burgerhartstraat 25
P.O. Box 211, 1000 AE Amsterdam
The Netherlands

ELSEVIER Inc.
525 B Street, Suite 1900
San Diego, CA 92101-4495
USA

ELSEVIER Ltd
The Boulevard, Langford Lane
Kidlington, Oxford OX5 1GB
UK

ELSEVIER Ltd
84 Theobalds Road
London WC1X 8RR
UK

First edition 2004

Library of Congress Cataloging in Publication Data
A catalog record is available from the Library of Congress.

British Library Cataloguing in Publication Data
A catalogue record is available from the British Library.

ISBN: 0-444-51373-6
ISSN: 0304-0208

∞ The paper used in this publication meets the requirements of ANSI/NISO Z39.48 -1992 (Permanence of Paper).
Printed in The Netherlands.

TABLE OF CONTENTS

PREFACE vii
LIST OF PARTICIPANTS ix

Stefan Banach and Lviv Mathematical School
Ya. G. Prytula 1

Existence of k-media and medians
E. Alvoni, P. L. Papini 5

On Optimization of approximate integration of multivariate periodic functions
V. F. Babenko, S. V. Borodachov 13

Topological spaces with the strong Skhorokhod property, II
T. Banakh, V. I. Bogachev, A. V. Kolesnikov 23

Fourth order spectral problem with eigenvalue in the boundary conditions
J. Ben Amara 49

Extending operators into $\mathcal{L}_\infty$ spaces under a twisted light
F. Cabello Sánchez, J. M. F. Castillo, Y. Moreno, D. T. Yost 59

Un théorème de point fixe pour les fonctions multivoques acycliques
R. Cauty 71

Almost periodic mappings to complex manifolds
S. Favorov, N. Parfyonova 81

Some new differential equations of the first-order in the spaces $M(1,3) \times R(u)$ and $M(1,4) \times R(u)$ with given symmetry groups
V. Fedorchuk, V. Fedorchuk 85

Inverse spectral problems for Sturm-Liouville operators with singular potentials, II. Reconstruction by two spectra
R. O. Hryniv, Ya. V. Mykytyuk 97

Incorrect nonlocal boundary value problem for partial differential equations
V. Il'kiv 115

Singular integral operators with flip and unbounded coefficients on rearrangement-invariant spaces
A. Karlovich 123

Spectrum as the support of functional calculus
V. V. Kisil 133

Generalized normality in topological products and related structures
A. P. Kombarov, A. N. Yakivchik 143

Nonstandard universe based on internal set theory
T. Kudryk, W. Lyantse, V. Neves 155

The Beurling theorem for entire functions of finite order
K. G. Malyutin, N. Sadyk 167

Dvoretzky's theorem by Gaussian method
I. Matsak, A. Plichko 171

On the inverse problem of scattering theory for a differential operator of the second order
K. R. Mamedov, H. Menken 185

On the automorphisms and multipliers related to generalized integration operators
S. Minczewa-Kamińska 195

On some classes of mappings preserving of measure
O. A. Ochakovskaya 205

Extreme problems and scales of the operator spaces
I. V. Orlov 209

On some problem of the convolution of bounded functions
K. Piejko, J. Sokół, J. Stankiewicz 229

Superstrictly singular and superstrictly cosingular operators
A. Plichko 239

Adic multivariate linear splines and their applications to Approximation Theory
A. Radyna 257

Change of variable in integrals over p-adic and Adelic domains
Ya. Radyna 267

Invariant subspaces of dissipative operators in Krein space and Sobolev problem on a rotating top
A. A. Shkalikov 273

Property of maximality for spectral components
A. Tikhonov 287

A General view on approximation ideals
N. Ţiţa 295

Inversion of the local Pompeiu transform
N. P. Volchkova 301

Space of probability measures and absolute extensors in the asymptotic category
M. M. Zarichnyi 311

F-Algebras: some results and open problems
W. Żelazko 317

PREFACE

The *International Conference on Functional Analysis and Applications dedicated to the 110th anniversary of Stefan Banach* took place on May 28 – May 31, 2002 at the Lviv National University in Lviv, Ukraine.

The Conference was organized under the auspices of Lviv Ivan Franko National University, Institute of Mathematics of Polish Academy of Sciences, Pidstryhach Institute for Applied Problems of Mechanics and Mathematics of Ukrainian NAS, Division of Mathematics of Ukrainian National Academy of Sciences, and National University "Lviv Polytechnika".

Program Committee of Conference: Yu. M. Berezanskii, Cz. Bessaga, B. Bojarski (Co-Chairman), M. L. Gorbachuk, G. S. Kit, A. A. Kondratyuk, O. V. Lopushansky, V. E. Lyantse, Yu. K. Rudavsky, M. M. Sheremeta, O. B. Skaskiv, I. V. Skrypnyk (Co-Chairman), I. O. Vakarchuk, M. M. Zarichnyi, W. Zelazko and A. V. Zagorodnyuk (secretary).

Organizing Committee of the Conference: T. O. Banakh, P. I. Kalenyuk, P. P. Kostrubiy (Co-Chairman), Kh. Yo. Kuchminska, V. M. Kyrylych (Co-Chairman), Ya. V. Mykytyuk, Ya. G. Prytula (Vice Chairman), B. Y. Ptashnyk, O. G. Storozh, M. V. Zabolotskyj, R. O. Hryniv (secretary).

The Rector of Lviv Ivan Franko National University Professor I. O. Vakarchuk and Co-Chairmen of the Program Committee Professors I. V. Skrypnyk and B. Bojarski took part in the Opening Ceremony. The Dean of the Faculty of Mechanics and Mathematics Ya. G. Prytula delivered a lecture on Stefan Banach and Lviv Mathematics school of the last century.

The program of the Conference consisted of plenary and sectional lectures. The plenary lectures were delivered by Cz. Bessaga (Warsaw) "Formulas for deleting sets from metric spaces, related to Lipschitz implicit function theorems", B. Bojarski (Warsaw) "The Fredholm Grassmanian", V. I. Bogachev (Moscow) "Extensions and approximations of mappings on Banach spaces with measures", J. Castillo (Badajor) "On the Lindenstrauss-Pelczynski theorem", R. Cauty (Paris) "Un théorème de point fixe pour les fonctions multivoques", V. V. Fedorchuk (Moscow) "On some topological properties of probability measures spaces", M. L. Gorbachuk (Kyiv) "On spaces of infinitely differentiable vectors of a closed operator on a Banach space and their applications", V. M. Kadets (Kharkiv) "Banach spaces with the Daugavet property", I. Polirakis (Athens) "Cones and Geometry of Banach Spaces", I. V. Skrypnyk (Kyiv) "The mapping index of nonlinear densely defined operators", I. A. Shevchuk (Kyiv) "Shape-preserving approximation by polynomials", A. A. Shkalikov (Moscow) "Abstract operator models for problems of elasticity and hydrodynamics", W. Żelazko (Warsaw) "F-algebras: some results and open problems".

There were six parallel sections: Banach Space Theory, Approximation Theory, Spectral Theory, Differential Equations, Theory of Functions and Topology. More than 150 participants from 21 countries (Ukraine, Poland, Russia, Moldova, Byelorus, Azerbaijan, Turkey, Bulgaria, Romania, Italy, Spain, Sweden, USA, France, Canada, Greece, the Netherlands, Czech Republic, Tunis, Norway, Algeria) took part in the work of sections and poster sessions. In particular some selected talks were: L. Maligranda "Type, cotype and convexity properties of some Banach spaces", A. Plichko "On the functional analytic approach to orthogonal systems", V. Babenko "Multivariate inequalities of Kolmogorov type", J. Janas "Discrete Levinson type results: application to Jacobi operators", M. Ivanchov "Free boundary problem for the diffusion equation", S. Ivashkovych "Some problems arising from the Jacobian conjecture", N. Zorii "On the theory of potential in locally compact spaces", P. Filevych "On the growth of entire functions defined by gap power series," A. Korenovskyy "Note on a theorem by Gurov-Reshetnyak," M. Choban "Mappings and Prohorov spaces", V. Maslyuchenko "Connection between separately and joint properties of several variables functions."

On behalf of the Organizers *Ya. G. Prytula, A. V. Zagorodnyuk*

LIST OF PARTICIPANTS

Abrahamsen Trond, Agder University College, Servicebox 422, Kristiansand, Norway 4604. E-mail: trond.a.abrahamsen@hia.no
Alexandrov Georgi, J.Bourchier Bld., 5, Sofia Bulgaria, 1164.
Alvoni Elisabetta, University of Bologna, Dipartimento di Matematica per le Scienze Economichee Sociali (University) Viale Fillopanti, 5 Bologna Italy 40126. E-mail: ealvoni@economia.unibo.it
Antonevich Anatoly, Belarussian State University, 4 F.Skaryna avenue, Minsk, Belarus 220039. E-mail: antonevich@bsu.by
Antonova Tamara, National University "Lvivska Politechnika", Lviv, Ukraine.
Arlinskii Yury, East Ukrainian National University, Lugansk Ukraine 91034. E-mail: yma@snu.edu.ua
Averbuch Vladimir, Silezian univercity in Opava, Bezrucovo nam. 13, Opava Czech Rep. 74601. E-mail: Vladimir.Averbuch@fpf.slu.cz
Babenko Yuliya, Dnepropetrovsk National University, Naukova, 13 Str., Dnepropetrovsk, Ukraine 49050. E-mail: babenko@mail.dnepr.net
Babenko Vladislav, Dnepropetrovsk National University, Naukova, 13 Str., Dnepropetrovsk, Ukraine 49050. E-mail: babenko@mail.dnepr.net
Banakh Taras, Ivan Franko Lviv National University, Lviv, Ukraine. E-mail: tbanakh@franko.lviv.ua
Barabash Galyna , Ivan Franko Lviv National University Universytets'ka, 1 Lviv Ukraine. 79000. E-mail: diffeq@uli2.franko.lviv.ua
Baran Oksana, Pidstryhach Institute for Applied Problems of Mechanics and Mathematics of NAS of Ukraine. Lviv, Ukraine.
Ben Amara Jamel, Faculte Des Sciences de Bizerte, Tunisia University. Tunis. E-mail: jamel.benamara@ipeigb.rnu.tn
Bereza Vitaly, Chernivtsi National University, Chernivtsy, Ukraine. E-mail: yasik@cv.ukrtel.net
Bessaga Czeslaw, Warsaw University, Warsaw Poland. bessaga@impan.gov.pl. E-mail: bessaga@mimuw.edu.pl
Bihun Oksana, Lviv Ivan Franko National University, Lviv, Ukraine. E-mail: obihun@ukr.net
Bodnar Dmytro, Ternopil Academy of National Economy, Department of ASP,

Ternopil, Ukraine. E-mail: dmytro_bodnar@hotmail.com
Bogachev Vladimir, Moscow State University, Dept. Mech. and Math. Moscow, Russia. E-mail: bogachev@vbogach.mccme.ru
Bojarski Bogdan, Mathematical Institute, Polish Academy of Sciences, Sniadeckich 8, P.O. Box 147, Warsaw, Poland.
Bordulyak Marta, L'viv Ivan Franko Lviv National University, 1 Universytets'ka str., Lviv Ukraine 79000. E-mail: tftj@franko.lviv.ua
Borodachov Sergiy, Department of Mechanics and Mathematics Dnipropetrovsk National University , Dnipropetrovsk, Ukraine. E-mail: borodach@ff.dsu.dp.ua
Boumaraf Fatima, Universite Badji Mokhtar, Deprtment de math'ematique de la facult'e des sciences.
Buhrii Oleh, Ivan Franko Lviv National University, 1 Universytets'ka str., Lviv Ukraine 79000. E-mail: diffeq@uli2.franko.lviv.ua
Castillo Jesus, Departamento de Matematicas, Universidad de Extremadura, Badajoz, Espana. E-mail: castillo@unex.es
Cauty Robert, Unversité Paris 6, Paris, France. E-mail: cauty@math.jussieu.fr
Cheremnikh Evgenii, Dept. of Higher Mathemativs, National University "Lvivska Polytechnika" , 12 Bandery str., Lviv, Ukraine. E-mail: echeremn@polynet.lviv.ua
Chmyr Oksana, Lviv Ivan Franko National University, 1 Universytets'ka str., Lviv Ukraine 79000. E-mail: diffeq@uli2.franko.lviv.ua
Choban Mitrofan, Tiraspol State University, Gh.Iablocichin 5, Kishinev, Republic of Moldova MD-2069.
Chyzhykov Igor, Ivan Franko Lviv National University, Lviv, Ukraine 79000. E-mail: tftj@franko.lviv.ua
Cobzas Stefan, Babes-Bolyai University, Faculty of Mathematics and Computer Science, Cluj-Napoca, Romania, RO-3400.
Dmytryshyn Maryan, Ternopil Academy of National Economy TANE, Department of ASP, Ternopil, Ukraine. E-mail: m_dmytryshyn@hotmail.com
Dobrowolski Tadeusz, Department of Mathmatics, Pittsburg State University, Pittsburg, KS U.S.A. 66762. E-mail: tdobrowo@mail.pittstate.edu
Dziok Jacek, University of Rzeszow ul. Rejtana 16 A, Rzeszow, Poland, 35-959.
Eremenko Alexandre, Department of Mathematics, Purdue University, West Lafayette, IN USA 47907. E-mail: eremenko@math.purdue.edu
Fedorchuk Vasyl, Pedagogical Academy, Institute of Mathematics, Krakow, Poland; Pidstryhach Institute for Applied Problems of Mechanics and Mathematics of NAS of Ukraine, Lviv, Ukraine. E-mail: 79053 vas_fedorchuk@yahoo.com, fedorchuk@wsp.krakow.pl
Fedorchuk Volodymyr, Franko Lviv National University, Lviv Ukraine. E-mail: fedorchukv@ukr.net
Fedorchuk Vitaly, Moscow State University, Moscow, Russia. E-mail: vitaly@gtopol.math.msu.su
Fedyk Mykhaylo, Lviv Ivan Franko National University, Lviv, Ukraine.
Fedynyak Stepan, Lviv Ivan Franko National university, Lviv, Ukraine 79000. E-mail: fedynyak@yahoo.com

Filevych Petro, Lviv Ivan Franko National university, Lviv, Ukraine, 79000. E-mail: tftj@franko.lviv.ua
Flyud Volodymyr, Lviv Ivan Franko National University, Lviv, Ukraine. E-mail: vmf@mail.lviv.ua, flyud@yahoo.com
Galibina Nadezhda, Donetsk National University 50 Gvardeyskoy, Donetsk, Ukraine. E-mail: gn1977@mail.ru
Goldstein Stanislaw, University of Lodz, Faculty of Mathematics, Lodz , Poland 90-238. E-mail: goldstei@math.uni.lodz.pl
Gorbachuk Myroslav, Institute of Mathematics of National Academy of Sciences of Ukraine, Kyiv, Ukraine. E-mail: imath@horbach.kiev.ua
Gorodnii Mykhaylo, Kyiv National Taras Shevchenko University Kyiv National Taras Shevchenko University, Dept. of Mathematics and Mechanics. E-mail: gorod@hotbox.ru
Grushka Yaroslav, Institute of Mathematics of Ukrainian National Academy of Sciences, Tereshchenkivska str. 3, Kyiv-4, Ukraine 1601. E-mail: grushka@imath.kiev.ua
Gutik Oleg, Pidstryhach Institute for Applied Problems of Machanics and Mathematics of Ukrainian NAS, Lviv Ukraine. E-mail: ogutik@iapmm.lviv.ua
Herasymchuk Vasyl', Chernivtsi National University.
Horbachuk Omelyan, Lviv Ivan Frnako National University 1 Universytets'ka str. Lviv, Ukraine 79000. E-mail: topos@franko.lviv.ua
Hryniv Rostyslav, Institute for Applied Problems of Mechanics and Mathematics, 3b Naukova str., Lviv, Ukraine 79060. E-mail: hryniv@mebm.lviv.ua
Ilkiv Volodymyr, National University "Lvivska Politechnika" , Lviv, Ukraine 79013.
Ivanchov Mykola, Lviv Ivan Franko National University, Universitetska 1 str. Lviv, Ukraine 79000. E-mail: ivanchov@franko.lviv.ua
Ivashkovych Serhiy, Pidstryhach Institute for Applied Problems of Machanics and Mathematics of Ukrainian NAS. E-mail: ivachkov@agat.univ-lille1.fr
Janas Jan, Krakow, Poland. E-mail: najanas@cyf-kr.edu.pl
Kadets Vladimir, Dept. of Mechanics and Mathematics, Kharkov National University, Pl. Svobody 4, Kharkov, Ukraine 61077.
Kalenyuk Petro, National University "Lvivska Politechnika", Lviv, Ukraine. E-mail: kalenyuk@polynet.lviv.ua
Kamiński Andrzej, Institute of Mathamatics, University of Rzeszów, Rejtana 16 A, Rzeszów, Poland, 35-310.
Karlova Olena, Chernivtsy National University, Chernivtsy, Ukraine. E-mail: alexolga@chv.ukrpack.net
Karlovich Alexei, Departamento de Matematica, Instituto Superior Tecnico, Av. Rovisco Pais 1, Lisbon, Portugal, 1049-001. E-mail: akarlov@math.ist.utl.pt
Kinash Orest Dept of Mech and Math, Lviv Ivan Frnako National University 1 Universytets'ka str., Lviv, Ukraine.
Kisil Vladimir, Odessa University and University of Leeds School of Mathematics, University of Leeds Leeds, UK LS2 9JT. E-mail: kisilv@maths.leeds.ac.uk
Kogut Peter Illich, Dnipropetrovsk, Ukraine. E-mail: kogut@a-teleport.com

Kohut Ihor, National University "Lvivska Politechnika", Lviv, Ukraine.
Kolodyazhny Volodymyr Institute for Problems of Machinery, Kharkiv, Ukraine. E-mail: kolodyazhny@univer.kharkov.ua
Kolysnyk Ruslana, Chernivtsy National University, Chernivtsy, Ukraine.
Kombarov Anatoly, Moscow State University Faculty of Mech. and Math., Moscow Russia 119899. E-mail: a@kombarov.mccme.ru
Kondratyuk Andrij, Lviv Ivan Franko National University, Lviv, Ukraine. E-mail: kond@franko.lviv.ua
Konstantinov Oleksiy, Department of Mechanics and Mathematics, Kyiv National Taras Shevchenko University, Ukraine. E-mail: konst@faust.kiev.ua
Kopchuk-Kashectky Andriy, National University "Lvivska Politechnika", Lviv, Ukraine.
Kopytko Bohdan, Dept of Mech and Math, Lviv Ivan Frnako National University, 1 Universytets'ka str., Lviv, Ukraine, 79000. E-mail: mstat@franko.lviv.ua
Korovina Natalia, Moscow State University MGU, Department of Mechanics and Mathematics, Chair of Differential Geometry and Applications, Vorobievy Gory, Moscow, Russia, 119899. E-mail: nathalie@mccme.ru
Kozhukar Olena, Chernivtsy National University, Chernivtsy, Ukraine.
Kshanovskyy Ivan, Lviv Ivan Franko National University, Lviv, Ukraine.
Kuchmin'ka Lubov, National University "Lvivska Politechnika", Lviv, Ukraine. E-mail: kuchmin@lms.lviv.ua
Kuchmin'ka Khrystyna, National University "Lvivska Politechnika", Lviv, Ukraine. E-mail: kuchmin@lms.lviv.ua
Kudryk Taras, Lviv National University, Lviv Ukraine, 79000. E-mail: kudryk@mail.lviv.ua
Lavrenyuk Serhiy, Ivan Franko Lviv National University, Lviv, Ukraine. E-mail: diffeq@uli2.franko.lviv.ua
Ligun Anatoly, Dniprodzerginsk University of Technology, Dnipropetrovsk, Ukraine. E-mail: gf@alicetele.com
Lopushanska Halyna, Lviv Ivan Frnako National University 1 Universytets'ka str., Lviv, Ukraine, 79000. E-mail: diffeq@uli2.franko.lviv.ua
Lopushanskyy Andriy, Pidstryhach Institute for Applied Problems of Mechanics and Mathematics of Ukrainian NAS, 3b Naukova Str., Lviv, Ukraine, 79060. E-mail: diffeq@uli2.franko.lviv.ua
Lozynska Vira, Pidstryhach Institute for Applied Problems of Machanics and Mathematics of Ukrainian NAS, Lviv, Ukraine.
Lyantse Wladyslaw, Lviv National University, Lviv, Ukraine, 79000.
Maligranda Lech, Department of Mathematics Lulea University of Technology, Lulea Sweden S-971 87. E-mail: lech@sm.luth.se
Malyutin Konstantin, Ukrainian Academy of Banking, Petropavlivska Street, 59 Sumy, Ukraine, 40030. E-mail: kgm@academy.sumy.ua
Mamedov Kh.R., Baku State Uviversity-Mersin University, Baku, Azerbajdjan. E-mail: hanlar@aport.ru
Manzij Lesia, National University "Lvivska Politechnika", Lviv, Ukraine. E-mail: Lesly@ukr.net

Maslyuchenko Volodymyr, Chernivtsy National University, Chernivtsy, Ukraine. E-mail: vmasl@math.chnu.cv.ua
Maslyuchenko Oleksandr, Chernivtsy National University, Chernivtsy, Ukraine. E-mail: slymak@math.chnu.cv.ua
Menken Hamza, Mersin University, Mersin, Turkey. E-mail: hanlar@aport.ru
Mincewa-Kamińska Swietlana, Institute of Mathematics University of Rzeszów, Rejtana 16 A, Rzeszów, Poland 35-310.
Moreno Salguero Yolanda, Departamento de Matemáticas, Universidad de Extremadura, Avda de Elvas s/n, Badajoz, Spain 6071. E-mail: ymoreno@unex.es
Mykhaylyuk Volodymyr, Chernivtsy National University, Chernivtsy, Ukraine. E-mail: mathan@math.chnu.cv.ua
Myjak Jozef, University of L'Aquila, Department of Mathematics, Via Vetoio L'Aquila, Italy, 67100.
Mykytyuk Yaroslav, Lviv Ivan Franko National University, Lviv, Ukraine. E-mail: mykytyuk@email.lviv.ua
Mytrofanov Mykhalo, Pidstryhach Institute for Applied Problems of Mechanics and Mathematics of Ukrainian NAS, 3b Naukova Str., Lviv, Ukraine, 79060.
Nygaard Olav , Agder University College Servicebox 422, Kristiansand, Norway 4604. E-mail: olav.nygaard@hia.no
Nykyforchyn Oleg, Ivano-Frankivsk State University, Ivano-Frankivsk, Ukraine. E-mail: nick@pu.if.ua
Nytrebych Zinoviy, National University "Lvivska Politechnika", Lviv, Ukraine.
Ochakovskaya Oksana, Institute of Applied Mathematics and Mechanics NAS Ukraine, Donetsk, Ukraine.
Orlov Igor, Taurida V. Vernadsky National University, Simferopol, Ukraine. E-mail: old@tnu.crimea.ua
Papini Pier Luigi, University of Bologna, Dipartimento di Matematica, Piazza Porta S. Donato 5, Bologna. Italy 40126. E-mail: papini@dm.unibo.it
Parfyonova Natalia, Karazin Kharkiv National University, Department of Mechanics and Mathematics, Kharkiv, Ukraine. E-mail: Natalia.D.Parfyonova@univer.kharkov.ua
Pekala Agnieszka Faculty of Mathematics and Natural Sciences, Rzeszow University. Rzeszow, Poland 35-326. aga_pekala@poczta.onet.pl
Pipa Hanna, Lviv Ivan Franko National University, Lviv, Ukraine.
Pirkovskii Alexandr, Moscow State University, Moscow, Russia. E-mail : pirkosha@online.ru
Plichko Anatolij, Kirovograd State Pedagogical University, Dept. of Mathematics, Kirovograd, Ukraine. E-mail: aplichko@kspu.kr.ua
Polyrakis Ioannis, Department of Mathematics, National Technical University of Athens, Zografou Campus, Athens, Greece, 157 80. E-mail: ypoly@math.ntua.gr
Popov Mykhailo, Chernivtsi National University, Chernivtsy, Ukraine. E-mail: popov@chv.ukrpack.net
Povstenko Yuriy, Pidstryhach Institute for Applied Problems of Mechanics and Mathematics of Ukrainian NAS, Naukova 3b, Lviv, Ukraine 79053. E-mail: povstenk@iapmm.lviv.ua

Protasov Igor, Kyiv National University, Kyiv, Ukraine.
Protsakh Natalia, Ivan Franko Lviv National University, Lviv, Ukraine. E-mail: diffeq@uli2.franko.lviv.ua
Protsyk Yuriy, Ivan Franko Lviv National University, Lviv, Ukraine. E-mail: kond@franko.lviv ua
Prytula Mykola, Ivan Franko Lviv National University, Lviv, Ukraine.
Prytula Yaroslav, Ivan Franko Lviv National University, Lviv, Ukraine.
Ptashnyk Bohdan, Pidstryhach Institute for Applied Problems of Machanics and Mathematics of Ukrainian NAS, Lviv, Ukraine. E-mail: ptashnyk@lms.lviv.ua
Puzyrova Iryna, Pidstryhach Institute for Applied Problems of Machanics and Mathematics of Ukrainian NAS, Lviv, Ukraine.
Radyna Aliaksandr, Belarusian State University, Minsk, Belarus'. E-mail: Yauhen_Radyna@tut.by
Radyna Yauhen, Belarusian State University, Minsk, Belarus'. E-mail: Radyna@bsu.by
Revenko Victor, Pidstryhach Institute for Applied Problems of Machanics and Mathematics of Ukrainian NAS, Lviv, Ukraine.
Rofe-Beketov Fedir, Mathematical Division of B.Verkin Institute for Low Temperature Physics, Kharkiv, Ukraine. E-mail: rofebeketov@ilt.kharkov.ua
Sarantopoulos Yannis, National Technical University of Athens, Faculty of Applied Sciences, Department of Mathematics, Zografou Campus, Athens, Greece 15780. E-mail: ysarant@math.ntua.gr
Schlomiuk Norbert, Departement de mathematiques et de statistique, Universite de Montreal, C.P. 6128, Succursale Centre-Ville, Montreal (Quebec), Canada H3C 3J7. E-mail: dasch@dms.umontreal.ca
Shevchuk Igor, Kyiv National University, Kyiv, Ukraine.
Sheremeta Myroslav, Ivan Franko Lviv National University, Lviv, Ukraine. E-mail: tftj@franko.lviv.ua
Shkalikov Andrey, Dept. of Mechanics and Mathematics Moscow State University, Moscow, Russia 119899. E-mail: shkal@mech.math.msu.su
Shyshyna Ol'ha, Chernivtsy National University, Chernivtsy, Ukraine. E-mail: mathan@math.chnu.cv.ua
Sidorenko Yuri, Lviv Ivan Franko National University, Universytetska, 1, Lviv, Ukraine, 79000.
Sipacheva Ol'ga, Department of General Topology and Geometry, Mechanics and Mathematics Faculty, Moscow State University, Vorobjovy gory, Moscow, Russia, 119992. E-mail: sipa@sipa.mccme.ru
Skaskiv Oleg, Lviv Ivan Franko National University, Lviv, Ukraine.
Skrypnik Igor Institute of Applied Mathematics and Mechanics NAS Ukraine, Donetsk, Ukraine.
Sobchuk Oleksandr, Chernivtsy National University, Chernivtsy, Ukraine. E-mail: osobchuk@math.chnu.cv.ua
Sokhan Petro, National University "Lviv Polytechnic", Lviv, Ukraine 79013.
Stankiewicz Jan, Uniwersytet Rzeszowski and Politechnika Rzeszowska Katedra Matematyki Politechniki Rzeszowskiej, Ul, W. Pola 2, Rzeszow, Poland, 35-959.

E-mail: jstan@man.rzeszow.pl
Starkov Pavel, Taurida National V. Vernadsky University, Simferopol, Ukraine. E-mail: pstarksci@aport.ru
Storozh Oleg, Lviv Ivan Franko National University, Lviv, Ukraine. E-mail: storozh@franko.lviv.ua
Sus' Olga, Pidstryhach Institute for Applied Problems of Machanics and Mathematics of Ukrainian NAS, Lviv, Ukraine. E-mail: sus@lms.lviv.ua
Syroid Ihor-Petro, Pidstryhach Institute for Applied Problems of Machanics and Mathematics of Ukrainian NAS, Lviv, Ukraine. E-mail: igor@iapmm.lviv.ua
Tabor Jacek, Institute of Mathematics, Jagiellonian University, Reymonta 4, Krakow, Poland, 30-059. E-mail: tabor@im.uj.edu.pl
Tabor Jozef, University of Rzeszow, Rejtana 16a, Rzeszow, Poland, 35-310. E-mail: tabor@wsp.krakow.pl, tabor@atena.rzeszow.pl
Tikhonov Alexey, Taurida National University, Simferopol, Ukraine. E-mail: alex@sunbay.crimea.ua
Tita Nicolae, University of Brasov, Bd.Eroilor 29, Brasov, Romania. 2200. E-mail: n.tita@info.unitbv.ro
Tocki Jerzy, University of Rzeszow, UL. Rejtana 16C Rzeszow, Poland, 35-959. E-mail: jjtocki@univ.rzeszow.pl
Tsvetkov Denis, Taurida National V. Vernadsky University, Simferopol, Ukraine. E-mail: old@tnu.crimea.ua
Tur Eduard, State Optical Institute, St.Petersburg, Russia. E-mail: Teduard@cards.lanck.net
Vasyl'kiv Yaroslav, Lviv Ivan Franko National University, Lviv, Ukraine. E-mail: kond@franko.lviv ua
Vatkina Elena, Belarussian State University, Minsk, Belarus'. E-mail: pugach@bsu.by
Vlasenko Larisa, Kharkov National University, Kharkov, Ukraine. E-mail: Larisa.A.Vlasenko@univer.kharkov.ua
Volchkov Valery, Donetsk National University, Universitetskaya 24 Str, Donetsk, Ukraine 83055. E-mail: volchkov@univ.donetsk.ua
Volchkov Vitaly, Donetsk National University, Universitetskaya 24 Str, Donetsk, Ukraine 83055.
Volchkova Natalia, Donetsk National University, Universitetskaya 24 Str, Donetsk, Ukraine 83055. E-mail: volchkov@univ.donetsk.ua
Vorobets Marija, National University "Lviv Polytechnika", Lviv, Ukraine. E-mail: vorobets@complex.lviv.ua
Vorobets Yaroslav, Lviv Ivan Franko National University, Lviv, Ukraine. E-mail: ya_vorobets@yahoo.com
Yakivchik Andrey, Moscow State University, Faculty of Mech. and Math., Moscow, Russia, 119992. E-mail: anyak@mech.math.msu.su
Yarka Ulyana, National University "Lviv Polytechnika", Lviv, Ukraine.
Yasinska Oksana, Chernivtsi National University, Chernivtsy, Ukraine. E-mail: yasik@cv.ukrtel.net
Yasinsky Volodymyr, Chernivtsi National University, Chernivtsy, Ukraine.

E-mail: yasik@cv.ukrtel.net
Yasinsky Lubov, Chernivtsi National University, Chernivtsy, Ukraine. E-mail: yasik@cv.ukrtel.net
Zagorodnyuk Andriy, Pidstryhach Institute for Applied Problems of Machanics and Mathematics of Ukrainian NAS, Lviv, Ukraine. E-mail: andriy@mebm.litech.net
Zarichnyj Mykhailo, Lviv Ivan Franko National University, Lviv, Ukraine. E-mail: mzar@litech.lviv.ua
Żelazko Wiesław, Mathematical Institute, Polish Academy of Sciences, Sniadeckich 8, P.O. Box 147, Warszawa, Poland 00-950. E-mail: zelazko@impan.gov.pl
Zorii Natalia, Institute of Mathematics of National Academy of Sciences of Ukraine, Tereshchenkivska str. 3, Kyiv Ukraine, 1601. E-mail: zorii@uprotel.net.ua
Zvozdetskyi Taras, Chernivtsy National University, Chernivtsy, Ukraine. E-mail: mathan@math.chnu.cv.ua

V. Kadets and W. Żelazko

Stefan Banach and Lviv Mathematical School

Yaroslav Prytula

Department of Mechanics and Mathematics, Lviv National University,
Universytetska 1, 79000 Lviv, Ukraine

The detailed biography of Stefan Banach (1892–1945), one of the greatest mathematicians of the twentieth century, is described in numerous publications and memoirs. In this note we would like to describe some events and persons connected with Lviv mathematics of that period.

One of the first Lviv mathematicians having a substantial influence on the development of mathematics in Lviv was Jozef Puzyna (1856–1919). He studied in Lviv University (1875–1882) obtaining his Ph.D. Degree in 1883. Puzyna continued his education in Berlin University (1883–1885) where he attended lectures of K. Weierstrass, L. Kronecker and O. Fuchs. These lectures determined his scientific interests: the theory of analytic functions, algebra and the theory of algebraic curves. Puzyna worked in Lviv University from 1885 to 1919. From 1892 he was a professor and headed a chair. Puzyna taught nearly 30 different courses. The masterpiece of his life was the monograph "Theory of analytic functions" ("Teorya funkcyj analitycznych") (1898–1900) in two volumes which included not only the most important investigations of Cauchy, Riemann and Weierstrass but also information from set theory, topology (Analysis situs), theory of groups and permutations. Later this book was highly praised by S. Saks and A. Zigmund. Puzyna organized two seminars (lower and higher) that played an important role in development of the Lviv school. Among the participants of those seminars were W. Lewicki, S. Ruzewich, O. Nikodym, A. Lomniecki and an number of other famous mathematicians. Puzyna helped create a second mathematical chair headed in 1903–1906 by Jan Rajewski. At the beginning of 1908 Puzyna proposed W. Sierpinski to do his habilitation in Lviv University and to head of the second chair of mathematics. In autumn of 1908 Sierpinski started to teach at the university and that year he gave one of the first university courses in set theory in the world. Till 1914 he taught, in particular, the following courses: "Measure of point-sets", "Lebesgue Integral", "The modern investigations in the theory of functions of a real variable". Puzyna charged W. Sierpinski with supervising one of the seminars (lower). In 1913, C. Mazurkiewicz, a student of Sierpinski, defended his Ph.D. In 1912 that Sierpinski proposed a position in the university to the young topologist Z. Janiszewski (1888–1920), who got his Ph.D. Degree (1911) in Paris. In 1913–14 Janiszewski started to work at the university as a private-docent. Later (in 1918) Janiszewski formulated the main principles of organization of the past-war Polish mathematical school. This is a picture of Lviv mathematics between 1910 and 1914.

Stefan Banach was a student of engineering Lviv Polytechnica. At that time in Lviv Polytechnica there were two mathematical chairs. One of them was headed (since 1898) by P. Dziwinski (1851–1936), who worked in geometry and history of mathematics. The other chair was headed by Z. Krysowski (1872–1955) whose main scientific results were related to the theory of elliptic and hyperbolic functions. In contrast to the University, the mathematicians of the Polytechnica of that time (till 1914) were not distinguished by prominent scientific results.

While of the Polytechnica Banach audited to the lectures of both Dziwinski and Krysowski professors. Except for his educational program we have little information about Banach's education in Lviv. Banach had come to Lviv from Krakow after graduating from the gymnasium there in 1910. There is a conjecture that for some time he audited lectures of Professor S. Zaremba in Jagellon University. In spring of 1914 Banach successfully passed the state exam concluding his two-year study in the Polytechnica. Of the First World War, in July of 1914 Banach returned to Krakow where he delivered private lessons to gymnasium pupils and study mathematics by himself. He never liked the work at school. Probably of that time he decided to make serious mathematical investigations.

In 1916 in Krakow Banach first met Hugo Steinhaus. That meeting of two great personalities and talents resulted in a collaboration which continued for many years. Banach's first scientific paper was published jointly with Steinhaus. After graduating from the gymnasium in Jasle, Steinhaus (1887–1972) studied mathematics and philosophy in Lviv University (1905–1906) and then continued in Göttingen (1906–1911), where under the supervision of D. Hilbert he obtained his Ph.D. in 1911. In 1917 Steinhaus made his habilitation in Lviv University and after the transfer of Sierpinski to Warsaw University he became an extraordinary professor (1920) and then an ordinary professor (1923). In 1920 A. Lomnicki now a professor of Lviv Polytechnica by the recommendation of Steinhaus invited Banach to be an assistant in the chair of mathematics. A. Lomnicki (1881–1941) graduated from the Lviv University and got his Ph.D. in 1903. He studied in Göttingen (1906–1907), made a habilitation in Lviv Polytechnica in 1919 and from 1920 headed the second mathematical chair. He was one of the first mathematicians who interpreted probability as a measure (1923).

In 1920 Banach defended his Ph.D. in Lviv University under the supervision of Lomnicki. In 1922 he made his habilitation in Lviv University. In the same year he became an extraordinary professor and in 1924 an ordinary professor of Lviv University. In this way Lviv University, taking into account the extraordinary talent and scientific value of his works, presented to Banach the highest scientific degrees, despite the fact that he had not formally obtained any higher education. Banach became the fourth professor of mathematics on the mathematics-natural sciences faculty. The other three professor were Steinhaus, E. Zylinski (1889–1954) and S. Ruziewicz (1889–1941). In 1928 Banach and Steinhaus founded a journal "Studia Mathematica" whose first volume appeared in 1929 and the tenth in 1940.

One of the most important books of classical analysis, S. Banach's monograph "Theorie des operations lineaires" was published in 1932. It had been published in the Polish language in 1931 and translated into Ukrainian in 1948. During those years in Lviv was formed a group of actively working mathematicians, who were mainly students of Banach and H. Steinhaus: P.-Ju. Schauder (1898–1943), S. Kaczmarz (1895–1939), S. Mazur (1905–1981), W. Orlicz (1903–1990), H. Auerbach (1901–1942) and W. Niklibore (1899–1948). In Lviv Polytechnica the chairs were headed by A. Lomnicki, W. Stozek (1883–1941) and K. Kuratowski (1896–1980). The study on the department supervised by K. Kuratowski was equivalent to university studies. One of the most famous students and doctorants of K. Kuratowski (1932) was S. Ulam (1909–1984). Meetings of the Lviv section of the Polish Mathematical Society were held each Saturday afternoon. Discussions often continued in down town cafe. One such, the famous "Scottish Cafe" became a meeting place of Lviv mathematicians and gave the name "Scottish Book" to the famous collection of mathematical problems. The first problem in this book was inscribed by Banach on 17 July 1935. In total Banach posed 14 problems (plus 11 others jointly with S. Mazur and S. Ulam). The last record in the book was made by Steinhaus on 31 May 1941.

In 1939 after the Soviet army entered Lviv, Lviv University was reorganized and the Faculty of Physics and Mathematics was created. In December 1939 S. Banach was appointed the head of the chair of analysis. During 1940–1941 he was a Dean of the Faculty. Many mathematicians from Poland, occupied by German troops, came to Lviv. In particular there were Professors: S. Saks, B. Knaster, M. Jakob, Docents: E. Szpilrain, O. Pepis, Assistants: I. Herzberg, M. Sperling, M. Stark, L. Sternbach, J. Mosler.

The work of Lviv University was interrupted by the Second World War and was renewed in 1944. During the German occupation Banach worked in an institute producing anti-typhus vaccine. That helped him to survive the war. He died on 31 September 1945.

Functional Analysis was the special area of Banach's investigations and his results in this direction gave him world fame. Yet, he has substantial achievements in other areas of mathematics. In particularly in: the theory of functions, theory of orthogonal series, measure theory, set theory. Lviv mathematical school, whose leader, without any doubt, was Banach, made a considerable contribution to world mathematics. In each mathematical area Lviv scientists had their own style.

The main achievements of the Lviv school are: creation of the fundamentals of functional analysis; introducing topological methods; interpreting probability as a measure and application of probability methods to Fourier series. A characteristic feature of Lviv mathematicians was masterful manipulation with non-constructive methods: the axiom of choice, Baire categories, and Lebesgue integral.

Functional Analysis and its Applications
V. Kadets and W. Żelazko

Existence of k-Media and Medians

Elisabetta Alvoni [a] **and Pier Luigi Papini** [b]

[a]Dipartimento di Matematica Applicata alle Scienze Economiche e Sociali Viale Filopanti, 40126 Bologna, Italy
[b]Dipartimento di Matematica, Piazza Porta S. Donato, 5, 40126 Bologna, Italy

Abstract

The notion of median, or Fermat point, of a finite set, has been recently generalized in two ways (see P.L. Papini and J. Puerto, preprint 2002; E. Alvoni, preprint 2002). Here we study conditions on the underlying space related to the existence of solutions concerning the second generalization.

Let X be a real Banach space; consider a finite subset A of X containing n elements and let k be an integer between 2 and n. For x in X, consider the distances among x and k points of A nearest to x; set $\mu_k(A,x)$ =average of these numbers. We want to minimize $\mu_k(A,x)$ (for $x \in X$) : a solution of this problem, if it exists, will be called a k-medium of A. The function $\mu_k(A,x)$is neither convex (or quasi-convex), nor concave; therefore the existence of solutions does not follow from standard results on convex functions.

Here we shall discuss existence of k- media; we will show that if the space X is such that every finite set A has a median (also called a Fermat point), then the same is true for k-media; in particular, in reflexive spaces, as well as in several classical spaces, k-media always exist.

Key words: Average distance, Fermat point, median
2000 MSC: 46B20, 46B99, 90B85

Email addresses: ealvoni@economia.unibo.it (Elisabetta Alvoni), papini@dm.unibo.it (Pier Luigi Papini).

1. Introduction and notation

Let X be a real Banach space. Consider a finite set $A = \{a_1, \ldots, a_n\} \subset X$, $n \geq 2$: we always assume $a_i \neq a_j$ for i$\neq j$; we shall also write $\#A = n$.

Given $x \in X$, we say that x *determines a permutation* $\Pi_0 = (\pi_1(x), \ldots, \pi_n(x))$ if $||x - a_{\pi_1(x)}|| \leq \ldots \leq ||x - a_{\pi_n(x)}||$.

Take $k \leq n$; according to [1], if x determines a permutation Π_0 we set:

$$\mu_k(A, x) = \frac{1}{k}\sum_{i=1}^{k} ||x - a_{\pi_i(x)}||.$$

Also, we set

$$\mu_k(A) = \inf_{x \in X} \mu_k(A, x).$$

We say that $x \in X$ is a *k-medium* of A if $\mu_k(A) = \mu_k(A, x)$; an n-medium of A ($\#A = n$) is usually called a *median*, or a *Fermat point* of A.

As shown in [1], the function $\mu_k(A, .)$ is neither convex (or quasi-convex), nor concave; so existence of k-media, also in "nice" Banach spaces, does not follow directly from standard results on convex functions (while it is so for medians).

Here we indicate a simple way to obtain existence results, by using simple subdivisions of the space into a finite number of regions.

2. k-Regions associated with a finite set

Given $A = \{a_1, \ldots, a_n\} \subset X$ ($\#A = n$), let $k \leq n$ ($k \geq 1$). Given a combination $\pi = (\pi_1, \ldots, \pi_k)$ of k numbers chosen in $\{1, \ldots, n\}$, we set:

$$M_\pi(A) = M(a_{\pi_1}, \ldots, a_{\pi_k}) = \{x \in X;\ \sup\{||x - a_{\pi_1}||, \ldots, ||x - a_{\pi_k}||\} \leq \leq \inf\{||x - a_{\pi_j}||;\ \pi_j \notin \pi\}\}. \tag{2.1}$$

We say that $M_\pi(A)$ is the *region determined by* $a_{\pi_1}, \ldots, a_{\pi_k}$.

The meaning is the following: x is in $M_\pi(A) = M(a_{\pi_1}, \ldots, a_{\pi_k})$ if and only if the k elements which "determine" that region form a set of k nearest points to x from A.

If the elements of the combination π form a set $I \subset \{1, 2, \ldots, n\}$, we shall also write M_I instead of M_π. For a combination of order $k = 1$, consisting of the index $\hat{\imath}$, we shall also write $M_{\hat{\imath}}$ for the corresponding region (determined by $a_{\hat{\imath}}$); similarly, for $k = 2$, if the combination consists of the indexes $\hat{\imath}$, $\hat{\jmath}$, we shall also write $M_{\hat{\imath}\hat{\jmath}}$ to denote the corresponding region (determined by $a_{\hat{\imath}}$, $a_{\hat{\jmath}}$).

When the underlying space is the euclidean plane, the regions determined by combinations like π create subdivisions of the plane which are called *order-k Voronoi diagrams* and have been extensively studied (see e.g. [3], p.357).

We indicate here some properties of the regions $M_\pi(A)$ considered here, for X any Banach space. Clearly, these regions (if we consider all combinations π) cover X.

1. Given A containing n elements and k between 1 and n, there are at most $\binom{n}{k}$ nonempty regions determined by subsets of A with k elements.

2. Two such regions may overlap; more precisely, given A and k, assume that two regions, say M_π and $M_{\pi'}$, have a common point x: then, if x determines a permutation $\Pi_0 = (\pi_1(x), \dots, \pi_n(x))$, we have $||x - a_{\pi_k(x)}|| = ||x - a_{\pi_{k+1}(x)}||$. But in case X is strictly convex, the overlaps cannot contain interior points.

3. For some π the corresponding region may be empty; for example, consider, in the euclidean plane: A={(0,2); (0,-2); (-1,0); (1,0)}, k=2, and the region determined by {(0,2); (0,-2)}.

4. There is no relation between $M(a_{\pi_1}, \dots, a_{\pi_k})$ and the convex hull of the same points, $conv\{a_{\pi_1}, \dots, a_{\pi_k}\}$.
In fact: of course, for any A, some of the regions are unbounded, while the convex hull of a finite set is bounded.
Conversely: let, for example, X be the euclidean plane; a_1=(-1,0); a_2=(1,0); a_3= (0,2); a_4=(0,-1); set $A = \{a_1, a_2, a_3, a_4\}$. We denote by θ the origin; then $\theta \in conv\{a_1, a_2, a_3\}$, but $\theta \notin M(a_1, a_2, a_3)$.

5. A median of a proper subset F of A (if it exists) does not necessarily belong to the region determined by F, also if X is euclidean: for example, the median of $\{a_1, a_2, a_3\}$ in the example indicated in 4. is the origin, but $\theta \notin M(a_1, a_2, a_3)$.

6. If X is an inner product space, then given two points a_1, a_2, the set

$$\{y \in X;\ ||y - a_1|| \leq ||y - a_2||\}$$

is convex; also, this property characterizes such spaces (see [2, p.26]).

Note that $M_\pi(A)$ as defined in (2.1) can be described in this way (with the notations used there): let, for h between 1 and k, $M_h(A) = \{x \in X;\ ||x - a_{\pi_h}|| \leq ||x - a||$ for $a \in A - \{a_{\pi_1}, \dots, a_{\pi_k}\}\}$. Then $M_\pi(A) = \cap_{h=1}^{k} M_h(A)$. This shows that, according to 6., all these regions are convex in inner product spaces (but only in these spaces).

Theorem 2.1. *Given two subsets I and J of {1,2,...,n}, we have $M_I \cap M_J \subset M_{I \cup J}$.*

Proof. With the usual notations ($\#A = n$) let $x \in M_I \cap M_J$: this means that $||x - a_i|| \leq ||x - a_k||$ for $i \in I$ and k (between 1 and n) not in I, and also that $||x - a_j|| \leq ||x - a_k||$ for $j \in J$ and k (between 1 and n) not in J. Therefore, if we consider a_h where h (between 1 and n) is not in I and not in J, then we have:

$||x - a_i|| \leq ||x - a_h||$ for all i in I;

$||x - a_j|| \leq ||x - a_h||$ for all j in J.

So, all elements of A with indexes in $I \cup J$ are nearest then a_h to x; therefore $x \in M_{I \cup J}$: this concludes the proof.

Remark. In general, given two combinations π_1, π_2, we do not have $M_{\pi_1 \cup \pi_2} \subset M_{\pi_1} \cup M_{\pi_2}$.

Theorem 2.2. *Let $I \subset \{1,2,\ldots,n\}$ contain l elements and let $k \leq l$; if $J_1, \ldots, J_h$ are all the subsets of I containing k elements, then $M_I \subset \cup_{i=1}^h M_{J_i}$.*

Proof. Let $I = \{i_1, \ldots, i_l\}$. If $x \in M_I$, then $A' = \{x_{i_1}, \ldots, x_{i_l}\}$ forms a set of l points of A nearest to x; if we consider $k \leq l$, k elements of A' nearest to x form a set J of k points of A nearest to x, so $J \subset I$ is such that $x \in M_J$: this proves the thesis.

Theorem 2.3. *If $M_i \cap M_j = \emptyset$, then $M_{ij} = \emptyset$.*

Proof. We prove the contrapositive. Assume that $M_{ij} \neq \emptyset$; it is not restriction to assume $||x - x_i|| \leq ||x - x_j||$. Take x_t on the line $x + t(x_j - x)$, such that $||x_t - x_i|| = ||x_t - x_j||$ $(0 \leq t \leq 1)$: if $x_t \in M_i \cap M_j$ we are done. Assume that $x_t \notin M_i \cap M_j$, so there exists x_k, $k \neq i, j$, such that $||x_t - x_k|| < ||x_t - x_i|| = ||x_t - x_j||$. This implies $||x - x_k|| \leq ||x - x_t|| + ||x_t - x_k|| < ||x - x_t|| + ||x_t - x_j|| = ||x - x_j||$, against $x \in M_{ij}$; this contradiction proves that necessarily $x_t \in M_i \cap M_j$; so $M_i \cap M_j \neq \emptyset$, which concludes the proof.

Remark. Let $conv(A) = conv\{a_1, \ldots, a_k\}$, $k < n$; by setting $I = \{1, \ldots, k\}$ we have, apart from some special cases, $M_I = \emptyset$: in fact, the farthest point to any $x \in X$ is a vertex of $conv(A)$, so it belongs to $\{a_1, \ldots, a_k\}$.

3. Existence of k-media

Let $A = \{a_1, \ldots, a_n\} \subset X$; denote by B_{a_i} the ball centered at the point a_i of A (i arbitrary), of radius $\delta(A)$ (the diameter of A). For k between 1 and n, let
$\gamma_k(A) = \{x \in X;\ \mu_k(A) = \mu_k(A, x)\}$.

It is clear that $\gamma_k(A) \neq \emptyset$ for every A if $k = 1, 2$ (see [1]). Also, $\gamma_k(A) \neq \emptyset$ always if $\dim(X) < \infty$: in fact, taken any i between 1 and n, we can minimize $\mu_k(A, x)$ for $x \in B_{a_i}$, a compact set.

We can also prove the following:

Theorem 3.1. *Assume that $[A]$, the linear span of the finite set A, is the range of a norm-one projection P from X: then, for all k between 1 and n, $\gamma_k(A) \neq \emptyset$. In particular, this is true in the following cases:*

a) A is contained in a line;

b) X is a Hilbert space.

Proof. Let $P : X \to [A]$ (P a linear projection), $||Px|| \leq ||x||$ for all x; given $x \in X$, assume that it determines a permutation $\Pi_0 = (\pi_1(x), \ldots, \pi_n(x))$. Then we have $(1 \leq k \leq n = \#A)$:

$$\sum_{i=1}^{k} ||Px - a_{\pi_i(x)}|| \leq \sum_{i=1}^{k} ||x - a_{\pi_i(x)}||,$$

so $\mu_k(A, Px) \leq \mu_k(A, x)$;

by using a ball $B = B_{a_i}$ (any i) as indicated before, we have:

$\mu_k(A) = \min\{\mu_k(A, x);\ x \in [A] \cap B\}$; since $[A] \cap B$ is compact, this gives the conclusion.

Also, as known, in case a) or b) is true, then [A] (or, in the first case, the one-dimensional translate of $[A]$) is the range of norm-one projection.

The above result is not too useful, since norm-one projections are rather rare (see e.g. [6]).

But we can prove the following

Theorem 3.2. *Let X be a space where all k-point sets have a median; then $\gamma_k(A) \neq \emptyset$ for every finite set A (with $\#A \geq k$).*

Proof. Given A, consider all regions, say M_π, determined by combinations $\pi = (\pi_1, \ldots, \pi_k)$ of k among the n elements of A and which are nonempty (the set of all these combinations contains at most $\binom{n}{k}$ elements).

Let $\mu(M_\pi) = \inf_{x \in X} \frac{1}{k} \sum_{i=1}^{k} ||x - a_{\pi_i}||$, or also, by our assumption, $\mu(M_\pi) = \min_{x \in X} \frac{1}{k} \sum_{i=1}^{k} ||x - a_{\pi_i}||$. Set $\mu = \min_{\pi \in \Pi} \mu(M_\pi)$ and let this minimum be achieved for $\pi' = \{\pi'_1, \ldots, \pi'_k\}$; also, let $m \in X$ be such that $\mu = \mu(M_{\pi'}) = \frac{1}{k} \sum_{i=1}^{k} ||m - a_{\pi_i'}||$.

We have $\mu = \mu_k(A)$: in fact, clearly $\mu_k(A) \leq \mu$. Now take any $x \in X$: there exists a permutation π'' (eventually $\pi'' = \pi'$) such that $x \in M_{\pi''}$ and we obtain:

$\mu_k(A, x) \geq \mu_k(M_{\pi''}) \geq \mu_k(M_{\pi'}) = \mu_k(A, m) = \mu$, so $\mu_k(A) \geq \mu$, and then equality holds: this concludes the proof.

Remark. Note that if $x \in \gamma_k(A)$, and $x \in M_\pi$, $\pi = (\pi_1, \ldots, \pi_k)$, then x is a median of $\{a_{\pi_1, \ldots,} a_{\pi_k}\}$.

Recall that general results concerning space where medians always exist have been given in [7].

4. Concluding remerks and some questions

Clearly, the condition indicated in Theorem 3.2 is necessary and sufficient: i.e., if for some k all sets A with $\#A \geq k$ have k-media, then -in particular- all k-point subsets of X have a median.

We could ask if this is equivalent to the fact that all finite subsets of X have a median.

A similar question could be asked for centers; recall that $c \in X$ is a *center* of A if $\inf_{x \in X} \sup_{a \in A} ||x - a|| = \sup_{a \in A} ||c - a||$.

Note that, as shown in [7], three-point sets are sufficient to characterize several properties of spaces by using centers or medians of sets (see also [4]).

More precisely, we can ask the following. Given a Banach space, consider the following properties:

m) all finite sets have (at least) a median;

m') all three-point sets have (at least) a median;

c) all finite sets have (at least) a center;

c') all three-point sets have (at least) a center;

of course, $m) \Longrightarrow m')$; $c) \Longrightarrow c')$.

We can raise the following

Question: are the above four properties equivalent?

Or at least:

are the conditions $m)$ and $m')$, and/or the conditions $c)$ and $c')$ equivalent? See also [8].

We conclude by proving a simple property.

Proposition 4.1. *Let $A = \{a_1, \ldots, a_n\}$ be such that $\gamma_k(A) = \emptyset$ for some $k \geq 3$ ($k \leq n$); then for any $h \geq 1$ we can embed A into a larger set A' with $\#A' = n + h$ such that $\gamma_k(A') = \emptyset$.*

Proof. By induction, it is enough to prove the following: given $A = \{a_1, \ldots, a_n\}$ such that $\gamma_k(A) = \emptyset$ $(3 \leq k \leq n)$, we can embed A into A' containing one more element and such that $\gamma_k(A') = \emptyset$. Take a_{n+1} such that $||a_{n+1} - a_i|| > k\mu_k(A)$ for $i = 1, \ldots, n$ $(a_{n+1} \notin A)$; then set $A' = A \cup \{a_{n+1}\}$. Since $A \subset A'$, we have $\mu_k(A') \leq \mu_k(A)$.

Let $x \in X$; assume that among the k points of A' which are nearest to x, there are a_{n+1} and some points a_j (j between 1 and n); if $a_{\bar{j}}$ is one of these points, then we have: $k\mu_k(A', x) > ||a_{\bar{j}} - a_{n+1}|| > k\mu_k(A) \geq k\mu_k(A')$.

Now assume that a_{n+1} is not among the k points of A' nearest to x; then, since $x \notin \gamma_k(A)$, we have: $\mu_k(A', x) = \mu_k(A, x) > \mu_k(A) \geq \mu_k(A')$.

Therefore, in any case, $\mu_k(A', x) > \mu_k(A')$ for every $x \in X$, so $\gamma_k(A') = \emptyset$. This concludes the proof.

Corollary 4.2. *The following statements are equivalent:*

(i) *every set containing k elements has a median;*

(ii) *there exists $n \geq k$ such that for every set A containing n elements we have $\gamma_k(A) \neq \emptyset$;*

(iii) *we have $\gamma_k(A) \neq \emptyset$ for every finite set A.*

Proof. (i)→(iii): see Theorem 3.2.

(iii)→(ii): trivial;

(ii)→(i): it follows from Proposition 4.1; in fact, if there exists A without medians and such that $\#A = k$, then for any $n \geq k$ there is a set A' such that $\#A' = n$ and $\gamma_k(A) \neq \emptyset$.

References

1. E. Alvoni, On k-nearest points of a finite set in Banach spaces, preprint.
2. D. Amir, *Characterizations of inner product spaces*. Birkhäuser, Basel 1986.
3. F. Aurenhammer, Voronoi diagrams - A survey of a fundamental geometric data structure, *ACM Computing Surveys* **23** (1991) 345-405.
4. Benítez C., Fernández M. and Soriano M.L., Location of the Fermat-Torricelli medians of three points, *Trans. Amer. Math. Soc.* **354** (2002) 5027-5038.
5. P.L. Papini and J. Puerto, Averaging the k-largest distances among n: k-centra in Banach spaces, preprint.
6. B. Randrianantoanina, Norm-one projections in Banach spaces, *Taiwanese J. Math.* **5** (2001) 35-95.
7. Veselý L., Generalized centers of finite sets in Banach spaces, *Acta Math. Univ. Comenian. New Ser.* **66** (1997) 83-115.
8. Veselý L., A Banach space in which all compact sets, but not all bounded sets, admit Chebyshev centers, *Arch. Math.* to appear.

V. Kadets and W. Żelazko

On Optimization of Approximate Integration of Multivariate Periodic Functions

Vladislav F. Babenko

Department of Mechanics and Mathematics, Dnepropetrovsk National University, vul. Naukova, 13, 49625, Ukraine
Donetsk Institute of Applied Mathematics and Mechanics of NAS of Ukraine, 74 Luxembourg st., Donetsk, 340114, Ukraine

Sergiy V. Borodachov

Department of Mechanics and Mathematics, Dnepropetrovsk National University, vul. Naukova, 13, 49625, Ukraine

Abstract

On the base of known optimal quadrature formulae for certain classes of differentiable periodic functions we construct optimal cubature formulae using values of a function at points of rectangular nets, and an optimal formula using mean values over small parallelepipeds centered at nodes of such nets for certain classes of multivariate periodic functions.

Key words: Optimal cubature formula, class of functions, error over a class
2000 MSC: 26B15

The problem of optimization of approximate integration was considered for the first time in the works of A.N. Kolmogorov and S.M. Nikol'skiy in the late forties – early fifties. A large number of results have been obtained in this direction for classes of one and several variables. A review of many of them one can find in [15,24,11].

Email addresses: babenko@dnepr.net (Vladislav F. Babenko), borodach@ff.dsu.dp.ua (Sergiy V. Borodachoy).

Let C_d, $d \in \mathrm{N}$, be the space of continuous functions $f : \mathrm{R}^d \to \mathrm{R}$, 2π-periodic along each variable, with the norm $\|f\|_{C_d} = \max\{|f(x)| : x \in [0, 2\pi]^d\}$. Put $C = C_1$. For classes $Y_1, \dots, Y_d \subset C$ denote by $(Y_1, \dots, Y_d)$ the set of all functions $f \in C_d$, such that for any fixed $j = 1, \dots, d$ and $x_1, \dots, x_{j-1}, x_{j+1}, \dots, x_d \in \mathrm{R}$ we have $\chi(t) = f(x_1, \dots, x_{j-1}, t, x_{j+1}, \dots, x_d) \in Y_j$. In other words, $(Y_1, \dots, Y_d)$ is the set of functions from C_d which belong to Y_j as functions of the j-th variable, $j = 1, \dots, d$. Denote

$$I_d(f) = \int_0^{2\pi} \dots \int_0^{2\pi} f(t_1, \dots, t_d)\, dt_1 \dots dt_d, \quad f \in C_d.$$

Let $d \in \mathrm{N}$. Given a vector $\overline{n} = (n_1, \dots, n_d) \in \mathrm{N}^d$ and d partitions of the segment $[0, 2\pi]$ – $\Delta^1_{n_1} : 0 \le x^1_1 < \dots < x^1_{n_1} < 2\pi, \dots, \Delta^d_{n_d} : 0 \le x^d_1 < \dots < x^d_{n_d} < 2\pi$, denote

$$X_{\overline{n}} = X_{n_1,\dots,n_d} = \{(x^1_{i_1}, \dots, x^d_{i_d}) : 1 \le i_1 \le n_1, \dots, 1 \le i_d \le n_d\}. \tag{1}$$

Every set of the form (1) and a d-dimensional matrix of real coefficients $C_{\overline{n}} = \{c_{i_1,\dots,i_d}\}_{i_j=\overline{1,n_j},\ j=\overline{1,d}}$ generate a cubature formula (for approximate integration along $[0, 2\pi]^d$) of the form

$$S(f) = S(f; X_{\overline{n}}, C_{\overline{n}}) = \sum_{i_1=1}^{n_1} \dots \sum_{i_d=1}^{n_d} c_{i_1,\dots,i_d} f(x^1_{i_1}, \dots, x^d_{i_d}). \tag{2}$$

Denote by $Q^d_{\overline{n}}$ the collection of formulae of the form (2) with nodes at all possible sets $X_{\overline{n}}$ of the form (1) with given vector $\overline{n} \in \mathrm{N}^d$.

Let again $\overline{n} = (n_1, \dots, n_d) \in \mathrm{N}^d$ and $\overline{h} = (h_1, \dots, h_d) \in \left(0, \frac{\pi}{n_1}\right) \times \dots \times \left(0, \frac{\pi}{n_d}\right)$ (numbers h_j are chosen to be sufficiently small). Given a function $f \in C_d$ and a point $\overline{x} = (x_1, \dots, x_d) \in \mathrm{R}^d$, denote $\Delta[\overline{x}, \overline{h}] = [x_1 - h_1, x_1 + h_1] \times \dots \times [x_d - h_d, x_d + h_d]$ and

$$I_{\overline{h}}(f; \overline{x}) = I_{\overline{h}}(f; x_1, \dots, x_d) = \frac{1}{\mathrm{mes}\Delta[\overline{x}, \overline{h}]} \cdot \int_{\Delta[\overline{x}, \overline{h}]} f(t)dt.$$

Every set $X_{\overline{n}}$ of the form (1) and the d-dimensional real matrix $C_{\overline{n}}$ generate the cubature formula of the form

$$S(f) = S(f; \overline{h}, X_{\overline{n}}, C_{\overline{n}}) = \sum_{i_1=1}^{n_1} \dots \sum_{i_d=1}^{n_d} c_{i_1,\dots,i_d} I_{\overline{h}}(f; x^1_{i_1}, \dots, x^d_{i_d}). \tag{3}$$

Denote by $Q^d_{\overline{n},\overline{h}}$ the collection of formulae (3), corresponding to all possible sets $X_{\overline{n}}$ of the form (1) with the given $\overline{n}$. Let $Y \subset C_d$ be a class of functions, $Q = Q^d_{\overline{n}}$ or $Q^d_{\overline{n},\overline{h}}$, $S \in Q$. Put

$$R(Y; S) = \sup_{f \in Y} |I_d(f) - S(f)|, \quad R(Y; Q) = \inf_{S \in Q} R(Y; S).$$

It is required to find the value $R(Y; Q)$ and Q-optimal formulae for the class Y (i.e. formulae $S^* \in Q$ such that $R(Y; S^*) = R(Y; Q)$), if they exist.

We will consider this problem for $d > 1$. If $Q = Q^1_n$, we have the Nikolskiy's problem of optimization of quadrature formulae $\sum_{k=1}^n c_k g(x_k)$ by nodes and coefficients for classes from C. In the case $Q = Q^1_{n,h}$, the problem set above is the problem of optimization of the so called "interval" formulae $\sum_{k=1}^n c_k I_h(g; x_k)$.

Given a modulus of continuity $\omega(t)$, denote by H^ω the class of functions f, such that $|f(s_1) - f(s_2)| \le \omega(|s_1 - s_2|)$, $s_1, s_2 \in \mathrm{R}$. Put $W^r H^\omega = \{f \in C : f^{(r)} \in H^\omega\}$. Let L_p, $p \in [1, \infty]$ be the space of 2π-periodic univariate functions with the corresponding norm $\|\cdot\|_p$. Denote by W^r_p, $r \in \mathrm{N}$, $p \in [1, \infty]$, the class of functions $f \in C$ such that $f^{(r-1)}$ is absolutely continuous, and $\|f^{(r)}\|_p \le 1$.

The problem of optimization of approximate integration of classes mentioned above and their generalizations was considered in [14,6,8,10,9,23,4,17,24,1,13]. For a detailed review of the results in this direction see [15,24,11].

If $Q = Q^d_{\overline{n}}$, $d \ge 2$, we have a problem about the best cubature formula with nodes forming rectangular nets. For $Q = Q^d_{\overline{n},\overline{h}}$, $d \ge 2$, we are required to find an optimal cubature formula using functions' means taken along d-dimensional small parallelepipeds centered at nodes of rectangular nets.

In mathematical literature a more general problem of optimization of cubature formulae for multivariate functions with nodes at arbitrary sets of points was considered. For reviews and results in this direction see [21,22].

Problems on optimal cubature formulae with nodes forming rectangular nets were considered in [6,7] and other papers. A detailed review of results in this direction can be found in [15]. Close problems concerning tensor products of quadrature and interpolation formulae were considered in [19,16].

Given moduli of continuity $\omega_1, \dots, \omega_d$, denote by $H^{(\omega_1,\dots,\omega_d)}$ the class of functions $f \in C_d$, such that for all $x_1, \dots, x_d, y_1, \dots, y_d \in \mathrm{R}$

$$|f(x_1, \dots, x_d) - f(y_1, \dots, y_d)| \le \omega_1(|x_1 - y_1|) + \dots + \omega_d(|x_d - y_d|).$$

We will consider the problems, stated above from a general point of view. Denote by C^*_d, $d \in \mathrm{N}$, the space of linear bounded functionals on C_d, and by $(C^*)^n$, $n \in \mathrm{N}$, – the n-th Cartesian power of the C^*. At first, introduce auxiliary definitions in the one-dimensional case. Let a class $H \subset C$, a functional $\Phi \in C^*$ and a collection of sets of functionals $A_n \subset (C^*)^n$ be given. Every set of functionals $M_n = (\Phi_1, \dots, \Phi_n) \in A_n$ and a set of coefficients $C_n = (c_1, \dots, c_n) \in \mathrm{R^n}$ generate a functional of the form $S(f) = S(f; M_n, C_n) = \sum_{k=1}^n c_k \Phi_k(f)$. Let $Q^1(A_n)$, $n \in \mathrm{N}$, be the collection of all functionals $S(\cdot; M_n, C_n)$ with $M_n \in A_n$, $C_n \in \mathrm{R^n}$. Given $S \in Q^1(A_n)$, put

$$R(H; \Phi, S) = \sup_{f \in H} |\Phi(f) - S(f)|, \quad R(H; \Phi; A_n) = \inf_{S \in Q^1(A_n)} R(H; \Phi, S).$$

A functional $S^* \in Q^1(A_n)$ is called an optimal one for approximation of the functional Φ on the class H (or $(A_n;\ \Phi;\ H)$–optimal), if $R(H; \Phi, S^*) = R(H; \Phi; A_n)$.

A set $U \subset C$ is called convex, if for all $f, g \in U$, $\alpha \in [0, 1]$ we have $\alpha f + (1-\alpha) g \in U$. A set $U \subset C$ is called centrally symmetric if $f \in U$ implies $-f \in U$. Given a functional $g \in C^*$, let $\widetilde{g}$ be a such function that $\bigvee_0^{2\pi} \widetilde{g} = \|g\|$ and $g(f) = \int_0^{2\pi} f(t) d\widetilde{g}$.

Given classes $Y_1, \ldots, Y_d \subset C$, $d > 1$, denote by $Y_1 + \ldots + Y_d$ the set of all functions $f \in C_d$, representable in the form $f(t_1, \ldots, t_d) = \chi_1(t_1) + \ldots + \chi_d(t_d)$, where $\chi_j \in Y_j$, $j = 1, \ldots, d$. The inclusion $(Y_1, \ldots, Y_d) \supset Y_1 + \ldots + Y_d$ holds, if $Y_1, \ldots, Y_d$ are invariant with respect to adding a constant. For functionals $F_1, \ldots, F_d \in C^*$ put $p_2(x_2, \ldots, x_d; f) = F_1[f(\cdot, x_2, \ldots, x_d)]$, $p_3(x_3, \ldots, x_d; f) = F_2[p_2(\cdot, x_3, \ldots, x_d; f)]$, $\ldots$, $p_d(x_d; f) = F_{d-1}[p_{d-1}(\cdot, x_d; f)]$. Let $F(f) = F_d[p_d(\cdot; f)]$. A functional F is representable as a d-fold Lebesgue-Stieltjes integral

$$F(f) = \int_0^{2\pi} \ldots \int_0^{2\pi} f(t_1, \ldots, t_d)\, d\widetilde{F}_1 \ldots d\widetilde{F}_d.$$

Since, by the Fubini theorem $F(f)$ does not depend on the order of applying of F_j, we will denote $F(f) = \langle F_1, \ldots, F_d; f \rangle$, writing the functional, applied to the j-th variable, on the j-th place. Note, that for all $f \in C_d$, $I_d(f) = \langle I_1, \ldots, I_1; f \rangle$, $\langle \Psi_1, \ldots, \Psi_d; f \rangle = f(x^1, \ldots, x^d)$, $f \in C_d$, $x^1, \ldots, x^d \in \mathrm{R}$, where $\Psi_j(g) = g(x^j)$, $j = 1, \ldots, d$, and $\langle I_{h_1}(\cdot; x_1), \ldots, I_{h_d}(\cdot; x_d); f \rangle = I_{\overline{h}}(f; x_1, \ldots, x_d)$. For all $f \in C_d$ and $j = 1, \ldots, d$ we have

$$|\langle F_1, \ldots, F_d; f \rangle| \le \|F_1\| \cdot \max_{t_1 \in [0, 2\pi]} |\langle F_2, \ldots, F_d; f(t_1, \cdot, \ldots, \cdot) \rangle| \le$$

$$\le \prod_{\substack{s=1 \\ s \ne j}}^{d} \|F_s\| \cdot \max_{(t_1, \ldots, t_{j-1}, t_{j+1}, \ldots, t_d) \in [0, 2\pi]^{d-1}} |F_j(f(t_1, \ldots, t_{j-1}, \cdot, t_{j+1}, \ldots, t_d))|. \quad (4)$$

Let a class $Y \subset C_d$, $d > 1$, a vector $(n_1, \ldots, n_d) \in \mathrm{N}^{\mathrm{d}}$ and a collection of non-empty sets $A^j_{n_j} \subset (C^*)^{n_j}$, $j = 1, ..., d$, be given. Each collection

$$\left(g^1_1, \ldots, g^1_{n_1}\right) \in A^1_{n_1}, \; \ldots, \; \left(g^d_1, \ldots, g^d_{n_d}\right) \in A^d_{n_d}, \quad (5)$$

and the real matrix $C_{\overline{n}} = \{c_{i_1, \ldots, i_d}\}_{i_j = \overline{1, n_j}, j = \overline{1, d}}$ generate an approximation algorithm of the functional F on the class Y of the form

$$S(f) = \sum_{i_1=1}^{n_1} \cdots \sum_{i_d=1}^{n_d} c_{i_1, \ldots, i_d} \langle g^1_{i_1}, \ldots, g^d_{i_d}; f \rangle. \quad (6)$$

Denote by $Q^d\left(A^1_{n_1}, \ldots, A^d_{n_d}\right)$ the collection of all functionals (6) for which (5) is true. Given $f \in Y$, $S \in Q^d\left(A^1_{n_1}, \ldots, A^d_{n_d}\right)$, put

$$R(f; F, S) = F(f) - S(f), \quad R(Y; F, S) = \sup_{f \in Y} |F(f) - S(f)|; \quad (7)$$

$$R\left(Y; F; A^1_{n_1}, \ldots, A^d_{n_d}\right) = \inf\{R(Y; F, S) : S \in Q^d\left(A^1_{n_1}, \ldots, A^d_{n_d}\right)\}. \quad (8)$$

Functional $S^* \in Q^d\left(A^1_{n_1}, \ldots, A^d_{n_d}\right)$, delivering the infimum on the right-hand side of (8), is said to be an $(A^1_{n_1}, \ldots, A^d_{n_d}; F; Y)$–optimal algorithm.

Theorem 1. *Let $d \in \mathrm{N}$, $d > 1$, classes $Y_1, \ldots, Y_d \subset C$ be convex and centrally symmetric, a class $Y \subset C_d$ be such that $(Y_1, \ldots, Y_d) \supset Y \supset Y_1 + \ldots + Y_d$, a non-zero functional $F \in C_d^*$ be representable in the form $F(f) = \langle F_1, \ldots, F_d; f\rangle$, $f \in C_d$, where $F_1, \ldots, F_d \in C^*$, correspond to monotone functions $\widetilde{F}_1, \ldots, \widetilde{F}_d$.*

Let also $\overline{n} = (n_1, \ldots, n_d) \in \mathrm{N}^d$, nonvoid sets $A^1_{n_1} \subset (C^*)^{n_1}$,..., $A^d_{n_d} \subset (C^*)^{n_d}$ be given, and for every $j = 1, \ldots, d$ some functional $S_j^*(f) = \sum\limits_{k=1}^{n_j} a_k^j \varphi_k^j(f) \in Q^1\left(A^j_{n_j}\right)$ be the $(A^j_{n_j};F_j;Y_j)$-optimal, with the inequality $\|S_j^*\| \le \|F_j\|$ being true for $j = 1, \ldots, d-1$. Then algorithm

$$S^*(f) = \sum_{i_1=1}^{n_1} \cdots \sum_{i_d=1}^{n_d} a^1_{i_1} \cdot \ldots \cdot a^d_{i_d} \cdot \langle \varphi^1_{i_1}, \ldots, \varphi^d_{i_d}; f\rangle$$

is thr $(A^1_{n_1}, \ldots, A^d_{n_d};F;Y)$-optimal. In addition,

$$R\left(Y; F; A^1_{n_1}, \ldots, A^d_{n_d}\right) = \sum_{j=1}^{d} R\left(Y_j; F_j; A^j_{n_j}\right) \prod_{\substack{s=1\\ s\neq j}}^{d} \|F_s\|. \tag{9}$$

Remark 1. If $Y = H^{(\omega_1,\ldots,\omega_d)} = (H^{\omega_1}, \ldots, H^{\omega_d})$, $F = I_d$, $F_j = I_1$ and $A^j_{n_j}$ be the collection of all ordered sets of functionals $\left(\Phi_1, \ldots, \Phi_{n_j}\right) \in (C^*)^{n_j}$, such that $\Phi_k(g) = g(x_k)$, $k = 1, \ldots, n_j$ and $0 \le x_1 < \ldots < x_n < 2\pi$, $j = 1, \ldots, d$, then theorem 1 follows from the results of [6].

Remark 2. Result, analogous to theorem 1, is true in the non-periodic case.

Following the terminology of [19], we will call the functional S^* the tensor product of functionals $S_1^*, \ldots, S_d^*$. Note, that because of arbitrariness of sets of coefficients $C_{\overline{n}}$ some algorithms from $Q^d\left(A^1_{n_1}, \ldots, A^d_{n_d}\right)$ are not tensor products of algorithms from $Q^1(A^j_{n_j})$.

Proof. First, obtain the upper bound for the error over the class Y of the algorithm S^*. For an arbitrary $f \in Y$ we will get

$$|R(f; F, S^*)| = \left| F(f) - \sum_{i_1=1}^{n_1} \cdots \sum_{i_d=1}^{n_d} a^1_{i_1} \cdot \ldots \cdot a^d_{i_d} \cdot \langle \varphi^1_{i_1}, \ldots, \varphi^d_{i_d}; f\rangle \right| =$$

$$= \left| F(f) - \langle \sum_{i_1=1}^{n_1} a^1_{i_1}\varphi^1_{i_1}, \ldots, \sum_{i_d=1}^{n_d} a^d_{i_d}\varphi^d_{i_d}; f\rangle \right| = |\langle F_1, \ldots, F_d; f\rangle - \langle S_1^*, \ldots, S_d^*; f\rangle| =$$

$$= \left| \sum_{j=1}^{d} \langle S_1^*, \ldots, S_{j-1}^*, F_j - S_j^*, F_{j+1}, \ldots, F_d; f\rangle \right| \le \sum_{j=1}^{d} \left|\langle S_1^*, \ldots, S_{j-1}^*, F_j - S_j^*, F_{j+1}, \ldots, F_d; f\rangle\right|.$$

Applying (4), we obtain

$$|R(f;F,S^*)| \le \sum_{j=1}^{d} \|S_1^*\|\cdot\ldots\cdot\|S_{j-1}^*\|\cdot\|F_{j+1}\|\cdot\ldots\cdot\|F_d\|\times$$

$$\times \max_{(t_1,\ldots,t_{j-1},t_{j+1},\ldots,t_d)\in[0,2\pi]^{d-1}} \left|(F_j - S_j^*)(f(t_1,\ldots,t_{j-1},\cdot,t_{j+1},\ldots,t_d))\right|.$$

For every $j = 1,...,d$ f is in Y_j as function of the j-th variable, and

$$\left|(F_j - S_j^*)(f(t_1,\ldots,t_{j-1},\cdot,t_{j+1},\ldots,t_d))\right| \le \sup_{g\in Y_j} \left|F_j(g) - S_j^*(g)\right| = R(Y_j;F_j,S_j^*),$$

$(t_1,\ldots,t_{j-1},t_{j+1},\ldots,t_d) \in [0,2\pi]^{d-1}$. Using the $(A^j_{n_j};F_j;Y_j)$–optimality of S_j^*, inequality $\|S_j^*\| \le \|F_j\|$, $j = 1,...,d-1$, for every $f \in Y$ we get

$$|R(f;F,S^*)| \le \sum_{j=1}^{d} \|S_1^*\|\cdot\ldots\cdot\|S_{j-1}^*\|\cdot R(Y_j;F_j,S_j^*)\cdot\|F_{j+1}\|\cdot\ldots\cdot\|F_d\| \le$$

$$\le \sum_{j=1}^{d} \|F_1\|\cdot\ldots\cdot\|F_{j-1}\|\cdot R(Y_j;F_j;A^j_{n_j})\cdot\|F_{j+1}\|\cdot\ldots\cdot\|F_d\|.$$

Then

$$R(Y;F,S^*) \le \sum_{j=1}^{d} \|F_1\|\cdot\ldots\cdot\|F_{j-1}\| R\left(Y_j;F_j;A^j_{n_j}\right) \|F_{j+1}\|\cdot\ldots\cdot\|F_d\|. \tag{10}$$

Show, that the error over the class Y of any algorithm from $Q^d\left(A^1_{n_1},\ldots,A^d_{n_d}\right)$ is at least the value on the right-hand side of the inequality (10). Let $T(f) = \sum_{i_1=1}^{n_1} \cdots \sum_{i_d=1}^{n_d} c_{i_1,\ldots,i_d} \langle \psi^1_{i_1},\ldots,\psi^d_{i_d};f\rangle \in Q^d\left(A^1_{n_1},\ldots,A^d_{n_d}\right)$ be an arbitrary algorithm. For $j = 1,\ldots,d$ put

$$\tau_j = \tau(\psi^j_1,\ldots,\psi^j_{n_j}) := \sup\{F_j(y) : y \in Y_j, \psi^j_k(y) = 0, k = 1,\ldots,n_j\},$$

$$R\left(Y_j;F_j,\psi^j_1,\ldots,\psi^j_{n_j}\right) := \inf_{\alpha_1,\ldots,\alpha_{n_j}\in\mathrm{R}} \sup_{y\in Y_j} \left|F_j(y) - \sum_{k=1}^{n_j} \alpha_k \psi^j_k(y)\right|. \tag{11}$$

Since $Y_1,\ldots,Y_d$ are central symmetric and convex, by Smolyak's lemma [20] (one can also find its proof in [3])

$$R\left(Y_j;F_j,\psi^j_1,\ldots,\psi^j_{n_j}\right) = \tau(\psi^j_1,\ldots,\psi^j_{n_j}),\quad j = 1,\ldots,d. \tag{12}$$

By the central symmetry of Y_j, for every $j = 1,\ldots,d$ there exists a sequence $\{y_{j,m}\}_{m=1}^{+\infty} \subset Y_j$, such that $\psi^j_k(y_{j,m}) = 0$, $k = 1,\ldots,n_j$, $m \in \mathrm{N}$ and

$$\lim_{m\to+\infty} F_j(y_{j,m}) = \tau_j \cdot \mathrm{sgn}\left(\widetilde{F}_j(2\pi) - \widetilde{F}_j(0)\right). \tag{13}$$

Denote $z_m(t_1,\ldots,t_d) = y_{1,m}(t_1) + \ldots + y_{d,m}(t_d)$, $m \in \mathrm{N}$. Obviously, $z_m \in Y_1 + \ldots + Y_d \subset Y$. Then,

$$T(z_m) = \sum_{i_1=1}^{n_1}\cdots\sum_{i_d=1}^{n_d} c_{i_1,\ldots,i_d} \int\limits_{[0,2\pi)^d} (y_{1,m}(t_1) + \ldots + y_{d,m}(t_d))\, d\widetilde{\psi}^1_{i_1}\ldots d\widetilde{\psi}^d_{i_d} =$$

$$= \sum_{i_1=1}^{n_1} \cdots \sum_{i_d=1}^{n_d} c_{i_1,\ldots,i_d} \left(\sum_{j=1}^{d} \int_{[0,2\pi)^{d-1}} \int_0^{2\pi} y_{j,m}(t_j) d\widetilde{\psi}_{i_j}^j d\widetilde{\psi}_{i_1}^1 \ldots d\widetilde{\psi}_{i_{j-1}}^{j-1} d\widetilde{\psi}_{i_{j+1}}^{j+1} \ldots d\widetilde{\psi}_{i_d}^d \right).$$

By the choice of the sequences, $\psi_{i_j}^j (y_{j,m}) = \int_0^{2\pi} y_{j,m}(t) d\widetilde{\psi}_{i_j}^j = 0$ for all values of the indexes. Then, $T(z_m) = 0$ and

$$R(Y; F, T) \geq |F(z_m)| = \left| \sum_{j=1}^{d} \int_{[0,2\pi)^{d-1}} \int_0^{2\pi} y_{j,m}(t_j) d\widetilde{F}_j \ d\widetilde{F}_1 \ldots d\widetilde{F}_{j-1} d\widetilde{F}_{j+1} \ldots d\widetilde{F}_d \right|$$

$$= \left| \sum_{j=1}^{d} F_j(y_{j,m}) \cdot \prod_{\substack{s=1 \\ s \neq j}}^{d} \left(\widetilde{F}_s(2\pi) - \widetilde{F}_s(0) \right) \right|. \tag{14}$$

Passing to the limit as $m \to +\infty$ and taking into account (13), we get

$$R(Y; F, T) \geq$$

$$\geq \left| \sum_{j=1}^{d} \tau_j \mathrm{sgn} \left(\widetilde{F}_j(2\pi) - \widetilde{F}_j(0) \right) \prod_{\substack{s=1 \\ s \neq j}}^{d} \left(\widetilde{F}_s(2\pi) - \widetilde{F}_s(0) \right) \right| = \sum_{j=1}^{d} \tau_j \prod_{\substack{s=1 \\ s \neq j}}^{d} \left| \widetilde{F}_s(2\pi) - \widetilde{F}_s(0) \right|.$$

Taking into account (11), (12) and the fact, that $\sum_{k=1}^{n_j} \alpha_k \psi_k^j \in Q^1(A_{n_j}^j)$, $\alpha_1, \ldots, \alpha_{n_j} \in$ R, we will obtain the inequality $\tau_j = R\left(Y_j; F_j, \psi_1^j, \ldots, \psi_{n_j}^j\right) \geq R\left(Y_j; F_j; A_{n_j}^j\right)$. Then, using monotonicity of $\widetilde{F}_1, \ldots, \widetilde{F}_d$, we get

$$R(Y; F, T) \geq \sum_{j=1}^{d} \tau_j \prod_{\substack{s=1 \\ s \neq j}}^{d} \bigvee_0^{2\pi} \widetilde{F}_s \geq \sum_{j=1}^{d} R\left(Y_j; F_j; A_{n_j}^j\right) \prod_{\substack{s=1 \\ s \neq j}}^{d} \|F_s\|. \tag{15}$$

From (10) and (15) we conclude, that for any algorithm $T \in Q^d\left(A_{n_1}^1, \ldots, A_{n_d}^d\right)$ the inequality $R(Y; F, T) \geq R(Y; F, S^*)$ holds, that is S^* is $(A_{n_1}^1, \ldots, A_{n_d}^d; F; Y)$-optimal. From (10) and the fact that (15) holds for S^* as well, we obtain (9). Theorem 1 is proved.

From theorem 1 and known results on optimization of quadrature formulae for classes of univariate functions we get the following result. Let $r_1, \ldots, r_d \in$ N. Denote by $W_\infty^{r_1,\ldots,r_d}$ – the set of functions $f \in C_d$, having for $j = 1, ..., d$ generalized derivatives $\frac{\partial^{r_j} f}{\partial x_j^{r_j}}$, satisfying the following condition

$$\|\frac{\partial^{r_j} f}{\partial x_j^{r_j}}\| = \sup_{(x_1,\ldots,x_d) \in [0,2\pi]^d} \mathrm{vrai} \left| \frac{\partial^{r_j} f}{\partial x_j^{r_j}}(x_1, \ldots, x_d) \right| \leq 1.$$

Denote by $W^{r_1,\ldots,r_d} H^{\omega_1,\ldots,\omega_d}$ the class of functions $f \in C_d$ such that $\frac{\partial^{r_j} f}{\partial x_j^{r_j}} \in C_d$ and

$$\left|\frac{\partial^{r_j} f}{\partial x_j^{r_j}}(x) - \frac{\partial^{r_j} f}{\partial x_j^{r_j}}(y)\right| \le \omega_j\left(|x-y|\right), \quad x, y \in \mathrm{R}^{\mathrm{d}}, \quad \mathrm{j} = 1, \ldots, \mathrm{d}.$$

These classes are interseting because, as it is known, the existence of the derivatives $\frac{\partial^{r_1} f}{\partial x_1^{r_1}}, \ldots, \frac{\partial^{r_d} f}{\partial x_d^{r_d}}$ implies the existence of all mixed derivatives $\frac{\partial^{i_1+\ldots+i_d} f}{\partial x_1^{i_1} \ldots \partial x_d^{i_d}}$, where $i_j \ge 0$, $\frac{i_1}{r_1} + \ldots + \frac{i_d}{r_d} < 1$.

Theorem 2. *Let* $d \in \mathrm{N}$, $d > 1$, $\overline{n} = (n_1, \ldots, n_d) \in \mathrm{N}^{\mathrm{d}}$. *The cubature formula* $q_{\overline{n}}^*(f) = (2\pi)^d (\prod_{s=1}^d n_s^{-1}) \sum_{i_1=0}^{n_1-1} \ldots \sum_{i_d=0}^{n_d-1} f\left(2\pi i_1 n_1^{-1}, \ldots, 2\pi i_d n_d^{-1}\right)$ *is* $Q_{\overline{n}}^d$*-optimal for the class* $W_\infty^{r_1,\ldots,r_d}$, $r_1, \ldots, r_d \in \mathrm{N}$, *and for the class* $W^{r_1,\ldots,r_d} H^{\omega_1,\ldots,\omega_d}$, *where* r_j *is odd,* ω_j *is convex,* $j = 1, \ldots, d$. *Let also* $\overline{h} = (h_1, \ldots, h_d) \in \left(0, \pi n_1^{-1}\right) \times \ldots \times \left(0, \pi n_d^{-1}\right)$. *The formula* $q_{\overline{n},\overline{h}}^*(f) = (2\pi)^d (\prod_{s=1}^d n_s^{-1}) \sum_{i_1=0}^{n_1-1} \ldots \sum_{i_d=0}^{n_d-1} I_{\overline{h}}\left(f; 2\pi i_1 n_1^{-1}, \ldots, 2\pi i_d n_d^{-1}\right)$ *is* $Q_{\overline{n},\overline{h}}^d$*-optimal for the classes* $W_\infty^{r_1,\ldots,r_d}$, $r_1, \ldots, r_d \in \mathrm{N}$, *and for the classes* $H^{(\omega_1,\ldots,\omega_d)}$.

For the class $W_\infty^{1,\ldots,1}$ $Q_{\overline{n}}^d$-optimality of $q_{\overline{n}}^*$ was in particular proved in [6].

Theorem 1 can be used for extending results on optimization of one-dimensional quadrature formulae with non-overlapping intervals or fixed nodes to the case of cubature formulae with non-overlapping node parallelepipeds or nodes at a fixed net. One can also use it to obtain optimal cubature formulae for some classes of the form $(Y_1, \ldots, Y_d)$ where optimal quadrature formulae are known for classes Y_j, $j = 1, \ldots, d$.

Let $n \in \mathrm{N}$, $h \in (0, \pi n^{-1})$, $q_n^*(f) = 2\pi n^{-1} \sum_{k=0}^{n-1} f\left(2\pi k n^{-1}\right)$, $q_{n,h}^*(f) = 2\pi n^{-1} \sum_{k=0}^{n-1} I_h\left(f; 2\pi k n^{-1}\right)$. Q_n^1-optimality of q_n^* was proved in [8,10] for the classes W_∞^r and in [10] – for $W^r H^\omega$, r is odd, ω is convex. $Q_{n,h}^1$-optimality of $q_{n,h}^*$ was proved for the classes W_1^1 in [18], for W_∞^r, $r > 1$, – in [12] and for H^ω – in [5]. Theorem 2 follows from these results and theorem 1. To prove Theorem 2 it is sufficient to check all the conditions of theorem 1 for the proof.

Results, obtained in this paper were announced in [2].

References

1. V.F. Babenko, Approximations, widths and optimal quadrature formulae for classes of periodic functions with rearrangement invariant sets of derivatives, *Anal. Math.* **13** (4) (1987) 281–306.
2. V. F. Babenko, S.V. Borodachov, On optimization of approximate integration of certain classes of multivariate periodic functions, Abstract of talk. *International Conference on Functional Analysis and Its Applications, dedicated to the 110-th anniversary of Stefan Banach.* – Lviv, Ukraine. (2002) 17–18.

3. N.S. Bakhvalov, On the optimality of linear methods of operator approximation in convex classes of functions, *USSR Comput. Math. Math. Phys.* **11** (1971) 244–249.
4. B.D. Bojanov, Characterization and existence of optimal quadrature formulae for one class of differentiable functions, Dokl. Akad. Nauk SSSR **232** (6) (1977) 1233–1236 (Russian).
5. S.V. Borodachov, On optimization of "interval" quadrature formulae for certain classes of absolutely continuous functions, *Vestnik of the Dnepropetrovsk Univ.* **5** (2000) 28–34 (Russian).
6. N.P. Korneychuk, Best cubature formulae for certain classes of functions of several variables, *Mat. Zametki* **3** (5) (1968) 565–576 (Russian). Translared in *Math Notes* **3** (1968) 360–367.
7. M.I. Levin, Yu.M. Girshovich, Extremal problems for cubature formulae, *Dokl. Akad. Nauk SSSR* **236** (6) (1977) (Russian).
8. N.E. Lushpay, Best quadrature formulae for classes of differentiable periodic functions, *Mat. Zametki* **6** (4) (1969) 475–481. Translated in *Math. Notes* **6** (1969) 740–744.
9. A.A. Lygun, Exact inequalities for spline-functions and best quadrature formulae for some classes of functions, *Mat. Zametki* **19** 6 (1976) 913–926 (Russian).
10. V.P. Motornyi, On the best quadrature formula of the form $\sum_{k=1}^{n} p_k f(x_k)$ for some classes of differentiable periodic functions, *Izv. Akad. Nauk SSSR Ser. Mat.* **38** (3) (1974) 583–614 Translated in *Math. USSR Izv.* **8** (1974) 591–620.
11. V.P. Motornyi, Investigations of Dnepropetrovsk Mathematicians on the Optimization of Quadrature Formulas, *Ukr. Mat. Zhurn.* **42** (1) (1990) 18–33 (Russian). Translated in *Ukrainian Math. J.* **42** (1) (1990) 13–27.
12. V.P. Motornyi, On the best interval quadrature formula in the class of functions with bounded r-th derivative, *East J. Approx.* **4** (4) (1998) 459–478.
13. T.T. Hoa Nguen, Best quadrature formulae and methods of recovery of functions, defined with the help of kernels not increasing the oscillation, *Mat. Sbornik* **130** (1) (1986) 105–119 (Russian).
14. S.M. Nikol'skiy, To the question on estimates of approximation by quadrature formulae, *Uspekhi Mat. Nauk* **5** 2(36) (1950) 165–177 (Russian).
15. S.M. Nikol'skiy, *Quadrature Formulae*. 4th ed., Nauka, Moscow, 1988 (Russian).
16. E. Novak, K. Ritter, High dimensional integration of smooth functions over cubes, *Numer. Math.* **75** (1) (1996) 79–97.
17. K.I. Oskolkov, On optimality of the quadrature formula with equidistant nodes on classes of periodic functions, *Dokl. Akad. Nauk SSSR* **249** (1) (1979) 49-52 (Russian).
18. R.N. Sharipov, *Best interval quadrature formulae for Lipschitz classes.* In: Constructive Function Theory and Functional Analysis, Kazan State University, Kazan, 1983, Issue 4, 124–132 (Russian).
19. S.A. Smolyak, Quadrature and interpolation formulae for tenzor products of certain classes of functions, *Dokl. Akad. Nauk SSSR* **148** (5) (1963) 1042–1045 (Russian).

20. S.A. Smolyak, *On the optimal recovery of functions and functionals of them*, Kand. Dissertation, Moscow State University, Moscow, 1965 (Russian).
21. S.L. Sobolev, *Introduction into the theory of cubature formulae*. Nauka, Moscow, 1974 (Russian).
22. V.N. Temlyakov, Approximate recovery of periodic functions of several variables, *Mat. Sbornik* **128** (2) (1985) 256–268 (Russian).
23. A.A. Zhensykbaev, Best quadrature formula for certain classes of periodic differentiable functions, *Izv. Akad. Nauk SSSR, Ser. Mat.* **41** (5) (1977) 1110–1124 (Russian).
24. A.A. Zhensykbayev, Monosplines of minimal norm and best quadrature formulae, *Uspekhi Mat. Nauk* **36** 4(220) (1981) 107–159.

Functional Analysis and its Applications
V. Kadets and W. Żelazko

Topological spaces with the strong Skhorokhod property, II

Taras Banakh

Lviv National University, Department of Mechanics and Mathematics, Universytetska 1, 79000, Ukraine

Vladimir I. Bogachev

Moscow State University, Department of Mechanics and Mathematics, 119899 Moscow, Russia

Alexander .V. Kolesnikov

Moscow State University, Department of Mechanics and Mathematics, 119899 Moscow, Russia

Abstract

A topological space X has the (uniformly tight) strong Skorokhod property if each probability Radon measure μ on X can be expressed as the image of Lebesgue measure under a Borel map $\xi_\mu : [0,1] \to X$ so that the correspondence $\mu \mapsto \xi_\mu$ takes (uniformly tight) weakly convergent sequences (μ_n) of measures on X to function sequences (χ_{μ_n}) that converge almost surely. It is shown that a generalized metric space X possessing a suitable algebraic structure has the uniformly tight strong Skorokhod property if and only if X is almost metrizable in the sense that X is the image of a metric space under a continuous bijective proper map. In particular, a strict inductive limit L of metrizable linear topological spaces has the (uniformly tight) strong Skorokhod property if and only if L is metrizable. The same equivalence holds for bornological locally convex spaces possessing a countable fundamental system of bounded sets. A Banach space X endowed with the weak topology has the (uniformly tight) strong Skorokhod property if and only if X is finite-dimensional (has the Shur property).

Key words: Weak convergence of measures, Skorokhod parametrization, metric space, topological group, locally convex space
2000 MSC: 30C45 28A33, 28C10, 28C15, 54D55, 54E35, 54H11, 60B05, 60B10, 60B11, 60B15

Introduction

This work can be viewed as a continuation of the papers [8] and [9] devoted to studying the strong Skorokhod property in non-metrizable topological spaces. Informally, the (uniformly tight) strong Skorokhod property allows to parametrize probability Radon Borel measures μ on a topological space X by Borel mappings $\xi_\mu : [0,1] \to X$ so that μ are images of Lebesgue measure under ξ_μ, and the correspondence $\mu \mapsto \xi_\mu$ takes (uniformly tight) weakly convergent sequences (μ_n) of measures on X to functions sequences (ξ_{μ_n}) that converge almost surely.

The study of the strong Skorokhod property was initiated by A.V. Skorokhod in his seminal paper [39] and then continued by many authors, see [13], [15], [16], [19], [21], [24], [35]. It was shown in [15] that each metrizable space has the strong Skorokhod property. In [8] this result was extended to the class of so-called almost metrizable spaces (i.e., images of metric spaces under bijective continuous proper maps). In this paper, we shall show that in certain classes of generalized metric spaces endowed with a suitable algebraic structure the almost metrizability is equivalent to the uniformly tight strong Skorokhod property.

To prove such an equivalence, in Section 2 we define a relatively simple space, called the Cantor fan, and prove that spaces containing a copy of the Cantor fan fail to have the uniformly tight strong Skorokhod property. The Cantor fan is a subspace of the product $2^\omega \times S(\omega)$ of the Cantor cube 2^ω and the Fréchet–Urysohn fan $S(\omega)$ and contains copies of these spaces. For topological groups or convex subsets in linear topological spaces (more generally, for spaces admitting a sequentially continuous binary operation with sequentially continuous cancellation) the converse statement is true: such a space contains a copy of the Cantor fan if and only if it contains copies of the Cantor cube and the Fréchet–Urysohn fan. Finally, using the fact that in certain classes of generalized metric spaces endowed with a suitable algebraic structure the almost metrizability is equivalent to the absence of a copy of the Fréchet–Urysohn fan, we obtain the required equivalence of the almost metrizability and the uniformly tight strong Skorokhod property.

The paper is organized as follows. Section 1 contains some preliminaries. In Section 2, we introduce the Cantor fan, characterize the topological spaces containing a copy of the Cantor fan, and prove that such spaces fail to have the strong Sko-

---

Email addresses: tbanakh@franko.lviv.ua (Taras Banakh), vbogach@mech.math.msu.su (Vladimir I. Bogachev).

rokhod property for uniformly tight weakly convergent sequences of measures with compact support.

In Section 3, we prove that a topological space endowed with a sequentially continuous binary operation with sequentially continuous cancellation contains a copy of the Cantor fan if and only if it contains copies of the Cantor cube and the Fréchet–Urysohn fan. This result is then applied to study strong Skorokhod properties in topological groups and convex (symmetric) subsets of linear topological spaces.

Finally, in Section 4 we study strong Skorokhod properties in locally convex spaces. A special attention is paid to the strict inductive limits and spaces possessing a countable fundamental system of bounded subsets. Also we study the strong Skorokhod property of Banach spaces endowed with the weak topology and show that the weak topology of a Banach space X has the (uniformly tight) strong Skorokhod property if and only if the Banach space is finite-dimensional (has the Shur property).

1. Preliminaries

All spaces considered in this paper are Tychonoff and all maps (unlike to functions) are continuous. For a subset A of a topological space X let $cl_X(A)$ or $\bar{A}$ denote the closure of A in X. We identify cardinals with initial ordinals; ω and ω_1 denote the smallest infinite and uncountable ordinals, respectively.

Given a topological space X let $\mathcal{P}(X)$ denote the space of all probability Radon Borel measures on X (that is, σ-additive measures defined on the σ-algebra of Borel subsets of X). Recall that a Borel measure μ on X is *Radon* if for every Borel subset $B \subset X$ and every $\epsilon > 0$ there is a compact subset $K \subset B$ with $\mu(B \setminus K) < \epsilon$. The space $\mathcal{P}(X)$ is equipped with the topology generated by the sub-base consisting of the sets $\{\mu \in \mathcal{P}(X) : \mu(U) > a\}$ where $a \in [0,1]$ and U runs over open subsets of X. It is well-known that this topology coincides with the weak topology generated by the seminorms $\mu \mapsto \left|\int_X f\, d\mu\right|$, where f is a bounded continuous function on X. Given a Borel function $f : X \to Y$ between topological spaces and a Borel measure μ on X let $\mu \circ f^{-1}$ be the Borel measure on Y defined by $\mu \circ f^{-1} = \mu(f^{-1}(B))$ for any Borel subset $B \subset X$.

A family $\mathcal{M}$ of non-negative Borel measures on a topological space X is called *uniformly tight* if, for every $\varepsilon > 0$, there exists a compact set $K_\varepsilon \subset X$ such that $\mu(X \backslash K_\varepsilon) < \varepsilon$ for all $\mu \in \mathcal{M}$. A topological space X is defined to be *sequentially Prokhorov* if every sequence of Radon probability measures on X that converges weakly to a Radon measure is uniformly tight.

We shall say that a family $\mathcal{M}$ of Borel probability measures on a topological space X has the *strong Skorokhod property* if, to every measure $\mu \in \mathcal{M}$, one can associate a Borel function $\xi_\mu \colon [0,1] \to X$ with $\lambda \circ \xi_\mu^{-1} = \mu$, where λ is Lebesgue measure, such that if a sequence of measures $\mu_n \in \mathcal{M}$ converges weakly to a measure $\mu \in \mathcal{M}$, then

$$\lim_{n\to\infty} \xi_{\mu_n}(t) = \xi_\mu(t) \quad \text{for almost all } t \in [0,1]. \tag{1.1}$$

If (1.1) holds under the additional assumption that $\{\mu_n\}$ is uniformly tight, then $\mathcal{M}$ is said to have the *uniformly tight strong Skorokhod property.*

We shall say that a topological space X has *the (uniformly tight) strong Skorokhod property for Radon measures* if the family $\mathcal{P}_r(X)$ of all Radon measures has that property. If each uniformly tight collection of Radon probability measures on X has the strong Skorokhod property, then we shall say that the space X has the *strong Skorokhod property for uniformly tight families of Radon measures.* In a similar manner we define also *the strong Skorokhod property for uniformly tight weakly convergent sequences of measures with compact support.* We shall use the terms *Skorokhod parameterization* and *Skorokhod representation* for mappings $\mu \mapsto \xi_\mu$ of the type described in this definition.

The following fundamental theorem was proven in [15, 3.1].

Theorem 1.1. Each metrizable space has the strong Skorokhod property for Radon measures.

In particular, each metrizable linear topological space has the strong Skorokhod property for Radon measures. This theorem fails beyond the class of metrizable spaces: the locally convex inductive limit $\mathbb{R}_0^\infty = \operatorname{ind}\mathbb{R}^n$ of finite-dimensional Euclidean spaces (in some sense, the simplest non-metrizable locally convex space) fails to have even the strong Skorokhod property for uniformly tight weakly convergent sequences of measures with compact support, [15, 4.1]. On the other hand, the space $\mathbb{R}_0^\infty$ has a weaker property called the weak Skorokhod property in [15] and [10]. A topological space X is defined to have the *weak Skorokhod property* if to each measure $\mu \in \mathcal{P}(X)$ one can associate a Borel function $\xi_\mu : [0,1] \to X$ such that $\mu = \lambda \circ \xi_\mu^{-1}$ and for every uniformly tight sequence of measures $(\mu_n) \subset \mathcal{P}(X)$ the function sequence (ξ_{μ_n}) contains a subsequence that converges almost surely.

It was shown in [10] that the weak Skorokhod property is preserved by monotone compact-covering maps. A map $f : X \to Y$ between topological spaces is called *monotone compact-covering* if to each compact subset K of Y one can assign a compact subset $l(K)$ of X so that $f(l(K)) = K$ and the function $l : K \mapsto l(K)$ is monotone in the sense that $l(K) \subset l(K')$ for any compact subsets $K \subset K'$ of Y.

A topological space X is defined to be *monotone $\aleph$-space* if X is the image of a metric space under a continuous monotone compact-covering map. The class of monotone $\aleph$-spaces is closed under many topological operations, see [10]. For such spaces we have the following result proven in [10], see also [24].

Theorem 1.2. Each monotone $\aleph$-space has the weak Skorokhod property. Moreover, for each uniformly tight sequence $(\mu_n) \subset \mathcal{P}(X)$ of Radon measures weakly convergent to a Radon measure $\mu_0 \in \mathcal{P}(X)$ there is a sequence $\xi_n : [0,1] \to X$,

$n \geq 0$, of Borel functions such that $\mu_n = \lambda \circ \xi_n$ for each $n \geq 0$ and the function sequence (ξ_n) contains a subsequence that converges almost surely to the function ξ_0.

Observe that the property described in the second assertion of Theorem 1.2 is a weakening of the strong Skorokhod property for uniformly tight weakly convergent sequences of Radon measures.

It is clear that the above defined Skorokhod properties have a sequential nature. So we recall some concepts related to sequences and sequentiality.

A subset U of a topological space X is called a *sequential neighborhood* of a point $x \in X$ if for each sequence $(x_n) \subset X$ convergent to x there is $n \in \mathbb{N}$ such that $x_m \in U$ for all $m \geq n$ (in [25] sequential neighborhoods are called *sequential barriers*). A subset U of X is *sequentially open* if U is a sequential neighborhood of every point $x \in U$. Passing to complements, we obtain the definition of a sequentially closed subset. Equivalently, a sequentially closed set can be defined as a subset $F \subset X$ containing all limit points of sequences $(x_n) \subset F$, convergent in X. The collection of all sequentially open subsets of a topological space X form a topology. The space X endowed with this topology is called the *sequential coreflection* of X and is denoted by σX.

A topological space X is called a *Fréchet–Urysohn* space if for each subset $A \subset X$ and each point $a \in \bar{A}$, there is a sequence $(a_n) \subset A$ convergent to a. It is easy to see that a space X is Fréchet–Urysohn if and only if each sequential neighborhood of a point $x \in X$ is a neighborhood of x.

If each sequentially open subset of a space X is open in X, then X is called a *sequential* space. It is clear that a space X is sequential if and only if $X = \sigma X$.

A topological space X is called a *strong Fréchet space* (or else *a countably bisequential space* [25]) if for each decreasing sequence (A_n) of subsets of X and a point $a \in \bigcap_{n=1}^{\infty} \bar{A}_n$ there is a sequence (a_n) convergent to a and such that $a_n \in A_n$ for all n, see [30].

Following [2] and [3] we say that a topological space X is an α_4*-space* if for any countable family $\{\xi_n\}$ of sequences convergent to a point $x \in X$, there is a sequence η that converges to x and meets infinitely many sequences ξ_n.

Lemma 1.3 ([25, p.191]). A topological space is strong Fréchet if and only if it is a Fréchet–Urysohn α_4-space.

A function $f : X \to Y$ between topological spaces is *sequentially continuous* if for each convergent sequence $(x_n) \subset X$ the sequence $(f(x_n))$ is convergent and $\lim f(x_n) = f(\lim x_n)$. It is clear that each sequentially continuous function defined on a sequential space is continuous. A bijective function $f : X \to Y$ is *a sequential homeomorphism* if both f and f^{-1} are sequentially continuous. Two spaces X and Y are sequentially homeomorphic if there is a sequential homeomorphism $h : X \to Y$. For example, any topological space X is sequentially homeomorphic to its sequential coreflection σX. Moreover, two topological spaces are sequentially homeomorphic if and only if their sequential coreflections σX and σY are homeomorphic.

Two sequential spaces — the Fréchet–Urysohn fan $S(\omega)$ and the Arens fan S_2 — are of crucial importance in the theory of generalized metric spaces, see [25], [26], [31], [32]. These fans are defined as follows. Given a cardinal τ, take any point $\infty \notin \tau \times \omega$ and consider the space $S(\tau) = \{\infty\} \cup (\tau \times \omega)$ with a unique non-isolated point ∞ having neighborhoods of the form $\{\infty\} \cup \{(\alpha, n) : n > f(\alpha),\ \alpha \in \tau\}$ where $f : \tau \to \omega$ is a function. Equivalently, $S(\tau)$ can be defined as a quotient space of the discrete sum of τ convergent sequences whose limit points are glued together. It can be shown that for each compact subset $K \subset S(\tau)$ there is a finite subset $F \subset \tau$ with $K \subset S(F) \subset S(\tau)$. The space $S(\omega)$ is called the *Fréchet–Urysohn fan* (or else, the *sequential fan*), see [6].

The *Arens fan* S_2 is the subset $S_2 = \{(0,0), (\frac{1}{i}, 0), (\frac{1}{i}, \frac{1}{j}) : i, j \in \mathbb{N}\}$ of the plane equipped with the strongest topology inducing the original (plane) topology on each compact set $K_n = \{(0,0), (\frac{1}{k}, 0), (\frac{1}{i}, \frac{1}{j}) : k, j \in \mathbb{N},\ i \le n\}$, $n \in \mathbb{N}$, see [1]. The Fréchet-Urysohn fan is an example of a countable Fréchet–Urysohn space that is not an α_4-space while the Arens fan is a sequential α_4-space that is not Fréchet–Urysohn.

Lemma 1.4 ([25, 3.6]). A topological space X contains a subspace sequentially homeomorphic to the Fréchet–Urysohn fan $S(\omega)$ if and only if X is not an α_4-space. Moreover, if a space X is a subspace of a regular space Y with countable pseudocharacter, then X is not an α_4-space if and only if X contains a closed in Y subspace W, sequentially homeomorphic to the Fréchet–Urysohn fan.

Lemma 1.5 ([25, 3.7]). A topological space X is sequentially homeomorphic to a strong Fréchet space if and only if X contains no subspace sequentially homeomorphic to the Fréchet–Urysohn or Arens fans.

A collection $\mathcal{N}$ of subsets of a topological space X is called a *network* if for any open subset $U \subset X$ and any point $x \in U$ there is a subset $N \in \mathcal{N}$ such that $x \in N \subset U$. A collection $\mathcal{N}$ of subsets of X is called a cs*-*network* at a point $x \in X$ if for every neighborhood $U \subset X$ of x and every sequence $(x_n) \subset X$ convergent to x there is an element $N \in \mathcal{N}$ such that $x \in N \subset U$ and $x_n \in N$ for infinitely many n. The smallest size $|\mathcal{N}|$ of a cs*-network $\mathcal{N}$ at a point $x \in X$ is called the cs*-*character of* X *at* x and is denoted by $\mathrm{cs}^*(X, x)$. The cardinal $\mathrm{cs}^*(X) = \sup_{x \in X} \mathrm{cs}^*(X, x)$ is called *the* cs*-*character* of X. It is shown in [11] that each non-metrizable sequential topological group with countable cs*-character has the countable pseudocharacter and contains an open k_ω-subgroup.

Lemma 1.6 ([25, 2.8 and 3.18]). A topological space X is sequentially homeomorphic to a first countable space if and only if X has countable cs*-character and contains no subspace sequentially homeomorphic to the Fréchet–Urysohn or Arens fans.

Monotone $\aleph$-spaces often have the countable cs*-character.

Lemma 1.7 ([10, 3.18]). A monotone $\aleph$-space X has the countable cs*-character provided X contains no sequentially closed subspace sequentially homeomorphic to $S(\omega_1)$.

Following [15] and [8] we say that a topological space X is *almost metrizable* if there is a bijective continuous proper map $f : M \to X$ from a metrizable space M onto X. If the space M is discrete, then X is called an *almost discrete space.* We remind that a map $f : X \to Y$ is *proper* if the preimage of each compact subset is compact. It is easy to see that a topological space X is almost metrizable if and only if each compact subset of X is sequentially compact and X is sequentially homeomorphic to a metrizable space. A space X is almost discrete if and only if X contains no infinite compact subset.

We recall that a topological space is *sequentially compact* if each sequence $(x_n) \subset X$ contains a convergent subsequence. It is well-known that each compact sequential space is sequentially compact. In the sequel we shall often use the following well known result, see [28, 6.2].

Lemma 1.8. Each uncountable compact metrizable space contains a subspace homeomorphic to the Cantor cube $\{0,1\}^\omega$.

Generalizing Theorem 1.1 the authors proved the following

Theorem 1.9 ([8]). Each almost metrizable space has the uniformly tight strong Skorokhod property for Radon measures. Moreover, an almost metrizable space has the strong Skorokhod property for Radon measures if and only if it is sequentially Prokhorov.

2. The failure of the strong Skorokhod property for the Cantor fan

In this section we define a relatively simple space CV called *the Cantor fan* and shall prove that it fails to have the strong Skorokhod property for uniformly tight weakly convergent sequences of measures with compact support. Since CV is a closed subspace of the product $2^\omega \times S(\omega)$ of the Cantor cube and the Fréchet–Urysohn fan we can conclude that the strong Skorokhod property for uniformly tight families of Radon measures (or even for uniformly tight weakly convergent sequences of measures with compact support) is not multiplicative.

We identify the number 2 with the discrete two-point space $\{0,1\}$. Let $T = \bigcup_{0\le n\le\omega} 2^n$ be the full binary tree, i.e., the set of all (finite and infinite) binary sequences, endowed with the natural partial order: $(a_i)_{i\le n} \le (b_i)_{i\le m}$ if and only if $n \le m$ and $a_i = b_i$ for each $i \le n$. Hence $a \le b$ if a coincides with an initial segment of b. It is clear that any two points $a, b \in T$ have the maximal lower bound $\min(a,b)$ (namely, their largest common initial segment which can be empty) which will be denoted by $a \wedge b$. For a sequence $a \in T$ by $|a|$ we denote its length. Let 2^ω and $2^{<\omega}$ be the sets of all infinite and finite binary sequences, respectively.

On the tree T we consider the interval topology generated by the sub-basis consisting of open rays $(\leftarrow, a) = \{x \in T : x < a\}$ and $(a, \rightarrow) = \{x \in T : x > a\}$, where $a \in T$. It is clear that T endowed with this topology is a metrizable compactum, each point $a \in 2^{<\omega}$ is isolated, 2^ω is a closed nowhere dense subset of T, and the topology on 2^ω coincides with the standard product topology of the Cantor cube. To each sequence $a \in T$ assign a unique infinite sequence $\bar{a} \in 2^\omega$ such that $\bar{a}_i = a_i$ for $i < |a|$ and $\bar{a}_i = 0$ for $i \ge |a|$. It is clear that the map $r : T \to 2^\omega$ defined by $r(a) = \bar{a}$ is a continuous retraction of T onto 2^ω and the image $r(2^{<\omega})$ coincides with the set 2^ω_0 of all eventually zero sequences in 2^ω.

The Cantor fan CV is the quotient space of the product $T \times \omega$ under the equivalence relation: $(a,n) \sim (b,m)$ if and only if $(a,n) = (b,m)$ or $a = b \in 2^\omega$. Thus CV looks like a book whose pages $T \times \{n\}$, $n \in \omega$, are glued together by the common Cantor set 2^ω. From a more formal point of view the Cantor fan can be written as the union $CV = 2^\omega \cup (2^{<\omega} \times \omega)$ where for every $n \in \omega$ the set $CV_n = 2^\omega \cup (2^{<\omega} \times \{0,\dots,n\})$ is a metrizable compactum, and a subset $U \subset CV$ is open in CV if and only if $U \cap CV_n$ is open in CV_n for every n. This implies, in particular, that CV is a sequential space and each compact subset $K \subset CV$ is contained in some CV_n (of course depending on K).

Sometimes it will be convenient to think of the Cantor fan CV as a closed subspace of the product $2^\omega \times S(\omega)$ of the Cantor cube and the Fréchet–Urysohn fan. A closed embedding $e : CV \to 2^\omega \times S(\omega)$ can be defined by the formula

$$e(x) = \begin{cases} (x, \infty) & \text{if } x \in 2^\omega \subset CV; \\ (\bar{a}, (n, |a|)) & \text{if } x = (a,n) \in 2^{<\omega} \times \omega \subset CV. \end{cases}$$

In the sequel we shall identify the Cantor fan CV with its image $e(CV)$ in $2^\omega \times S(\omega)$. By $\mathrm{pr} : CV \to 2^\omega$ we denote the restriction on CV of the projection $2^\omega \times V \to 2^\omega$ onto the first factor. Observe that $\mathrm{pr}^{-1}(a) = \{a\}$ if $a \in 2^\omega \setminus 2^\omega_0$ and $\mathrm{pr}^{-1}(a)$ is homeomorphic to the Fréchet–Urysohn fan for every $a \in 2^\omega_0$. Thus the Cantor fan consists of the Cantor set 2^ω and countably many closed topological copies of the Fréchet–Urysohn fan.

In Proposition 5.12 of [8] it was proved that the Fréchet–Urysohn fan has the strong Skorokhod property for uniformly tight families of Radon measures. In spite of some resemblance between the Cantor and Fréchet–Urysohn fans, their Skorokhod properties differ substantially.

Theorem 2.1. The Cantor fan fails to have the strong Skorokhod property for uniformly tight weakly convergent sequences of measures with compact support.

Proof. Notice that a sequence $(x_n)_{n\in\mathbb{N}}$ of a topological space X converges to a point $x_0 \in X$ if and only if for each neighborhood $O(x_0) \subset X$ of x_0 there is a finite subset $F \subset \mathbb{N}$ such that $x_n \in O(x_0)$ for each $n \in \mathbb{N} \setminus F$. This simple observation allows us to index sequences by an arbitrary countable set (not necessarily by $\mathbb{N}$).

The convergent sequence of measures on the Cantor fan without the Skorokhod parametrization which we are going to construct will be parametrized by the index set $\mathcal{I} = \{(a,i) \in 2^{<\omega} \times \omega : |a| < i\}$. To each $(a,i) \in \mathcal{I}$ assign the measure $\mu_{(a,i)} = \sum_{b\in 2^i} \frac{1}{2^i}\delta_{(b,|a\wedge b|)}$ with finite support $\operatorname{supp}(\mu_{(a,i)}) = \{(b,|a\wedge b|) : b \in 2^i\}$ on $CV_{|a|} \subset CV$.

Claim A. *The set $\{\mu_{(a,i)} : (a,i) \in \mathcal{I}\}$ is uniformly tight.*

This is so because $\mu_{(a,i)}(CV_n) \geq 1 - \frac{1}{2^n}$ for every $n \in \mathbb{N}$ and every $(a,i) \in \mathcal{I}$ (to prove this consider separately two cases: $|a| \leq n$ and $|a| > n$).

Claim B. *The sequence $(\mu_{(a,i)})_{(a,i)\in\mathcal{I}}$ converges to the standard product measure μ on 2^ω.*

It suffices to verify that for every open set $U \subset CV$ and every $n \in \mathbb{N}$ the inequality $\mu_{(a,i)}(U) > \mu(U) - \frac{2}{2^n}$ holds for all but finitely many indices $(a,i) \in \mathcal{I}$.

By the regularity of the measure μ one can find a compact subset $K \subset U$ with $\mu(K) > \mu(U) - \frac{1}{2^n}$. Since the Cantor cube 2^ω carries the product topology, there is $m > n$ and $A \subset 2^m$ such that $K \subset \operatorname{pr}_m^{-1}(A) \subset U$, where $\operatorname{pr}_m : 2^\omega \to 2^m$ stands for the projection onto the first m factors. Since μ is the product measure, $\mu(\operatorname{pr}_m^{-1}(A)) = \frac{|A|}{2^m}$.

Using the compactness of the sets $\operatorname{pr}_m^{-1}(A) \subset CV_n$, we may find a number $k > m$ such that $(b,i) \in U$ for any $(b,i) \in 2^{<\omega} \times \omega$ with $i \leq n$, $|b| \geq k$ and $\bar{b} \in \operatorname{pr}_m^{-1}(A)$.

Consider the finite subset $F = \{(a,i) \in \mathcal{I} : i \leq k\}$. We claim that $\mu_{(a,i)}(U) > \mu(U) - \frac{1}{2^{n-1}}$ for every $(a,i) \in \mathcal{I} \setminus F$. Indeed, for any given $(a,i) \in \mathcal{I} \setminus F$ we have $i > k > m$. Let $B = \{b \in 2^i : \bar{b} \in \operatorname{pr}_m^{-1}(A)\}$ and $B' = \{b \in B : |a \wedge b| \leq n\}$. It is clear that $\frac{|B|}{2^i} = \frac{|A|}{2^m}$ and $|B'| \geq |B| - |B \setminus B'| \geq \frac{2^i}{2^m}|A| - |\{b \in 2^i : |a \wedge b| > n\}| \geq \frac{2^i}{2^m}|A| - \frac{2^i}{2^n}$. Consequently,

$$\mu_{(a,i)}(U) = \frac{1}{2^i}\left|\{b \in 2^i : (b,|a\wedge b|) \in U\}\right| \geq \frac{|B'|}{2^i} \geq \frac{|A|}{2^m} - \frac{1}{2^n} = \mu(\operatorname{pr}_m^{-1}(A)) - \frac{1}{2^n} \geq \mu(K) - \frac{1}{2^n} > \mu(U) - \frac{2}{2^n}. \quad (2.2)$$

Claim C. *Suppose $(\Omega, \mathcal{F}, \lambda)$ is a probability space and $\xi_{(a,i)} : \Omega \to CV$, $(a,i) \in \mathcal{I}$, are λ-measurable maps such that $\mu_{(a,i)} = \lambda \circ \xi_{(a,i)}^{-1}$ for each $(a,i) \in \mathcal{I}$. Then the set D of all $t \in \Omega$ for which the sequence $(\xi_{(a,i)}(t))_{(a,i)\in\mathcal{I}}$ diverges has positive λ-measure.*

Suppose that $\lambda(D)=0$. Redefining the functions $\xi_{(a,i)}$ on the null set D we may assume that the sequence $(\xi_{(a,i)}(t))_{(a,i)\in\mathcal{I}}\subset CV$ converges to a point $\xi(t)\in CV$ for every $t\in\Omega$, and $\xi_{(a,i)}(\Omega)=\{(b,|a\wedge b|):b\in 2^i\}=\mathrm{supp}(\mu_{(a,i)})$ for each $(a,i)\in\mathcal{I}$. It can be shown that the function $\xi:\Omega\to CV$ is λ-measurable and $\mu=\lambda\circ\xi^{-1}$.

Let $A_n=\{t\in\Omega:\xi_{(a,i)}(t)\in CV_{n-1}$ for all $(a,i)\in\mathcal{I}\}$. Since each convergent sequence $S\subset CV$ lies in some CV_{n-1}, we get that $\Omega=\bigcup_{n\in\mathbb{N}}A_n$. Then $\lambda(A_m)>0$ for some $m\in\mathbb{N}$. Let λ_0 be the restriction of the measure λ on the set A_m, i.e., $\lambda_0(B)=\lambda(B\cap A_m)$ for any $B\in\mathcal{F}$.

Let $\tau=\lambda_0\circ\xi^{-1}$ and $\tau_{(a,i)}=\lambda_0\circ\xi^{-1}_{(a,i)}$ for $(a,i)\in\mathcal{I}$. It is clear that $\tau\le\mu$ and $\tau_{(a,i)}\le\mu_{(a,i)}$ for every $(a,i)\in\mathcal{I}$. Since the function sequence $(\xi_{(a,i)})_{(a,i)\in\mathcal{I}}$ converges to ξ on the set A_m, the sequence $(\tau_{(a,i)})_{(a,i)\in\mathcal{I}}$ converges to the measure τ. It is clear that $\tau(2^\omega)=\lambda(A_m)>0$. Consequently, there is $c\in 2^m$ such that $\tau(\mathrm{pr}_m^{-1}(c))>\frac{1}{2^n}$ for some $n\in\mathbb{N}$.

Observe that the set $U=\{x\in CV:pr_m\circ\mathrm{pr}(x)=c\}$ is an open neighborhood of $\mathrm{pr}_m^{-1}(c)$ in CV. Since the sequence $(\tau_{(a,i)})$ converges to the measure τ, there is a finite set $F\subset\mathcal{I}$ such that $\tau_{(a,i)}(U)>\tau(U)-\frac{1}{2^{n+1}}$ for each $(a,i)\in\mathcal{I}\setminus F$.

Now pick $i>m$ so that $(c,i)\in\mathcal{I}\setminus F$. Since $\tau_{(c,i)}(CV_{m-1})=1$ we have

$$\begin{aligned}\tau_{(c,i)}(CV_{m-1}\cap U)=&\tau_{(c,i)}(U)>\tau(U)-\frac{1}{2^{n+1}}\ge\\ &\ge\tau(\mathrm{pr}_m^{-1}(c))-\frac{1}{2^{n+1}}>\frac{1}{2^n}-\frac{1}{2^{n+1}}=\frac{1}{2^{n+1}}.\end{aligned}$$

Next we estimate the measure $\mu_{(c,i)}(CV_{m-1}\cap U)=\frac{1}{2^i}|\{b\in 2^i:(b,|c\wedge b|)\in CV_{m-1}\cap U\}|$. Pick any $b\in 2^i$. If $|c\wedge b|\ge m$, then $(b,|c\wedge b|)\notin CV_{m-1}$. If $|c\wedge b|<m$, then $\bar{b}\notin\mathrm{pr}_m^{-1}(c)$ and $(b,|c\wedge b|)\notin U$.

Therefore $x_{(c,b)}\notin CV_{m-1}\cap U$ for each $b\in 2^i$ which yields

$$\tau_{(c,b)}(CV_{m-1}\cap U)\le\mu_{(c,b)}(CV_{m-1}\cap U)=0,$$

a contradiction.

Our first application of Theorem 2.1 is the following important statement which can be easily derived from Theorem 2.1 and the sequentiality of the Cantor fan.

Corollary 2.2. If a space X contains a subspace sequentially homeomorphic to the Cantor fan, then X fails to have the strong Skorokhod property for uniformly tight weakly convergent sequences of measures with compact support.

Corollary 2.2 implies that the strong Skorokhod property for uniformly tight families of Radon measures as well as the strong Skorokhod property for uniformly tight weakly convergent sequences of measures with compact support are not multiplicative.

Corollary 2.3. The interval $[0,1]$ and the Fréchet–Urysohn fan $S(\omega)$ have the strong Skorokhod property for uniformly tight families of Radon measures but their product $[0,1] \times S(\omega)$ fails to have the strong Skorokhod property for uniformly tight weakly convergent sequences of measures with compact support.

Proof. It is well known that the segment $[0,1]$ has the strong Skorokhod property for Radon measures, see [15, 3.1] or [37, Ch. III. §8]. According to Proposition 2.15 of [8] the Fréchet–Urysohn fan $S(\omega)$ has the strong Skorokhod property for uniformly tight families of Radon measures. Finally, Corollary 2.2 implies that the product $[0,1] \times S(\omega)$ fails to have the strong Skorokhod property for uniformly tight weakly convergent sequences of measures with compact support.

Question 1. Does the Fréchet–Urysohn fan have the strong Skorokhod property for Radon measures?

Question 2. Suppose X, Y are topological spaces with the strong Skorokhod property for Radon measures. Does the product $X \times Y$ have that property?

Let us remark that a positive answer to the first question implies a negative answer to the second and vice versa.

In light of Corollary 2.2 it is important to detect spaces containing a subspace sequentially homeomorphic to the Cantor fan.

Theorem 2.4. A topological space X contains a subspace sequentially homeomorphic to the Cantor fan if and only if there is a countable family $\mathcal{M}$ of metrizable subspaces whose intersection $\cap\mathcal{M}$ contains an uncountable compactum $K \subset \cap\mathcal{M}$ such that for every point $x \in K$ the following two conditions are satisfied:

1) for each sequence $(x_n)_{n=1}^{\infty} \subset \cup\mathcal{M}$ convergent to x there are $M \in \mathcal{M}$ and $n_0 \in \mathbb{N}$ such that $x_n \in M$ for all $n \geq n_0$;

2) for every $M \in \mathcal{M}$ there is a sequence $(x_n)_{n=1}^{\infty} \subset \cup\mathcal{M} \setminus M$ convergent to x.

Proof. The "only if" part is trivial. To prove the "if" part, suppose that a collection $\mathcal{M}$ satisfies the conditions of the theorem.

For each $k \in \mathbb{N}$ fix a metric ρ_k generating the topology of M_k. Next, for every $p \in \mathbb{N}$ find a neighborhood $O_p(x)$ of x in $\cup\mathcal{M}$ such that $O_p(x) \cap M_k \subset \{y \in M_k : \rho_k(y,x) < \frac{1}{p}\}$ for all $k \leq p$.

Claim. *For each $x \in K$ and $n \in \mathbb{N}$ there are a number $m > n$ and a sequence $(x_k)_{k=1}^{\infty} \subset M_m \setminus \bigcup_{k=1}^{n} M_k$, convergent to x.*

Let $\alpha : \mathbb{N} \to \mathbb{N}$ be any function such that $\alpha^{-1}(n)$ is infinite for every $n \in \mathbb{N}$. We claim that there is $p \in \mathbb{N}$ such that $O_p(x) \cap (\bigcup_{k=1}^{n} M_k) \subset M_{\alpha(p)}$. Assuming the converse, for every $p \in \mathbb{N}$ find a point $y_p \in O_p(x) \cap (\bigcup_{k=1}^{n} M_k)$ with $y_p \notin M_{\alpha(p)}$. Then the sequence $(y_p)_{p=1}^{\infty}$ converges to x. By Condition 1, there are $q \in \mathbb{N}$ and

$p_0 \in \mathbb{N}$ such that $y_p \in M_q$ for all $p \geq p_0$. Since the set $\alpha^{-1}(q)$ is infinite, there is a number $p_1 \in \alpha^{-1}(q)$ with $p_1 \geq p_0$. Then $y_{p_1} \in M_q = M_{\alpha(p_1)}$ which contradicts our choice of the points y_p.

Therefore, $O_p(x) \cap (\bigcup_{k=1}^n M_k) \subset M_{\alpha(p)}$ for some $p \in \mathbb{N}$. By Condition 2, there is a sequence $(x_i)_{i=1}^\infty \subset \bigcup \mathcal{M} \setminus M_{\alpha(p)}$ convergent to x. By Condition 1, there are $m \in \mathbb{N}$ and $i_0 \in \mathbb{N}$ such that $x_i \in M_m$ for all $i \geq i_0$. It is clear that $m > n$. We can take i_0 so large that $x_i \in O_p(x)$ for all $i \geq i_0$. Then $(x_i)_{i=i_0}^\infty \subset (M_m \cap O_p(x)) \setminus M_{\alpha(p)} \subset M_m \setminus \bigcup_{k=1}^n M_k$ and the claim is proved. □

Given $x \in K$ and $n \in \mathbb{N}$ let $f(x,n)$ be the minimal number m with the property described in the Claim. It is easy to verify that for every $n, m \in \mathbb{N}$ the set $K_{n,m} = \{x \in K : f(x,n) \leq m\}$ is closed in K.

The uncountable metrizable compact space K contains a topological copy of the Cantor cube [28, 6.2] and consequently admits a probability measure μ such that $\mu(C) = 0$ for every countable set $C \subset K$. Since $K = \bigcup_{m=1}^\infty K_{n,m}$, for every $n \in \mathbb{N}$ by induction we can find a number $l(n) > l(n-1)$ such that $\mu(K_{n,l(n)}) > 1 - \frac{1}{2^{n+1}}$. Then $\mu(\bigcap_{n=1}^\infty K_{n,l(n)}) > 1 - \sum_{n=1}^\infty \frac{1}{2^{n+1}} = \frac{1}{2}$ and thus the closed subset $\bigcap_{n=1}^\infty K_{n,l(n)}$ is uncountable. Since each uncountable metrizable compact space contains a topological copy of the Cantor cube 2^ω, we can find a closed subset $X_0 \subset \bigcap_{n=1}^\infty K_{n,l(n)}$ homeomorphic to 2^ω. In the sequel we shall identify X_0 with the Cantor cube 2^ω.

Observe that for each point $x \in X_0$ and each $n \in \mathbb{N}$ there is a sequence $(x_k)_{k=1}^\infty \subset \bigcup_{k=1}^{l(n)} M_k \setminus \bigcup_{k=1}^n M_k$, convergent to x. By recursion let us define a function $g : \mathbb{N} \to \mathbb{N}$ letting $g(1) = l(1)$ and $g(n+1) = l(g(n))$ for $n \geq 1$.

Then, for each point $a \in 2^\omega = X_0$ and each $n \in \mathbb{N}$, there is a sequence $(x_k)_{k=1}^\infty \subset \bigcup_{k=1}^{g(n+1)} M_k \setminus \bigcup_{k=1}^{g(n)} M_k$, convergent to a.

This enables us to find, for every $(a,n) \in 2^{<\omega} \times \mathbb{N} \subset CV$, a point $x(a,n) \in \bigcup_{k=1}^{g(n+1)} M_k \setminus \bigcup_{k=1}^{g(n)} M_k$ such that $x(a,n) \in O_{|a|}(\bar{a})$. Without loss of generality, we may assume that $x(a,n) \neq x(b,m)$ if (a,n), (b,m) are distinct points of $2^{<\omega} \times \mathbb{N}$. Now consider the map $h : CV \to 2^\omega \cup \{x(a,n) : (a,n) \in 2^{<\omega} \times \mathbb{N}\} \subset X$ defined by $h(a) = a$ if $a \in 2^\omega$ and $h(a,n) = x(a,n)$ if $(a,n) \in 2^{<\omega} \times \mathbb{N}$. It can be shown that the map h is a sequential homeomorphism.

Theorem 2.4 implies the following useful fact.

Corollary 2.5. An uncountable image X of the Cantor fan under a continuous proper map $f : CV \to X$ contains a subspace sequentially homeomorphic to the Cantor fan.

Proof. Let $M_n = f(CV_n)$ for $n \in \mathbb{N}$. It is clear that the countable family $\{M_n\}_{n \in \mathbb{N}}$ consists of metrizable compacta and has the uncountable intersection $\bigcap_{n \in \mathbb{N}} M_n \supset f(2^\omega) \supset X \setminus f(CV \setminus 2^\omega)$. To complete the proof it suffices to verify Conditions 1 and 2 of Theorem 2.4.

3. The strong Skorokhod property in topological groups and spaces admitting a cancellative operation

It is clear that a topological space containing a subspace sequentially homeomorphic to the Cantor fan contains a subspace homeomorphic to the Cantor cube and a subspace sequentially homeomorphic to the Fréchet–Urysohn fan. We shall study the cases when the inverse statement is also true.

Under a *(sequentially) continuous binary operation* on a topological space X we understand any (sequentially) continuous function $o : X \times X \to X$. An operation $o : X \times X \to X$ is *cancellative* if $o(x, z) \neq o(y, z)$ and $o(z, x) \neq o(z, y)$ for any distinct points $x, y \in X$ and any $z \in X$. We say that $o : X \times X$ is *an operation with sequentially continuous cancellation* if for any sequences $(x_n), (y_n) \subset X$ with convergent product $(o(x_n, y_n))_{n=1}^{\infty}$ the sequence (x_n) converges in X if and only if so does the sequence (y_n).

Observe that each operation $o : X \times X \to X$ with sequentially continuous cancellation on a Hausdorff space X is cancellative. Indeed, assuming that $o(x, y) = o(x', y)$ for some $x \neq x'$, we can consider the sequences $(x_n), (y_n)$ such that $y_n = y$ for all n while $x_n = x$ for odd n and $x_n = x'$ for even n. Then the sequences (y_n) and $(o(x_n, y_n))$ converge in X while (x_n) does not. By analogy consider the case when $o(x, y) \neq o(x, y')$ for some $y \neq y'$.

Note that each space X, sequentially homeomorphic to a topological group, admits a sequentially continuous binary operation with sequentially continuous cancellation. But this operation does not need to be (and often even can not be) continuous, see [7].

Theorem 3.1. Suppose a regular topological space X admits a sequentially continuous operation $o : X \times X \to X$ with sequentially continuous cancellation. The space X contains a subspace sequentially homeomorphic to the Cantor fan if and only if X contains an uncountable compact metrizable subset K and a subspace W sequentially homeomorphic to the Fréchet–Urysohn fan.

Proof. Observe that the product $K \times o(K \times W)$ can be written as a countable union of metrizable compacta and thus has a countable network. Let $L = \{(x, z) \in K \times o(K \times W) : \exists y \in X \text{ such that } z = o(x, y)\}$. Since the operation o is cancellative we conclude that for any point $(x, z) \in L$ there is a unique point $y = l(x, z) \in X$ such that $z = o(x, y)$. Thus we define a function $l : L \to X$ (which can be thought as a left cancellation). Since o is an operation with sequentially continuous cancellation, the function l is sequentially continuous. Next, the space L, being a regular space with countable network, is the image of a separable metrizable space M under a surjective continuous map $\alpha : M \to L$, see [23, 4.9]. Then the composition $l \circ \alpha : M \to X$ is continuous and hence the image $l(L) = l \circ \alpha(M)$ has a countable network and, being regular, has the countable pseudocharacter.

By Lemma 1.4, the space $W \subset l(L)$ contains a subspace W', sequentially closed in $l(L)$ and sequentially homeomorphic to the Fréchet–Urysohn fan $S(\omega)$. Let $h : S(\omega) \to W'$ be a sequential homeomorphism. The metrizable compactum K, being uncountable, contains a topological copy $2^\omega \subset K$ of the Cantor cube. Now consider the map $f : 2^\omega \times S(\omega) \to X$ defined by $f(x, y) = o(x, h(y))$. Since the operation o is cancellative, the set $f(2^\omega \times \{\infty\})$ is uncountable. Next, we verify that the map $f : 2^\omega \times S(\omega) \to f(2^\omega \times S(\omega))$ is proper. Fix any compact subset $C \subset f(2^\omega \times S(\omega)) \subset o(K \times W)$. The compact space C has countable network and hence is metrizable [20, 3.1.19]. Assuming that the preimage $f^{-1}(C)$ is not compact, we would find an infinite closed discrete subset $\{(x_n, y_n) : n \in \mathbb{N}\} \subset f^{-1}(C)$. Passing to a suitable subsequence, we can assume that the sequences (x_n) and $(f(x_n, y_n))$ converge to some points $x_0 \in 2^\omega$ and $z_0 \in C$, respectively. By the sequential continuity of the cancellation, the sequence $(h(y_n))$ converges to some point $w_0 \in X$. Observe that $o(x_0, w_0) = o(\lim x_n, \lim h(y_n)) = \lim o(x_n, h(y_n)) = \lim f(x_n, y_n) = z_0$. Then $(x_0, z_0) \in L$ and $w_0 \in l(L)$. Since the set $W' = h(S(\omega))$ is sequentially closed in $l(L)$, we conclude that $w_0 \in W'$ and thus $w_0 = h(y_0)$ for some point $y_0 \in V$. By the sequential continuity of the inverse function h^{-1}, the sequence (y_n) converges to y_0. Then the sequence $\{(x_n, y_n)\}_{n=1}^\infty$ converges to (x_0, y_0) and can not be closed and discrete $2^\omega \times S(\omega)$. This contradiction proves that the map $f : 2^\omega \times S(\omega) \to f(2^\omega \times S(\omega))$ is proper.

Since the Cantor fan CV is a closed subspace of $2^\omega \times S(\omega)$ we can apply Corollary 2.5 to conclude that $X \supset f(2^\omega \times S(\omega)) \supset f(CV)$ contains a subspace sequentially homeomorphic to the Cantor fan.

We say that an operation $o : X \times X \to X$ on a space X is *sequentially invertible* if for every convergent sequence $(x_n) \subset X$ there are a point $e \in X$ and a sequence $(y_n) \subset X$ such that $o(x_n, y_n) = e$ for infinitely many numbers n.

It is clear that each topological group and each convex symmetric subset of a linear topological space admit a continuous sequentially invertible binary operation with sequentially continuous cancellation (in the case of a convex set X such an operation $o : X \times X \to X$ can be defined by $o(x, y) = \frac{1}{2}x + \frac{1}{2}y$). A convex subset X of a linear space is called *symmetric* if there is a point $e \in X$ such that for every $x \in X$ there is $y \in X$ with $e = \frac{1}{2}x + \frac{1}{2}y$.

The following proposition is well known for topological groups [31] and [32].

Proposition 3.2. Suppose that a Hausdorff space X admits a sequentially continuous sequentially invertible operation $o : X \times X \to X$ with a sequentially continuous cancellation. Then the space X contains a subspace sequentially homeomorphic to the Fréchet–Urysohn fan if and only if X contains a subspace sequentially homeomorphic to the Arens fan.

Proof. Assume that W is a subspace of X, sequentially homeomorphic to the Fréchet–Urysohn fan $S(\omega)$ and let $h : S(\omega) \to W$ be a sequential homeomorphism. Let $x_0 = h(\infty)$ and $x_{n,m} = h(n, m)$ for $(n, m) \in \mathbb{N} \times \mathbb{N} \subset S(\omega)$. Let $(y_n) \subset X$ be a sequence, convergent to y_0 and such that $y_n \neq y_m$ for any $n \neq m$ (for example, we can take $y_0 = x_0$ and $y_n = x_{1,n}$ for $n \geq 1$).

Next, consider the sequence $(z_n)_{n=0}^{\infty}$ where $z_n = o(x_0, y_n)$. By the sequential continuity and cancellativity of the operation o we obtain $\lim z_n = z_0$ and $z_n \neq z_m$ for any $n \neq m$. Using the Hausdorff axiom of X by induction we construct, for every $n \geq 1$, an open neighborhood U_n of z_n such that $U_n \cap U_m = \emptyset$ if $n \neq m$. By the sequential continuity of the operation o for every $n \geq 1$ find a number $m_n \in \mathbb{N}$ such that $o(x_{n,m}, y_n) \in U_n$ for all $m > m_n$. For every $n, m \in \mathbb{N}$ let $z_{n,m} = o(x_{n,m+m_n}, y_n)$. It follows from the construction and the cancellativity of the operation that $z_0 \neq z_{n,m} \neq z_{n',m'}$ for any $(n,m) \neq (n',m')$. Let $A = \{z_0, z_n, z_{n,m} : n, m \in \mathbb{N}\}$ and consider the sequential homeomorphism $h : S_2 \to A$ defined by $h(0,0) = z_0$, $h(\frac{1}{n}, 0) = z_n$, and $h(\frac{1}{n}, \frac{1}{m}) = z_{n,m}$ for $n, m \in \mathbb{N}$. Hence the subspace A of X is sequentially homeomorphic to the Arens fan. Next, assume that $h : S_2 \to W$ is a sequential homeomorphism of the Arens fan onto a subspace of X. In order to show that X contains a subspace sequentially homeomorphic to the Fréchet–Urysohn fan it suffices to verify that X is not an α_4-space, see Lemma 1.4. Let $x_0 = h(0,0)$, $x_n = h(\frac{1}{n}, 0)$, and $x_{n,m} = h(\frac{1}{n}, \frac{1}{m})$ for $n, m \in \mathbb{N}$. Since the operation o is sequentially invertible, there are $e \in X$ and a sequence $(y_n) \subset X$ such that $o(x_n, y_n) = e$ for infinitely many numbers n. Replacing W by a suitable subspace we can assume that $o(x_n, y_n) = e$ for all n. Since the sequences (x_n) and $(o(x_n, y_n))$ converge, so does the sequence (y_n). Let $y_0 = \lim y_n$ and $z_{n,m} = o(x_{n,m}, y_n)$ for $n, m \in \mathbb{N}$. By the sequential continuity of the operation o, the sequence $(z_{n,m})_{m=1}^{\infty}$ converges to e for each n. Assuming that X is an α_4-space, we can find a sequence $\{a_k\}_{k=1}^{\infty} \subset \{z_{n,m} : n, m \in \mathbb{N}\}$ converging to e and meeting infinitely many sequences $\{z_{n,m}\}_{m=1}^{\infty}$. For each k find numbers $n(k)$ and $m(k)$ such that $a_k = z_{n(k),m(k)}$. Replacing the sequence (a_k) by a suitable subsequence, we can assume that the number sequence $(n(k))_{k=1}^{\infty}$ is strictly increasing. Then the sequence $(y_{n(k)})_{k=1}^{\infty}$ converges to y_0 and consequently, the sequence $(x_{n(k),m(k)})_{k=1}^{\infty}$ converges to a point $x_\infty \in X$. Note that

$$o(x_\infty, y_0) = o(\lim_{k\to\infty} x_{n(k),m(k)}, \lim_{k\to\infty} y_{n(k)}) = \lim_{k\to\infty} o(x_{n(k),m(k)}, y_{n(k)}) =$$
$$= \lim_{k\to\infty} z_{n(k),m(k)} = \lim_{k\to\infty} a_k = e = \lim_{n\to\infty} o(x_n, y_n) = o(\lim_{n\to\infty} x_n, \lim_{n\to\infty} y_n) =$$
$$= o(x_0, y_0).$$

It follows from the cancellativity of the operation o that $x_\infty = x_0$. Hence the sequence $(x_{n(k),m(k)})_{k=1}^{\infty}$ converges to x_0 and by the sequential continuity of h^{-1} the sequence $(\frac{1}{n(k)}, \frac{1}{m(k)})_{k=1}^{\infty}$ converges to $(0,0)$ in the Arens fan, which is a contradiction. Thus X is not an α_4-space. By Lemma 1.4, X contains a subspace sequentially homeomorphic to the Fréchet–Urysohn fan.

Combining Theorem 3.1, Proposition 3.2, and Lemmas 1.3–1.7 we get

Theorem 3.3. Suppose a regular topological space contains an uncountable metrizable compact subset and has the strong Skorokhod property for uniformly tight weakly convergent sequences of measures with compact support. If X admits a sequentially continuous binary operation $o : X \times X \to X$ with a sequentially continuous cancellation, then X is an α_4-space. If the operation o is sequentially invertible (and the space X has countable cs*-network or is a monotone $\aleph$-space), then X is sequentially homeomorphic to a (first-countable) strong Fréchet space.

In the particular case of (symmetric) convex sets and topological groups we get

Corollary 3.4. If a convex subset X of a linear topological space has the strong Skorokhod property for uniformly tight weakly convergent sequences of measures with compact support, then X is an α_4-space. If X is symmetric (and has countable cs*-network or is a monotone $\aleph$-space), then X is sequentially homeomorphic to a (first countable) strong Fréchet space.

Theorem 3.5. If a topological group G contains an uncountable compact metrizable subspace and has the strong Skorokhod property for uniformly tight weakly convergent sequences of measures with compact support, then G is sequentially homeomorphic to a strong Fréchet space. If, in addition, G is a monotone $\aleph$-space, then the space G is almost metrizable and consequently, has the uniformly tight strong Skorokhod property for Radon measures.

Proof. The first assertion follows from Theorem 3.3. If G is a monotone $\aleph$-space, then by the second assertion of Theorem 3.3, G is sequentially homeomorphic to a first-countable space H. This implies that the sequential coreflection σG of G is first countable. Since the identity map $\sigma G \to G$ is a sequential homeomorphism and σG is first countable, the group operation is continuous on σG. Consequently, σG is a first countable topological group. By the classical Birkhoff–Kakutani Theorem [12], [27], the space σG is metrizable. Since G is a monotone $\aleph$-space, all compact subsets of G are metrizable and hence sequentially compact. This fact and the sequential continuity of the identity map $G \to \sigma G$ implies that the identity map $\sigma G \to G$ is proper and hence G is an almost metrizable space. By Theorem 1.9, the space G has the uniformly tight strong Skorokhod property for Radon measures.

By the same reasoning we can prove

Theorem 3.6. If a topological group G has the countable cs*-character, contains an uncountable metrizable compact subset, and has the strong Skorokhod property for uniformly tight weakly convergent sequences of measures with compact support, then G is sequentially homeomorphic to a metrizable space. If, in addition, every compact subset of G is sequentially compact, then the space G is almost metrizable and hence G has the uniformly tight strong Skorokhod property for Radon measures.

For sequential topological groups Theorem 3.5 implies

Corollary 3.7. If a sequential topological group G contains an uncountable metrizable compact subset and has the strong Skorokhod property for uniformly tight weakly convergent sequences of measures with compact support, then G is a strong Fréchet space.

It should be mentioned that separable strong Fréchet topological groups are rather exotic objects: all known examples of such groups require some additional set-theoretic assumptions (like $\mathfrak{p} = \mathfrak{b}$), see [33]. In fact, under $\mathfrak{p} = \mathfrak{b}$ there is even a non-metrizable separable linear topological space [33]. In contrast, if a strong Fréchet group has countable cs*-character, then it is metrizable, see [5], [11] or [36]. On the other hand, there are many natural examples of non-metrizable sequential groups with countable cs*-character (take for example, any free topological group over a metrizable compact space). According to [11] each non-metrizable sequential topological group with countable cs*-character has countable pseudocharacter and contains an open σ-compact subgroup. We use this fact to generalize Theorem 3.6 in the case when the topological group is sequential. A *dispersion character* of a topological space is the smallest cardinal τ such that any point $x \in X$ has a neighborhood $U \subset X$ of size $|U| \leq \tau$.

Theorem 3.8. For a sequential topological group G with countable cs*-character and uncountable dispersion character the following conditions are equivalent:

1) the space G has the strong Skorokhod property for Radon measures;

2) the space G has the strong Skorokhod property for uniformly tight weakly convergent sequences of measures with compact support;

3) the space G is metrizable.

Proof. The implication 1) $\Rightarrow$ 2) is trivial and 3) $\Rightarrow$ 1) follows from Theorem 1.1. To prove that 2) $\Rightarrow$ 3) assume that the space G has the strong Skorokhod property for uniformly tight weakly convergent sequences of measures with compact support but is not metrizable. By [11], G contains an open σ-compact subgroup H. Since the group G has countable pseudo-character, it admits a continuous metric [40, 2.2]. Consequently, each compact subset of G is metrizable. Since G has the uncountable dispersion character, the open σ-compact subgroup H is uncountable and thus contains an uncountable metrizable compact subset. Now it is legal to apply Theorem 3.6 and conclude that the space G is almost metrizable. Since each sequential almost metrizable space is metrizable, we arrive to a contradiction.

Now we apply the above results to studying the strong Skorokhod property in free topological groups. We recall that a *free topological (abelian) group* of a Tychonoff space X is a topological group $F(X)$ containing X so that every continuous map $f : X \to G$ into a topological (abelian) group G can be uniquely extended to a continuous group homomorphism $\bar{f} : F(X) \to G$. It is well-known that a free topological (abelian) group $F(K)$ over an infinite metric compact space is a non-metrizable sequential space with countable cs*-character, see [29] or [4].

Theorem 3.9. If a Tychonoff space X contains an uncountable metrizable compact subset K, then the free topological (abelian) group $F(X)$ of X fails to have the strong Skorokhod property for uniformly tight weakly convergent sequences of measures with compact support.

Proof. Let $h : F(K) \to F(X)$ be a unique continuous group homomorphism extending the natural inclusion $K \subset X \subset F(X)$. By [38], h is a topological embedding. By Theorem 3.8, the free topological (abelian) group $F(K)$ of the metrizable compactum K fails to have the strong Skorokhod property for uniformly tight weakly convergent sequences of measures with compact support, and so does the space $F(X) \supset h(F(K))$.

Question 3. Does there exist a non-discrete metrizable space X whose free topological (abelian) group $F(X)$ has the strong Skorokhod property for uniformly tight weakly convergent sequences of measures with compact support? In particular, does the free topological group $F(K)$ over a countable compactum K with a unique non-isolated point have the strong Skorokhod property for uniformly tight weakly convergent sequences of measures with compact support?

It is worth mentioning that non-discrete Tychonoff spaces whose free topological group has the uniformly tight strong Skorokhod property for Radon measures do exist.

Proposition 3.10. For any countable subspace $X \subset \beta\mathbb{N}$ the free topological group $F(X)$ of X is almost discrete and has the uniformly tight strong Skorokhod property for Radon measures.

Proof. Let $h : F(X) \to F(\beta\mathbb{N})$ be the injective continuous group homomorphism extending the embedding $X \subset \beta\mathbb{N}$. It suffices to show that each compact subset $K \subset F(X)$ is finite. Assuming that it is not so we can find a non-trivial convergent sequence $S \subset F(X)$. Then $h(S)$ is a non-trivial convergent sequence in $F(\beta\mathbb{N})$. But the group $F(\beta\mathbb{N})$ contains no non-trivial convergent sequence (see [18, Lemma 4.8]).

4. Strong Skorokhod properties in linear topological spaces

In this section, we study the strong Skorokhod property in linear topological spaces. It is clear that all results proven in the preceding sections for topological groups or convex sets apply also to linear topological spaces. In particular, we have

Theorem 4.1. If a linear topological space X contains a subspace sequentially homeomorphic to the Fréchet–Urysohn fan, then the space X contains a subspace sequentially homeomorphic to the Cantor fan and fails to have the strong Skorokhod property for uniformly tight weakly convergent sequences of measures with compact support.

A (locally convex) linear topological space L is called the *(locally convex) strict inductive limit* of a sequence $L_1 \subset L_2 \subset \dots$ of linear subspaces of L if $L = \bigcup_{n\in\mathbb{N}} L_n$ and L carries the strongest (locally convex) linear topology inducing the original topology on each space L_n, $n \in \mathbb{N}$.

Proposition 4.2. If a (locally convex) linear topological space L is a (locally convex) strict inductive limit of a sequence $L_1 \subset L_2 \subset \dots$ of linear subspaces of L and L has the strong Skorokhod property for uniformly tight weakly convergent sequences of measures with compact support, then one of the spaces L_n is dense in L.

Proof. Assuming that none of L_n is dense in L, we can pass to a subsequence of the sequence (L_n) and assume that each L_n is nowhere dense in L_{n+1}. For every $n \in \mathbb{N}$, let us pick a point $x_n \in L_{n+1} \setminus \bar{L}_n$. It can be easily shown that the subspace $W = \{0, \frac{1}{m}x_n : n, m \in \mathbb{N}\}$ of L is sequentially homeomorphic to the Fréchet–Urysohn fan. Then by Theorem 4.1, the space L fails to have the strong Skorokhod property for uniformly tight weakly convergent sequences of measures with compact support.

Corollary 4.3. Suppose that a (locally convex) linear topological space L is a (locally convex) strict inductive limit of a sequence $L_1 \subset L_2 \subset \dots$ of metrizable subspaces of L. The following conditions are equivalent:

1) the space L has the strong Skorokhod property for Radon measures;

2) the space L has the strong Skorokhod property for uniformly tight weakly convergent sequences of measures with compact support;

3) the space L is metrizable.

Proof. The implication 1) $\Rightarrow$ 2) is trivial and 3) $\Rightarrow$ 1) follows from Theorem 1.1. To prove that 2) $\Rightarrow$ 3) assume that the space L has the strong Skorokhod property for uniformly tight weakly convergent sequences of measures with compact support. Then by Proposition 4.2, one of the spaces L_n is dense in L. Then the space L is metrizable because it contains a dense metrizable linear subspace.

A subset B of a linear topological space X is *bounded* if for every neighborhood $U \subset X$ of the origin there is $n \in \mathbb{N}$ such that $\frac{1}{n}B \subset U$. A family $\mathcal{B}$ of bounded subsets of a locally convex space X is called *fundamental* if every bounded subset $A \subset X$ lies in some $B \in \mathcal{B}$.

Theorem 4.4. If a locally convex space X possessing a countable fundamental family $\mathcal{B}$ of bounded subsets has the strong Skorokhod property for uniformly tight weakly convergent sequences of measures with compact support, then X is the image of a normed space N under a bijective linear continuous operator $T : N \to X$ with sequentially continuous inverse $T^{-1} : X \to N$.

Proof. Write $\mathcal{B} = \{B_n\}_{n\in\mathbb{N}}$. Without loss of generality, each set B_n is convex, symmetric and lies in B_{n+1}. We claim that some $B_n \in \mathcal{B}$ is a sequential neighborhood of the origin. If not, for every n, we can find a sequence $(x_{n,m})_{m\in\mathbb{N}} \subset X \setminus B_n$ convergent to zero. Then the subspace $W = \{0, x_{n,m} : n, m \in \mathbb{N}\}$ of X is sequentially homeomorphic to the Fréchet–Urysohn fan and by Theorem 4.1 the space X fails to have the strong Skorokhod property for uniformly tight weakly convergent sequences of measures with compact support, which is a contradiction. Thus some $B \in \mathcal{B}$ is a sequential neighborhood of the origin in X. Let $\|\cdot\|$ be the gauge functional of the set B. Then $N = (X, \|\cdot\|)$ is a normed space and the identity map $T : N \to X$ is a bijective continuous linear operator with sequentially continuous inverse T^{-1}.

A linear function $f : X \to \mathbb{R}$ on a locally convex space is called *bounded* if for any bounded subset $B \subset X$ its image $f(B)$ is bounded in $\mathbb{R}$.

Theorem 4.5. Let X be a locally convex space possessing a countable fundamental family of bounded subsets. Suppose that each compact subset of X is sequentially compact or each bounded linear functional $f : X \to \mathbb{R}$ is continuous on compact subsets of X. Then the following conditions are equivalent:

1) the space X has the uniformly tight strong Skorokhod property for Radon measures;

2) the space X has the strong Skorokhod property for uniformly tight weakly convergent sequences of measures with compact support;

3) the space X is the image of a normed space N under a bijective linear proper

continuous operator $T : N \to X$;

4) the space X is almost metrizable.

Proof. The implications 1) $\Rightarrow$ 2) and 3) $\Rightarrow$ 4) are trivial, while 4) $\Rightarrow$ 1) follows from Theorem 1.9. In order to prove that 2) $\Rightarrow$ 3), let us apply Theorem 4.4 to conclude that X is the image of a normed space N under a bijective continuous linear operator $T : N \to X$ with sequentially continuous inverse. To prove that the operator T is proper it suffices to verify that each compact subset K of X is sequentially compact. Assume that each bounded linear functional $f : X \to \mathbb{R}$ is continuous on compact subsets of X. The sequential continuity of the map T^{-1} implies that for each continuous linear functional $f : N \to \mathbb{R}$ the functional $f \circ T^{-1} : X \to \mathbb{R}$ is bounded and by our hypothesis is continuous on the compact set K. This yields that the map $T^{-1} : K \to (T^{-1}(K), \text{weak})$ of K onto $T^{-1}(K)$ endowed with the weak topology is continuous and because of the compactness of K, it is a homeomorphism. By the Eberlein Theorem [34, Ch. IV.§11] each compact subset of the space (N, weak) is sequentially compact. Consequently, the compact space K is sequentially compact.

We recall that a locally convex space X is *bornological* if each convex symmetric set $U \subset X$ absorbing bounded subsets of X is a neighborhood of the origin, see [34, Ch.II.§8].

Corollary 4.6. For a bornological locally convex space X possessing a countable fundamental family of bounded subsets the following conditions are equivalent:

1) the space X has the strong Skorokhod property for Radon measures;

2) the space X has the strong Skorokhod property for uniformly tight weakly convergent sequences of measures with compact support;

3) X is a normable linear topological space;

4) the space X is metrizable.

Proof. The implications 1) $\Rightarrow$ 2) and 3) $\Rightarrow$ 4) are trivial and 4) $\Rightarrow$ 1) follows from Theorem 1.1. To prove that 2) $\Rightarrow$ 3) assume that the space X has the strong Skorokhod property for uniformly tight weakly convergent sequences of measures with compact support. Then Theorem 4.4 implies that there are a normed space N and a bijective linear continuous operator $T : N \to X$ with a sequentially continuous inverse T^{-1}. To show that the space X is normable, it suffices to show that the image $T(B)$ of the unit ball B of the normed space N is a neighborhood of the origin in X. Since the space X is bornological it suffices to show that $T(B)$ absorbs bounded subsets of X. Let $A \subset X$ be any bounded subset of X. We have to find

$m \in \mathbb{N}$ such that $A \subset mT(B)$. Assuming that no such m exists we could construct a sequence $(a_n)_{n=1}^{\infty} \subset A$ such that $a_n \notin nT(B)$ for each n. Since the sequence (a_n) is bounded, the sequence $(\frac{1}{n}a_n)$ converges to the origin, see [34, I.5.3]. Then by the sequential continuity of the operator T^{-1} there is m such that $T^{-1}(\frac{1}{m}a_m) \in B$. This yields that $a_m \in mT(B)$, which contradicts the choice of the sequence (a_n).

Next, we investigate strong Skorokhod properties of the weak topology of normed spaces. For a linear topological space X by (X, weak) we denote X endowed with the weak topology. We recall that a normed space X has the *Shur property* if each weakly convergent sequence in X converges in norm. A standard example of a normed space with the Shur property is the Banach space l^1, see [17, p.85]. It is also known that each Banach space with the Shur property contains an isomorphic copy of l^1, see [17, p.212].

Theorem 4.7. For a normed space X the following conditions are equivalent:

1) the space (X, weak) has the uniformly tight strong Skorokhod property for Radon measures;

2) the space (X, weak) has the strong Skorokhod property for uniformly tight weakly convergent sequences of measures with compact support;

3) the space (X, weak) is almost metrizable;

4) the normed space X has the Shur property.

Proof. The implication 1) $\Rightarrow$ 2) is trivial, 4) $\Rightarrow$ 3) is proven in [8, 2.8] and 3) $\Rightarrow$ 1) follows from Theorem 1.9. Thus it remains to show that 2) $\Rightarrow$ 4). Assuming that the normed space X fails to have the Shur property, we can easily construct a subspace of (X, weak) sequentially homeomorphic to the Fréchet–Urysohn fan which by Theorem 4.1 implies that the space (X, weak) fails to have the strong Skorokhod property for uniformly tight weakly convergent sequences of measures with compact support.

Thus, there exist infinite-dimensional Banach spaces whose weak topologies have the uniformly tight strong Skorokhod property for Radon measures. Concerning the strong Skorokhod property for Radon measures we have the following negative result.

Theorem 4.8. If X is an infinite-dimensional Banach space, then the space (X, weak) fails to have the strong Skorokhod property for Radon measures.

Proof. Suppose that for some infinite-dimensional Banach space X the space (X, weak) has the strong Skorokhod property for Radon measures. By Theorem 4.7, the space (X, weak) is almost metrizable and by Theorem 1.9, this space is sequentially Prokhorov, which contradicts [22].

Question 4. Does there exist a non-metrizable linear topological space with the strong Skorokhod property for Radon measures?

All known examples of topological spaces with the strong Skorokhod property have first countable sequential coreflections.

Question 5. Suppose that a topological space X has the strong Skorokhod property for Radon measures. Does X have the countable cs*-character? Is X sequentially homeomorphic to a first countable space?

References

1. R. Arens, Note on convergence in topology, *Math. Mag.* **23** (1950), 229–234.
2. A.V. Arhangel'skii, The frequency spectrum of a topological space and the classification of spaces, *Soviet Math. Dokl.* **13** (1972), 265–268.
3. A.V. Arhangel'skii, The frequency spectrum of a topological space and the product operation, *Trans. Moscow Math. Soc.* **2** (1981), 163–200.
4. A.V. Arkhangel'kiĭ, Classes of topological groups, *Russian Math. Surveys* **36** (1981), 151–174. Russian original in: *Uspekhi Mat. Nauk* **36** (3) (1981), 127–146.
5. A.V. Arkhangel'skii, Topological properties in topological groups, in: *XVIII All Union Algebraic Conference*. Kishinev, 1985.
6. A.V. Arkhangel'skii, S. Franklin, Ordinal invariants for topological spaces, *Michigan Math. J.* **15** (1968), 313–320.
7. T. Banakh, On topological groups containing a Fréchet–Urysohn fan, *Matem. Studii* **9** (2) (1998), 149–154.
8. T.O. Banakh, V.I. Bogachev, A.V. Kolesnikov, Topological spaces with the strong Skorokhod property, *Georgian Math. J.* **8** (2) (2001), 201–220.
9. T.O. Banakh, V.I. Bogachev, A.V. Kolesnikov, Topological spaces with the Prokhorov and Skorokhod properties, *Doklady Math.* **64** (2) (2001), 244–247; Russian original in: *Doklady RAN.* **380** (6) (2001), 727–730.
10. T.O. Banakh, V.I. Bogachev, A.V. Kolesnikov, Monotone compact-covering maps, monotone $\aleph$-spaces, and their applications in probability theory, *preprint*, 2001.
11. T. Banakh, L. Zdomskyĭ, Topological groups with countablecs*-character, *preprint*, 2002.
12. G. Birkhoff, A note on topological groups, *Comput. Math.* **3** (1936), 427–430.
13. D. Blackwell, L.E. Dubins, An extension of Skorohod's almost sure representation theorem, *Proc. Amer. Math. Soc.* **89** (4) (1983), 691–692.

14. V.I. Bogachev, Measures on topological spaces, *J. Math.Sci. (NY)* **91** (4) (1998), 3033–3156.
15. V.I. Bogachev, A.V. Kolesnikov, Open mappings of probability measures and the Skorohod representation theorem, *Theory Probab. Appl.* **46** (1) (2001), 1–21.
16. J.A. Cuesta-Albertos, C. Matrán-Bea, Stochastic convergence through Skorohod representation theorems and Wasserstein distances, In: *First International Conference on Stochastic Geometry, Convex Bodies and Empirical Measures* (Palermo, 1993). *Rend. Circ. Mat. Palermo (2), Suppl. No.* **35** (1994), 89–113.
17. J. Diestel, *Sequences and series in Banach spaces*. Springer, New York, 1984.
18. D. Dikranjan, M. Tkačenko, Weakly complete free topological groups, *Topology Appl.* **112** (3) (2001), 259–287.
19. R.M. Dudley, Distances of probability measures and random variables, *Ann. Math. Stat.* **39** (1967), 1563–1572.
20. R. Engelking, *General Topology*. PWN, Warsaw, 1977.
21. X. Fernique, Un modèle presque sûr pour la convergence enloi, *C. R. Acad. Sci. Paris, Sér. 1* **306** (1988), 335–338.
22. D. Fremlin, D. Garling, R. Haydon, Bounded measures ontopological spaces, *Proc. London Math. Soc.* **25** (1972),115–136.
23. G. Gruenhage, Generalized metric spaces, In: K.Kunen and J.Vaughan (eds.) *Handbook of Set-Theoretic Topology*. Elsevier Sci., 1984, 423–501.
24. A. Jakubowski, The almost sure Skorokhod representation for subsequences in nonmetric spaces, *Teor. Veroyatn. i Primenen.* **42** (1) (1997), 209–217; English transl.: *Theory Probab. Appl.* **42** (1) (1997), 167–174.
25. S. Lin, A note on the Arens' space and sequential fan, *Topology Appl.* **81** (1997), 185–196.
26. C. Liu, M. Dai, g-metrizability and S_ω, *Topology Appl.* **60** (1994), 185–189.
27. S. Kakutani, Über die Metrization der topologischen Gruppen, *Proc. Imp. Acad. Tokyo.* **12** (1936), 82–84.
28. A.S. Kechris, *Classical descriptive set theory*. Springer-Verlag, Berlin, New York, 1995.
29. A.A. Markov, On free topological groups, *Topology and Topological Algebra*, Translations Series 1. American Mathematical Society **8** (1962), 195–272. Russian original in: *Izvestiya Akad. Nauk SSSR. Ser. Mat.* **9** (1945), 3–64.
30. E. Michael, A quintuple quotient quest, *General Topology Appl.* **2** (1972), 91–138.
31. T. Nogura, D. Shakhmatov, Y. Tanaka, Metrizability of topological groups having weak topologies with respect to good covers, *Topology Appl.* **54** (1993), 203–212.
32. T. Nogura, Y. Tanaka, Spaces which contain a copy of S_ω or S_2 and their applications, *Topology Appl.* **30** (1988), 51–62.
33. P.J. Nyikos, Subsets of ${}^\omega\omega$ and the Fréchet–Urysohn and α_i-properties, *Topology Appl.* **48** (1992), 91–116.

34. H.H. Schaefer, *Topological vector spaces*. The Macmillan Company. NY, London, 1966.
35. A. Schief, Almost surely convergent random variables with given laws, *Probab. Theory Relat. Fields* **81** (1989), 559–567.
36. A. Shibakov, Metrizability of sequential topological groups with point-countable k-networks, *Proc. Amer. Math. Soc.* **126** (1998), 943–947.
37. A.N. Shiryaev, *Probability*. Nauka, Moscow, 1989; English transl.: Springer, 1990.
38. O.V. Sipacheva, Free topological groups of spaces and their subspaces, *Topology Appl.* **101** (2000), 181–212.
39. A.V. Skorohod, Limit theorems for stochastic processes, *Teor. veroyatn. i Primen.* **1** (3) (1956), 289–319; English transl.: *Theor. Probab. Appl.* **1** (3) (1956), 261–290.
40. M. Tkačenko, Introduction to topological groups, *Topology Appl.* **86** (1998), 179–231.

Functional Analysis and its Applications
V. Kadets and W. Żelazko

Fourth Order Spectral Problem with Eigenvalue in the Boundary Conditions

Jamel Ben Amara

Faculté des Sciences de Bizerte (Tunisia)

Abstract

In this paper we consider a fourth order eigenvalue problem containing a spectral parameter both in the equation and in the boundary conditions. We associate this problem with a self-adjoint operator in a Hilbert or in a Pontryagin space. Using this operator-theoretic formulation and analytic methods, we investigate locations, multiplicities and the asymptotic behaviour of the eigenvalues.

Key words: Eigenvalue problem, spectral parameter in boundary conditions, Pontryagin space
2000 MSC: 47A75, 47A25

1. Introduction

We shall investigate the following eigenvalue problem

$$l(y) = (py'')'' - (qy')' = \lambda ry, \qquad ' := \frac{d}{dx} \tag{1}$$

$$\begin{cases} y(0) = y'(0) = y''(\pi) = 0 \\ [(py'')' - qy']_{|x=\pi} = -m\lambda y(\pi), \end{cases} \tag{2}$$

on the interval $[0, \pi]$. We assume that p, q and r are positive functions, $r, 1/r \in L_1[0, \pi]$, $p \in W_1^2[0, \pi]$, $q \in W_1^1[0, \pi]$, $m \in \mathbf{R} \setminus \{0\}$ and λ is a spectral parameter,

Email address: `jamel.benamara@fsb.rnu.tn` (Jamel Ben Amara).

$W_p^k[0,\pi]$ is the Sobolev function space having a generalized (in the sense of distributions) k-th derivative in $L_p[0,\pi]$. We intentionally restrict ourselves to special boundary conditions in order to avoid nonessential complications.

Problem (1)–(2) is derived from a wave equation which describes the vibration of a nonhomogeneous rod or beam fixed at one end and subjected to a follower force at the other. In particular, in our situation for $m > 0$, there is a massive load at the end π (e.g.; see [5, Ch. 1–3], [10, Ch. 4], [13]). In a particular case where $p = r \equiv 1$ and $q \equiv x$, Problem (1)–(2) has partially been investigated in [1]. A large body of literature has built up, over the last years, on problems of Sturm-Liouville type where the boundary conditions depend on the spectral parameter λ (e.g., see [3], [4], [6], [15]). Note that there are many distinctions between the second and the fourth order differential equations and between the associated eigenvalue problems. For example, the eigenvalues of the vibrating rod are, in general, not necessarily simple; e.g., the eigenvalue $\lambda = 0$ of the free end eigenvalue problem (i.e., Equation (1) with $q \equiv 0$ and the boundary conditions $y''(0) = y'''(0) = y''(\pi) = y'''(\pi) = 0$) has an algebraic multiplicity 2.

The sign of m plays an essential role. In the case $m > 0$, we associate with Problem (1)–(2) a self-adjoint operator in the Hilbert space $L_2\big((0,\pi),r\big)\times\mathbf{C}$ with an appropriate inner product. Using this fact and extending analytic methods (which are usually used for the second order boundary value problems) to fourth order problems, we investigate locations, multiplicities and an asymptotic behaviour of the eigenvalues λ_n for large n. More precisely, we show that all λ_n are positive, simple and interlace with eigenvalues ρ_n of the standard problem with the boundary conditions $y(0) = y'(0) = y(\pi) = y''(\pi) = 0$. This leads to an asymptotic estimate of the form $\lambda_n = (\frac{\pi n}{\gamma})^4 + o(n^4)$, as $n \to \infty$. Thereafter, we shall give a better estimate

$$\lambda_n - \rho_{n-1} = O(n^2).$$

In the case of Sturm–Liouville problems, similar but slightly different asymptotics was established in [4] using Prüfer transformation.

For $m < 0$, Problem (1)–(2) can be interpreted as an eigenvalue problem for a self-adjoint operator in a Pontryagin space Π_1, and so it can be rewritten in the form

$$A(\cdot) = \lambda G(\cdot),$$

where $A = A^*$, $A > 0$ and $G = G^*$ in the Hilbert space $L_2\big((0,\pi),r\big) \times \mathbf{C}$. It is proved bellow that all the eigenvalues are real and simple, such that the first eigenvalue is negative and is of negative type in the sense of Definition 7, while all other eigenvalues are positive and are of positive type. The asymptotic formulae obtained for $m > 0$ remain valid in this case.

2. Spectral Analysis

Let $H = L_2\big((0,\pi),r\big) \times \mathbf{C}$ be a Hilbert space equipped with the inner product

$$(\hat{y}_1, \hat{y}_2)_{\mathrm{H}} = \int_0^{\pi} r(t) y_1(t) \overline{y_2(t)}\, dt + |m| p(\pi) a \bar{b}, \tag{3}$$

where

$$\hat{y}_1 = \begin{pmatrix} y_1(x) \\ a \end{pmatrix} \in \mathrm{H}, \; \hat{y}_2 = \begin{pmatrix} y_2(x) \\ b \end{pmatrix} \in \mathrm{H}.$$

We define in H an operator

$$\hat{L}_m \hat{y} = \hat{L}_m \begin{pmatrix} y(x) \\ y(\pi) \end{pmatrix} = \begin{pmatrix} r^{-1}(x)((py'')''(x) - (qy')'(x)) \\ -\dfrac{[(py'')' + qy']_{|x=\pi}}{m} \end{pmatrix} \tag{4}$$

with the domain

$$\mathcal{D}(\hat{L}_m) = \left\{ \hat{y} = \begin{pmatrix} y(\cdot) \\ y(\pi) \end{pmatrix} \in \mathrm{H}, y \in W_2^4[0,\pi], y(0) = y'(0) = y''(\pi) = 0 \right\}.$$

Obviously, the operator $\hat{L}_m$ is well defined. By immediate verification we conclude that Problem (1)-(2) is equivalent to the following spectral problem

$$\hat{L}_m \hat{y} = \lambda \hat{y}, \qquad \hat{y} \in D(\hat{L}_m); \tag{5}$$

i.e., the eigenvalues λ_k of Problem (1)-(2) and those of Problem (5) coincide; moreover, there exists a correspondence between the eigenfunctions and the adjoint functions of the two problems,

$$\begin{pmatrix} y(x) \\ y(\pi) \end{pmatrix} \leftrightarrow y(x).$$

Lemma 1. The domain of $\hat{L}_m$ is dense in H.

Proof. For a more general situation, this result is proved in [11]. □

Theorem 2. The operator $\hat{L}_m$ has for $m \in \mathbf{R} \setminus \{0\}$ a discrete spectrum. Moreover, for $m > 0$ the operator $\hat{L}_m$ is positive and self-adjoint in H, its spectrum consists of positive eigenvalues with finite multiplicities tending to $+\infty$.

Proof. Problem (1)–(2) has regular boundary conditions in the sense of [11] for all $m \in \mathbf{R}$; in particular, it has a discrete spectrum. For $m > 0$, integrating by parts we obtain

$$(\hat{L}_m \hat{y}, \hat{y})_H = (py'', y'')_{L_2[0,\pi]} + (qy', y')_{L_2[0,\pi]},$$

where $\hat{y} \in D(\hat{L}_m)$. Thus, for $m > 0$, $\hat{L}_m$ is a positive operator with a discrete spectrum and therefore it is self-adjoint in H. □

Our next aim is to show that all the eigenvalues $\{\lambda_n\}_1^{\infty}$ are algebraically simple. The following result will play a fundamental role in the sequel.

Lemma 3. Let $\lambda > 0$ and E be the space of all solutions of the following boundary value problem

$$\begin{cases} (py'')'' - (qy')' = \lambda ry \\ y(0) = y'(0) = y''(\pi) = 0. \end{cases} \tag{6}$$

Then $\dim \mathrm{E} = 1$.

For the proof of Lemma 3 we need the following result which is a particular case of Lemma 2.1 of [8].

Lemma 4. Let $P \in W_1^2[0,\pi]$, $Q \in L_1[0,\pi]$, $P > 0$ and $Q > 0$. If y is a solution of the equation

$$(Py'')'' - Qy = 0, \tag{7}$$

satisfying the initial conditions

$$y(0) = y'(0) = y''(0) = 0, (Py'')'(0) = 1$$

then the functions $y(x), y'(x), y''(x)$ and $(py'')'(x)$ are positive for $x \in (0,\pi]$.

We give now the proof of Lemma 3.

Proof of Lemma 3. Let h be the solution of the Cauchy problem

$$\begin{cases} (py')' - qy = 0 \\ y(\pi) = 1, y'(\pi) = 0. \end{cases} \tag{8}$$

Then, by Sturm comparison theorem, $h > 0$ on $[0,\pi]$. Using the following change of variables,

$$t(x) := \frac{\pi \int_0^x h(s)ds}{\int_0^\pi h(s)ds}; \tag{9}$$

Problem (6) can be transformed into the following problem

$$\begin{cases} (ph^3\ddot{y})^{\cdot\cdot} = \lambda h^{-1} ry, & \cdot := \dfrac{d}{dt} \\ y(0) = \dot{y}(0) = \ddot{y}(\pi) = 0. \end{cases} \tag{10}$$

Suppose that there exist two linear independent solutions y_1, y_2 of Problem (10). Both $\ddot{y}_1(0)$ and $\ddot{y}_2(0)$ must be different from zero since otherwise it would follow from Lemma 4 that $\ddot{y}_1$ and $\ddot{y}_2$ could not vanish at $x = \pi$. In view of the assumptions about y_1 and y_2, the solution

$$w(x) = \ddot{y}_2(0)y_1(x) - \ddot{y}_1(0)y_2(x)$$

will satisfy $w(0) = \dot{w}(0) = \ddot{w}(0) = \ddot{w}(\pi) = 0$, and this again contradicts Lemma 4 unless $w(x) \equiv 0$. The proof of the lemma is complete. □

Theorem 5. Each of the eigenvalues $\{\lambda_n\}_1^\infty$ of Problem (1)–(2) is algebraically simple.

Proof. From Lemma 3 it follows that each eigenvalue of Problem (1)–(2) (or of $\hat{L}_m$) has a geometric multiplicity equal to 1. By Theorem 2, $\hat{L}_m$ is self-adjoint in H. Therefore, all its eigenvalues are algebraically simple. □

We turn now to the case $m < 0$. It is easy to verify that for $m < 0$, the operator $\hat{L}_m$ is not self-adjoint in H. Let

$$\hat{G} = \begin{pmatrix} I & 0 \\ 0 & -1 \end{pmatrix}$$

where I is the identity operator on the space $L_2[0, \pi]$. The operator $\hat{G}$ is unitary and symmetric on H, its spectrum consists of two eigenvalues -1 with multiplicity 1 and $+1$ with infinite multiplicity. Hence, it generates a Pontryagin space Π_1. It follows from

$$(\hat{G}\hat{L}_m\hat{y}, \hat{y})_{\mathrm{H}} = \int_0^\pi |y''(x)|^2 dx + \int_0^\pi q(x)|y'(x)|^2 dx, \tag{11}$$

where $\hat{y} \in \mathcal{D}(L_m)$, that $\hat{L}_m$ is selfadjoint in Π_1. Thus, Problem (1)-(2) can be rewritten in the form

$$\hat{A}\hat{y} = \hat{G}\hat{L}_m\hat{y} = \lambda\hat{G}\hat{y}, \qquad \hat{y} \in \mathcal{D}(\hat{A}) = \mathcal{D}(\hat{L}_m), \tag{12}$$

Proposition 6. The operator $\hat{A}$ is positive, self-adjoint in the Hilbert space H and has a discrete spectrum.

Proof. It is clear from (11) that $\hat{A} > 0$. The rest of the proof is similar to the proof of Theorem 2. □

Definition 7. We say that an eigenvalue λ is of "positive" (resp. "negative") type if $(\hat{G}y, y) > 0$ (< 0) for all nonzero $y \in \mathrm{Ker}(\hat{A} - \lambda\hat{G})$. If λ is not one of these types we say that λ is an eigenvalue of "neutral type". If λ is of positive or of negative type we say that it has a definite type.

Theorem 8. Let $m < 0$. Then the spectrum of Problem (1)-(2) consists of real and simple eigenvalues $\{\lambda_n\}_1^\infty$ tending to $+\infty$. Moreover, all the eigenvalues have a definite type and there exists exactly one negative eigenvalue $\lambda_1 < 0$ which is of negative type, and all the positive eigenvalues $\lambda_2, \lambda_3, \cdots$ are of positive type.

Proof. By Theorem 2, the operator $\hat{L}_m$ has for $m < 0$ a discrete spectrum. For the eigenpair $(\lambda_k, \hat{y_k})$ of Problem (12) we have

$$0 < (\hat{A}\hat{y}_k, \hat{y}_k)_{\mathrm{H}} = \lambda_k(\hat{G}\hat{y}_k, \hat{y}_k)_{\mathrm{H}}. \tag{13}$$

Hence $\hat{L}_m$ has only real eigenvalues. Since the operator $\hat{G}$ generates a Pontryagin space metric with one negative square, it follows that $\hat{L}_m$ has one eigenvalue of negative or neutral type (see [2], Ch. 3). By (13), the eigenvalue must be negative and have a negative type, while all others are positive and have a positive type. It is known that an eigenvalue of definite type is necessarily real and semisimple (i.e., the algebraic multiplicity coincides with the geometric multiplicity). Since in our case all the eigenvalues have a definite type and by Lemma 3 their geometric multiplicity is equal to 1, it follows that they are simple. □

3. Asymptotic distribution of the eigenvalues λ_n

In this section we establish the asymptotic formulae for the eigenvalues λ_n of Problem (1)–(2) for large n. First, we establish two Lemmas.

Lemma 9. Let ρ_n and ρ'_n, $n = 1, 2, \cdots$ be respectively the eigenvalues of the Problem

$$\begin{cases} (py'')'' - (qy')' = \lambda r y \\ y(0) = y'(0) = 0 \\ y''(\pi) = y(\pi) = 0, \end{cases} \tag{14}$$

and Problem (1)–(2) for $m = 0$. Then $\rho_n \neq \rho'_n, \quad n = 1, 2, \cdots$.

Proof. It is known [7] that ρ_n and ρ'_n, $n = 1, 2, \cdots$ are positive and simple eigenvalues. Suppose that there exists integers n_1 and n_2 such that $\rho_{n_1} = \rho'_{n_2} =: \mu$. Then by Lemma 3, for $\lambda = \mu$ Equation (1.1) has a non trivial solution $y = y(x, \mu)$ satisfying the boundary conditions

$$\begin{cases} y(0) = y'(0) = 0 \\ y''(\pi) = y(\pi) = [((py'')' - qy')]_{|x=\pi} = 0. \end{cases} \tag{15}$$

using the change of variables (9), the boundary value problem (1.1)-(15) can be transformed into the problem

$$(ph^3\ddot{y})^{\cdot\cdot} = \mu h^{-1} r y \tag{16}$$

$$\begin{cases} y(0) = \dot{y}(0) = 0 \\ y(\pi) = \ddot{y}(\pi) = 0 \\ y^{(3)}(\pi) = 0, \end{cases} \tag{17}$$

where $h > 0$ is the solution of Problem (8). By Lemma 4, this is impossible unless $y \equiv 0$. □

Consider the function

$$F(\lambda) = \frac{[(py''(x,\lambda))' - qy'(x,\lambda)]_{|x=\pi}}{y(\pi,\lambda)}, \tag{18}$$

where $y(x,\lambda)$ is a non trivial solution of the boundary value problem (6). Obviously, the eigenvalues of Problem (1)–(2) coincide with the roots of the equation

$$F(\lambda) = m\lambda. \tag{19}$$

Lemma 10. The function $F(\lambda)$ is meromorphic and increases on each interval $(-\infty,\rho_1)$, (ρ_n,ρ_{n+1}), $n = 1,2,\cdots$. In particular, it increases from $-\infty$ to $+\infty$ on the finite intervals.

Proof. By Lemma 3, $F(\lambda)$ is defined in a unique way. Moreover, since each solution of Problem (6) is analytic of λ, then $F(\lambda)$ is a finite-order meromorphic function. By virtue of Lemma 9, all the eigenvalues ρ_n and ρ'_n coincide respectively with all poles and zeros of $F(\lambda)$. Let $y(x,\lambda)$ and $y(x,\mu)$ be two solutions of Problem (6), where λ and μ ($\lambda \neq \mu$) both belong to one of the intervals $(-\infty,\rho_1)$, (ρ_n,ρ_{n+1}), $n = 1,2,\cdots$. Using the Lagrange formula [9, Ch. 1], we obtain

$$y(\pi,\mu)[(py''(x,\lambda))' - qy'(x,\lambda)]_{|x=\pi} - y(\pi,\lambda)[(py''(x,\mu))' - qy'(x,\mu)]_{|x=\pi} = (\lambda-\mu)\big(ry(\lambda),y(\mu)\big)_{L_2(0,\pi)}. \tag{20}$$

Thus,

$$(\lambda-\mu)\big(ry(\lambda),y(\mu)\big)_{L_2(0,\pi)} = y(\pi,\lambda)y(\pi,\mu)[F(\lambda)-F(\mu)]. \tag{21}$$

Dividing this equality by $(\lambda-\mu)$ and passing to the limit as $\mu \to \lambda$, we obtain

$$\int_0^\pi r(x)y^2(x,\lambda)dx = y^2(\pi,\lambda)\frac{\mathrm{d}F(\lambda)}{\mathrm{d}\lambda}. \tag{22}$$

Since $y(\pi,\lambda) \neq 0$ for all $\lambda \neq \rho_n, n = 1,2,\cdots$, then

$$\frac{\mathrm{d}F(\lambda)}{\mathrm{d}\lambda} > 0.$$

Therefore, the desired result follows. □

Lemmas 9, 10 lead to the following result.

Theorem 11. Let $m > 0$ and let λ_n, ρ_n, ρ'_n be respectively the eigenvalues of Problems (1)–(2), (14) and (1)–(2) for $m = 0$. Then

$$\rho'_{n-1} < \rho_{n-1} < \lambda_n < \rho'_n < \rho_n, \qquad n \geq 2. \tag{23}$$

Proof. From Lemmas 9, 10 we obtain the interlacing of ρ_n, ρ'_n, i.e.,

$$\rho'_n < \rho_n < \rho'_{n+1} < \rho_{n+1}, \qquad n \geq 1.$$

Thus and from the graphs of $F(\lambda)$ and of the straight line with negative slope $g(\lambda) = -m\lambda$, the required inequalities (23) follows. □

Corollary 12. The eigenvalues λ_n satisfy the asymptotic

$$\lambda_n = \left(\frac{\pi n}{\gamma}\right)^4 + \mathrm{o}(n^4), \quad \text{as} \quad n \to \infty, \tag{24}$$

where

$$\gamma = \int_0^{\pi} \left(\frac{r}{p}\right)^{1/4} dt.$$

Proof. This follows from Theorem 11 and the asymptotic (e.g; see [9], Chap. 2)

$$\rho_n = \left(\frac{\pi n}{\gamma}\right)^4 \left[1 + \mathrm{O}(\frac{1}{n})\right], \qquad \text{as} \qquad n \to \infty. \tag{25}$$

□

We shall now examine the relation between λ_n and ρ_n more carefully. The graphs of F and g might suggest that $(\lambda_n - \rho_{n-1}) \to 0$, as $n \to \infty$, but this is not true as shown in the following result.

Theorem 13. Let $m > 0$ and let p, q, r be positive and belong to $W_1^3[0,1]$. Then

$$\lambda_n - \rho_{n-1} \asymp n^2, \qquad \text{as} \qquad n \to \infty. \tag{26}$$

Proof. Let $G(\lambda) = (F(\lambda))^{-1}$. Since $F(\lambda)$ is meromorphic, then by (23), G is $\mathbf{C}^1$ on $I_n = [\rho_{n-1}, \lambda_n]$ (G has no poles in I_n). Thus we obtain

$$G(\lambda_n) - G(\rho_{n-1}) = (\lambda_n - \rho_{n-1})G'(\lambda_n) \tag{27}$$

for some $\lambda \in I_n$. From (19) and (24), it follows that the left side of (27) is of order n^{-4}. We study now the behaviour of $G'(\lambda)$ for $\lambda \in I_n$ and large n. From (18) and (22) we obtain

$$G'(\lambda) = \frac{\int_0^{\pi} ry^2(x,\lambda)dx}{\left([(py''(x,\lambda))' - qy'(x,\lambda)]_{|x=\pi}\right)^2}. \tag{28}$$

From (23) it follows that $[(py''(x,\lambda))' - qy'(x,\lambda)]_{|x=\pi} \neq 0$ for $\lambda \in I_n$, hence G' is a bounded function on I_n. Let $\{y_i(x,\lambda)\}_{i=1}^{i=4}$ be a system of fundamental solutions of Equation (1.1) satisfying the asymptotics (e.g., see [9], Chap 2)

$$y_i(x) = e^{\mu w_i X}[1 + \mathrm{O}(\mu^{-1})], \quad \text{as} \quad |\mu| \to \infty$$

and

$$y_i^{(k)}(x) = (r/p)^{k/4}(\mu w_i)^k e^{\mu w_i X}[1 + \mathrm{O}(\mu^{-1})], \quad k \in \{1,2,3\},$$

where $\mu^4 = \lambda$, ${w_i}^4 = 1$ and $X = \int_0^x (r/p)^{1/4} dt$. Without loss of generality, we can suppose that the solution $y(x,\lambda)$ of Problem (6) satisfies $y''(0) = 1$. Using the above asymptotics we obtain

$$\begin{aligned} &y(x,\lambda) \\ &= [(\cos\mu X - \cosh\mu X) - \frac{\cos\mu\gamma + \cosh\mu\gamma}{\sin\mu\gamma - \sinh\mu\gamma}(\sin\mu X - \sinh\mu X)][1 + \mathrm{O}(\mu^{-1})] \\ &\qquad = (\cos\mu X - \sin\mu X)[1 + \mathrm{O}(\mu^{-1})] \end{aligned} \tag{29}$$

as $|\mu| \to \infty$, where the constant γ is defined in Corollary 13. Moreover,

$$\begin{aligned}[(py''(x,\lambda))' - qy'(x,\lambda)]_{|x=\pi} &= \mu^3 \left(\frac{1+\cosh\mu\gamma\cos\mu\gamma}{\sin\mu\gamma - \sinh\mu\gamma}\right)[1+\mathrm{O}(\mu^{-1})] \\ &= \mu^3 \cos\mu\gamma[1+\mathrm{O}(\mu^{-1})], \quad \text{as} \quad |\mu| \to \infty.\end{aligned} \tag{30}$$

From (29), (30), a simple calculation gives

$$G'(\lambda_n) \asymp \mu_n^{-6} \asymp n^{-6}, \quad \text{as} \quad n \to \infty.$$

Therefore, from (27) the required result follows. □

Remark 14. In the case $m < 0$, the interlacing property (23) of the eigenvalues slightly changes, while the asymptotics of λ_n obtained for $m > 0$ remains valid. More deep studies of this and of the indefinite case will be done in a separate work.

The author thanks A.A.Vladimirov of the Moscow State University, for many helpful discussions. The author also thanks the referee for useful comments.

References

1. M. J. Ahn, *Quart. Appl. Math.* **39** (1) (1981) 109–117.
2. T.Ya. Azizov, I. S. Iokhvidov, *Linear operators in spaces with indefinite metric*. John Wieley, Chichester, 1989.
3. J.A. Ben Amara, A.A.Shkalikov, A Sturm-Liouville Problem With Physical and Spectral Parameters in Boundary Conditions, *Mathem. Notes* **66** (2) (1999) 163–172.
4. P.A. Binding, P. J. Browne, K. Seddighi, Sturm-Liouville problems with eigenparameter dependent boundary conditions, *Proc. Edinburgh Math. Soc.* **37** (1993), 57–72.
5. L. Collatz, *Eigenwertaufgaben mit Technischen Anwendungen*. Akademishe Verlag, Leipzig, 1963.
6. C.T. Fulton, Two-point boudary value problems with eigenvalue parameter contained in the boundary conditions, *Proc. Soc. Edinburgh* **Sect. A 77** (1977) 393–308.
7. F.P. Gantmaher, M. G. Krein, *Oscillating matrices and kernels and small oscillation of mechanic systems*. Moscow 1957.
8. W. Leighton, Z. Nehari, On the oscillation of solutions of self-adjoint linear differential equations of the fourth order, *Trans. Amer. Math. Soc.* **98** (1958) 325–377.
9. M.A. Naimark, *Linear differential operators*. Ungar, New York, 167.
10. F. Roseau, *Theory of vibration in mechanicals systems*. Springer-Verlag 1983.

11. A.A. Shkalikov, Boundary value problems for ordinary differential equations with parameter in the boundary conditions, *J. Soviet Math.* **33** (1986) 1311–1342.
12. A.A. Shkalikov, Operator pencils arising in elasticity and hydrodynamics: Instability index formula, *in the book "Operator theory: Advances and Applications"*, Vol. **87** 258–285, Birkhaüser Verlag, 1996.
13. S.P. Timoshenko, *Strength and vibrations of structural members*. Collection of paper (E. I. Grigolyuk, editor) in Russian, Nauka, Moscow, 1975.
14. A.G. Kostyuchenko, A.A. Shkalikov, On the theory of self-adjoint quadratic operator pencils, *Math. Bull. Moscow State Univ.* **38** (1983) 44–58.
15. J. Walter, Regular eigenvalue problems with eigenvalue parameter in the boundary conditions, *Math. Z.* **133** (1973) 301–312.

Functional Analysis and its Applications
V. Kadets and W. Żelazko

Extending Operators into $\mathcal{L}_\infty$ Spaces Under a Twisted Light

Félix Cabello Sánchez, Jesús M. F. Castillo, Yolanda Moreno

Departamento de Matemáticas, Universidad de Extremadura
06071 Badajoz, Spain

David T. Yost

Department of Mathematics, King Saud University
P.O. Box 2455, Riyadh 11451, Saudi Arabia

Abstract

We present a homological approach to the problem of extending operators which take values in $C(K)$ or $\mathcal{L}_\infty$ spaces. In this way we obtain unified simpler proofs of results of Lindenstrauss-Pełczyński and Johnson-Zippin.

Key words: Extension of linear operators, operator ideals, Banach spaces
2000 MSC: 46B20, 47L20

Introduction and preliminaries

The classical problem of extension of linear operators is: when can a given operator $Y \longrightarrow B$ from a subspace Y of a Banach space X be extended to an operator

Email addresses: fcabello@unex.es (Félix Cabello Sánchez), castillo@unex.es (Jesús M. F. Castillo), ymoreno@unex.es (Yolanda Moreno), dthoyost@ksu.edu.sa (David T. Yost).
[1] FCS, JMFC and YM where supported in part by the DGICYT project BFM 2001–0813. DTY'sde Extremadura and partly by KSU Research Centre Project Number Math/1420/25.

$X \longrightarrow B$? Equivalently, when is the restriction operator $\mathfrak{L}(X,B) \longrightarrow \mathfrak{L}(Y,B)$ surjective?

Linear continuous operators cannot, as a rule, be extended; but it is possible in many interesting situations. For instance, when either **i)** Y is complemented in X; or **ii)** B is injective; or **iii)** the operator is 2-summing (regardless of the spaces involved; see [10, Theorem 4.15]). We see that there are three variables involved here: the pair $Y \longrightarrow X$, the space B, and the class of operators $\mathfrak{A}$ we want to extend; and, accordingly, there are three different extension problems depending on which two of these three elements are chosen as the data.

Our main concern will be generalizations of **ii)** due to Lindenstrauss-Pełczyński and Johnson-Zippin. The fact that an injective Banach space has to be an $\mathcal{L}_\infty$-space attracts one's attention to $\mathcal{L}_\infty$-spaces. The fact that the identity operator $c_0 \longrightarrow c_0$ cannot be extended to ℓ_∞ sets the limits of what can be done: in general, even $C(K)$-valued operators cannot be extended.

The problem admits an algebraic formulation and treatment, and this is precisely the point of view and the language we adopt in this paper. For general information about $\mathcal{L}_\infty$-spaces, see [10, Chapter 3] or [2]. In section 3 we assume from the reader familiarity with the basic theory of exact sequences and extensions; in section 4, only, we moreover assume familiarity with the theory of quasi and zero-linear mappings. All this background can be found in [1] or [8]. Here is a brief description.

A short exact sequence (also called an extension of Z by Y) is a diagram

$$0 \longrightarrow Y \xrightarrow{\imath} X \xrightarrow{\pi} Z \longrightarrow 0 \qquad (1)$$

composed of Banach spaces and operators in which the kernel of each arrow coincides with the image of the preceding one. Two exact sequences $0 \longrightarrow Y_i \longrightarrow X_i \longrightarrow Z_i \longrightarrow 0$ $(i = 1, 2)$ are said to be **isomorphically equivalent** if there exist isomorphisms α, β, γ making the diagram

$$\begin{array}{ccccccccc} 0 & \longrightarrow & Y_1 & \longrightarrow & X_1 & \longrightarrow & Z_1 & \longrightarrow & 0 \\ & & {\scriptstyle\alpha}\downarrow & & {\scriptstyle\beta}\downarrow & & {\scriptstyle\gamma}\downarrow & & \\ 0 & \longrightarrow & Y_2 & \longrightarrow & X_2 & \longrightarrow & Z_2 & \longrightarrow & 0 \end{array}$$

commutative. This notion was introduced in [5,9] and clearly generalizes the older notion of equivalent sequences (where it is assumed that $Y_2 = Y_1$, $Z_2 = Z_1$ and α and γ are the identity). An exact sequence (1) is said to split if it is equivalent to the trivial sequence $0 \longrightarrow Y \longrightarrow Y \oplus Z \longrightarrow Z \longrightarrow 0$; a common shorthand for this is $\mathrm{Ext}(Z,Y) = 0$.

There is a correspondence between exact sequences of Banach spaces (1) and the so-called **zero-linear** maps, which are homogeneous maps $F : Z \longrightarrow Y$ satisfying an estimate $\|F(\sum_{i=1}^n x_i) - \sum_{i=1}^n F(x_i)\| \le \Delta_F \sum \|x_i\|$ for some constant Δ_F, each $n \in \mathbb{N}$ and all $x_i \in Z$.

With these tools let us return to the extension problem and consider it in full generality, replacing $\mathfrak{L}$ by other operator ideals. Since it is well known that given a surjective operator ideal $\mathfrak{A}$ (see [8, §2.5]), an exact sequence (1) and a fixed Banach space B the induced sequence

$$0 \longrightarrow \mathfrak{A}(Z, B) \xrightarrow{\pi^*} \mathfrak{A}(X, B) \xrightarrow{\imath^*} \mathfrak{A}(Y, B) \longrightarrow 0 \qquad (2)$$

is exact **up to** $\mathfrak{A}(X, B)$, the extension problem consists in deciding whether the whole diagram (2) is exact. If so, we shall say that the functor $\mathfrak{A}(\cdot, B)$ is exact at (1). A functor is called exact (see [8]) when it is exact at every exact sequence. For instance, the functor $\mathfrak{L}(\cdot, B)$ is exact if and only if the space B is injective.

A brief reminder of extension results

When $\mathfrak{A} = \mathfrak{K}$, the ideal of compact operators, an affirmative answer to the extension problem characterizes $\mathcal{L}_\infty$-spaces. Precisely:

Proposition 1 (Lindenstrauss [15]). A Banach space B is an $\mathcal{L}_\infty$ space if and only if the functor $\mathfrak{K}(\cdot, B)$ is exact.

Proof. Let B be an $\mathcal{L}_\infty$-space and let $T : Y \longrightarrow B$ be a finite rank operator. Since $T(Y)$ is λ-isomorphic to a subspace of some ℓ_∞^n-space, the operator T admits a finite rank extension $\tau : X \longrightarrow B$ with $\|\tau\| \leq \lambda\|T\|$. But $\mathcal{L}_\infty$ spaces have the (bounded) approximation property and so the finite rank operators are dense in $\mathfrak{K}(Y, B)$. Thus, the restriction $\mathfrak{K}(X, B) \longrightarrow \mathfrak{K}(Y, B)$ is almost open and therefore (open and) surjective.

To prove the converse just observe that the canonical inclusion $\delta : B \longrightarrow B^{**}$ extends to any superspace X by: $\Delta(x)(b^*) = \lim_{\mathfrak{U}}(\Delta_E x)(b^*)$, where E denotes a typical finite dimensional subspace of B, $\mathfrak{U}$ is an ultrafilter refining the Fréchet filter on the directed set of all such subspaces, and Δ_E is an extension (with norm at most λ) to X of the restriction of δ to E. It follows easily that B^{**} is λ-complemented in X^{**}, and by a well known characterization of $\mathcal{L}_\infty$-spaces, that is all that we need. □

The choice $\mathfrak{A} = \mathfrak{W}$, the ideal of weakly compact operators, was first published by Bourgain and Delbaen [3, Theorem 2.4]. Their real contribution was to give examples of spaces showing that the following result is not vacuous.

Proposition 2. A Banach space B is an $\mathcal{L}_\infty$ space with the Schur property if and only if the functor $\mathfrak{W}(\cdot, B)$ is exact.

The proof is as before taking into account that weakly compact operators into a Schur space are compact (only if); to get the if part, take a weakly null sequence in B. The sequence is contained in the image of some reflexive space; when extended to all the $C(K)$-space this weakly compact operator should be completely continuous.

Straeuli considered in [19, 3.1] the extension problem for different operator ideals. We quote:

Proposition 3. Let $\mathfrak{A}$ be an idempotent operator ideal with $\mathfrak{K} \subset \mathfrak{A} \subset \mathfrak{R}$, where $\mathfrak{R}$ denotes the Rosenthal operators. Then the functor $\mathfrak{A}(\cdot, B)$ is exact if and only if B is an $\mathcal{L}_\infty$-space such that $\mathfrak{A}(\cdot, B) = \mathfrak{K}(\cdot, B)$

The extension problem for the choices $\mathfrak{A} = \mathfrak{K}$ or $\mathfrak{A} = \mathfrak{W}$ but fixing the exact sequence was solved by Fakhoury [11, Théorème 3.1]:

Proposition 4. The functors $\mathfrak{K}(\cdot, B)$ —or $\mathfrak{W}(\cdot, B)$— are exact at (1) if and only if the dual sequence $0 \longrightarrow Z^* \longrightarrow X^* \longrightarrow Y^* \longrightarrow 0$ splits.

One then says that (1) **locally splits** or that Y is **locally complemented** in X. (Other authors use different terminologies.)

Returning to the general problem, there are two distinguished situations in which $C(K)$-valued operators extend: the Lindenstrauss-Pełczyński theorem [16] for subspaces of c_0 and the Johnson-Zippin theorem [12] for weakly* closed subspaces of ℓ_1.

Proposition 5 (Lindenstrauss-Pełczyński). Every $C(K)$-valued operator defined on a subspace of c_0 can be extended to all of c_0.

Their proof is of Hahn-Banach style, showing that if $H \subset c_0$, $T : H \longrightarrow C(K)$ is an operator, $\varepsilon > 0$ and $p \notin H$, then there exists an extension $T_p : H + [p] \longrightarrow C(K)$ with norm $\|T_p\| \leq \|T\| + \varepsilon$. Therefore, an operator $T : H \longrightarrow C(K)$ defined on a subspace H of c_0 can, for every $\varepsilon > 0$, be extended to an operator $\tau : c_0 \longrightarrow C(K)$ with $\|\tau\| \leq (1 + \varepsilon)\|T\|$. An equal norm extension (i.e., $\varepsilon = 0$) cannot be achieved as an example of Johnson and Zippin [13] shows. In this same paper Johnson and Zippin extend the validity of the extension result to subspaces of $c_0(\Gamma)$ using a different approach, which we will discuss now, that allows one to circumvent the ε in the original proof.

Given an inclusion $\imath : Y \longrightarrow X$, Lindenstrauss [15] approached the extension problem for $C(K)$-valued operators in different ways, one of them being the existence of a certain selection for the dual quotient map $\imath^*$. This was precisely formulated by Zippin in [20]. There, Y is said to be **λ-almost complemented** in X if each operator $T : Y \longrightarrow C(K)$ extends to X with norm at most $\lambda\|T\|$:

Lemma 1. Let $\imath : Y \longrightarrow X$ be an isomorphic embedding. Then Y is λ-almost complemented in Xif and only if there exists a weakly* continuous function ω : $\operatorname{Ball} Y^* \longrightarrow \lambda \operatorname{Ball} X^*$ such that $\imath^*\omega = \mathrm{id}$. (We shall call such a function a selector for $\imath^*$.)

Proof. As usual, we identify K with the carrier space of $C(K)$. If a weakly* continuous selector $\omega : \text{Ball}\, Y^* \longrightarrow \lambda\, \text{Ball}\, X^*$ exists for $\imath^*$ then $\tau(x)(k) = \omega(T^*k)(x)$ defines an operator $\tau : X \longrightarrow C(K)$ with $\|\tau\| \leq \lambda\|T\|$ and such that $\tau(\imath(y))(k) = \omega(T^*k)(\imath y) = \imath^*\omega(T^*k)(y) = T^*k(y) = Ty(k)$.

Conversely, if every $C(K)$-valued operator admits such an extension then consider the canonical operator $\delta : Y \longrightarrow C(\text{Ball}\, Y^*, \text{weak}^*)$ and let D be its extension to X with norm at most λ. The weakly* continuous selector for $\imath^*$ is $\omega(y^*)(x) = D(x)(y^*)$, since $\omega(y^*)(\imath(y)) = D(\imath(y))(y^*) = y^*(y)$. □

Let us say that (1) λ-almost splits if Y is λ-almost complemented in X through $\imath$. Observe that the notions of almost split and locally split sequences are independent. On one hand, the sequence

$$0 \longrightarrow X \xrightarrow{\ \delta\ } C(\text{Ball}\, X^*) \longrightarrow C(\text{Ball}\, X^*)/\delta(X) \longrightarrow 0$$

almost splits, as follows from the existence of the weakly* continuous selector $\text{Ball}\, X^* \longrightarrow \text{Ball}\, C(\text{Ball}\, X^*)^*$ for δ^* given by $D(x^*)(f) = f(x^*)$. Since $\|D(x^*)\| = \|x^*\|$ the sequence 1-almost splits. This sequence locally splits if and only if X is an $\mathcal{L}_\infty$-space since in that case X^{**} would be complemented in $C(\text{Ball}\, X^*)^{**}$. On the other hand, the sequence $0 \longrightarrow X \longrightarrow X^{**} \longrightarrow H(X) \longrightarrow 0$ locally splits, while it does not almost split when X is a non-injective $C(K)$-space.

With the aid of this approach Zippin obtained in [21,22] different proofs for the Lindenstrauss-Pełczyński theorem. At the end of the paper [21] Zippin poses three questions, one of which connects with the nature of the Ext functor: When does an exact sequence $0 \longrightarrow E \longrightarrow \ell_1 \longrightarrow Z \longrightarrow 0$ almost split? By, e.g. [7, Theorem 1.1], this is the same as asking: When is $\text{Ext}(Z, C(K)) = 0$? We will use this equivalence several times in the sequel. A substantial answer to this problem was obtained by Johnson and Zippin in [12]. (A partial converse to this in [14] shows that the hypothesis on the subspace cannot be weakened much.)

Proposition 6 (Johnson-Zippin). Every $\mathcal{L}_\infty$-valued operator defined on a $\sigma(\ell_1, c_0)$-closed subspace of ℓ_1 can be extended to ℓ_1.

The proof of Johnson and Zippin is rather technical and long and at first sight seems to have no common point with that of Lindenstrauss and Pełczyński. Nevertheless, the remainder of this paper will show that the two results share a common homological nature. The basic idea of both our proofs is the same: each result will first be established under the additional assumption of a finite dimensional decomposition, and then we appeal to the result that any separable Banach space X admits a short exact sequence $0 \longrightarrow Y \longrightarrow X \longrightarrow Z \longrightarrow 0$ with Y and Z both having FDDs. Of course the details differ a bit.

The Lindenstrauss-Pełczyński theorem

We reformulate Proposition 5.

Theorem 1. Every exact sequence $0 \longrightarrow H \longrightarrow c_0 \longrightarrow c_0/H \longrightarrow 0$ almost splits.

Proof. The first ingredient of our proof is a decomposition result of Johnson, Rosenthal and Zippin (see [18, Theorems 1.g.2 and 2.d.1]) asserting that every subspace H of c_0 admits an exact sequence $0 \longrightarrow c_0(A_n) \longrightarrow H \longrightarrow c_0(B_n) \longrightarrow 0$ in which A_n and B_n are finite dimensional spaces. Crossing the new sequence with the starting one and completing the diagram we get

$$\begin{array}{ccccccccc}
 & & 0 & & 0 & & & & \\
 & & \downarrow & & \downarrow & & & & \\
 & & c_0(A_n) & = & c_0(A_n) & & & & \\
 & & \downarrow & & \downarrow & & & & \\
0 & \longrightarrow & H & \longrightarrow & c_0 & \longrightarrow & c_0/H & \longrightarrow & 0 \\
 & & \downarrow & & \downarrow & & \| & & \\
0 & \longrightarrow & c_0(B_n) & \underset{w}{\longrightarrow} & c_0/c_0(A_n) & \longrightarrow & c_0/H & \longrightarrow & 0 \\
 & & \downarrow & & \downarrow & & & & \\
 & & 0 & & 0 & & & &
\end{array}$$

To control the vertical sequence centered at c_0 we use a theorem of Lindenstrauss and Rosenthal [17] which implies that any such sequence is isomorphically equivalent to a sequence

$$0 \longrightarrow c_0(A_n) \stackrel{\kappa}{\longrightarrow} c_0(\ell_\infty^{m(n)}) \longrightarrow c_0(\ell_\infty^{m(n)}/A_n) \longrightarrow 0,$$

where κ is defined by $\kappa((a_n)_n) = (\jmath_n(a_n))_n$, for suitable $(1+\varepsilon)$-into isomorphisms $\jmath_n : A_n \longrightarrow \ell_\infty^{m(n)}$. The Bartle-Graves selection principle (we use the stronger version appearing in [1, Proposition 1.19 (ii)]) yields a continuous selector ω_n : $\operatorname{Ball} A_n^* \longrightarrow (1+\varepsilon)\operatorname{Ball} \ell_1^{m(n)}$ for $\jmath_n^*$. Putting together all these maps, we obtain a selector $\omega : \operatorname{Ball}(\ell_1(A_n^*)) \longrightarrow (1+\varepsilon)\operatorname{Ball}(\ell_1(\ell_1^{m(n)}))$ for κ^* defined by $\omega((a_n^*)_n) = (\omega_n(a_n^*))_n$. This selector turns out to be weakly* continuous since its domain is metrizable and a bounded sequence in a space $\ell_1(F_n)$ with F_n finite-dimensional is weakly* null if and only if the norms of its projections into F_n are convergent to 0. In conclusion, every exact sequence $0 \longrightarrow c_0(A_n) \longrightarrow c_0 \longrightarrow Q \longrightarrow 0$ almost-splits.

We just finish the proof with a kind of 3-space property for the extension of operators. (Actually, one should be a bit more careful because the Lindenstrauss-Rosenthal result used above only gives an isomorphically equivalent sequence; we leave the reader to worry about these details.)

Proposition 7. If the functor $\mathfrak{L}(\cdot, \spadesuit)$ is exact at the sequences $0 \longrightarrow A \longrightarrow X \longrightarrow X/A \longrightarrow 0$ and $0 \longrightarrow Y/A \longrightarrow X/A \longrightarrow X/Y \longrightarrow 0$ then it is also exact at $0 \longrightarrow Y \longrightarrow X \longrightarrow X/Y \longrightarrow 0$.

Proof. This can be seen by drawing homology sequences but this time it is easier to go back to basics. So let $T \in \mathfrak{L}(Y, \spadesuit)$ be any operator. By our first assumption, $T|_A$ admits an extension $S : X \longrightarrow \spadesuit$. Since $S - T$ vanishes on A, its restriction to Y factorizes as RQ where $Q : Y \longrightarrow Y/A$ is the quotient map and $R \in \mathfrak{L}(Y/A, \spadesuit)$. Our second assumption guarantees that R has an extension (for which we use the same symbol) in $\mathfrak{L}(X/A, \spadesuit)$. Then $x \mapsto Sx - R(x + A)$ is the required extension of T. □

Our homological approach has some advantages and, of course, some disadvantages. The worst of these is the absence of a good estimate for the norm of the extension: we obtain $(1+\varepsilon)$-for subspaces of c_0 of the form $c_0(A_n)$ but the estimate for general subspaces blows up during the last part of the proof. This is precisely the benefit of Zippin's careful work [22]: he obtains $1+\varepsilon$. The advantages, on the other hand, are clear: our proof is clean and, as we shall see, by duality it provides a proof for the Johnson-Zippin theorem. We close this section with an example showing that the hypotheses of the Lindenstrauss-Pełczyński theorem are optimal, in the sense that $C(K)$-spaces cannot be replaced by arbitrary $\mathcal{L}_\infty$-spaces. Since we will refer to the following construction of Bourgain and Pisier [4] later on, we quote it at this stage.

Proposition 8. Any separable Banach space Y can be embedded in a separable $\mathcal{L}_\infty$ space $\mathcal{L}_\infty(Y)$ such that $\mathcal{L}_\infty(Y)/Y$ has the Schur property.

Example. Let $0 \longrightarrow H \longrightarrow c_0 \longrightarrow Q \longrightarrow 0$ be an exact sequence with $Q \neq c_0$. Then there is an operator from H into an $\mathcal{L}_\infty$-space that cannot be extended to c_0.

Proof. Let $0 \longrightarrow H \longrightarrow \mathcal{L}_\infty(H) \longrightarrow S \longrightarrow 0$ be the corresponding Bourgain-Pisier sequence. Pushing out and completing the diagram [8] we obtain

$$\begin{array}{ccccccccc}
 & & 0 & & 0 & & & & \\
 & & \downarrow & & \downarrow & & & & \\
0 & \longrightarrow & H & \longrightarrow & c_0 & \longrightarrow & c_0/H & \longrightarrow & 0 \\
 & & \downarrow & & \downarrow & & \| & & \\
0 & \longrightarrow & \mathcal{L}_\infty(H) & \longrightarrow & \mathrm{PO} & \longrightarrow & c_0/H & \longrightarrow & 0 \\
 & & \downarrow & & \downarrow & & & & \\
 & & S & = & \mathrm{PO}/c_0 & & & & \\
 & & \downarrow & & \downarrow & & & & \\
 & & 0 & & 0 & & & &
\end{array}$$

Now, if the inclusion $H \longrightarrow \mathcal{L}_\infty(H)$ extends to c_0 then the lower row splits and PO $= \mathcal{L}_\infty(H) \oplus c_0/H$. However, the middle vertical sequence certainly splits by Sobczyk's theorem, so that PO $= c_0 \oplus S$. Therefore $\mathcal{L}_\infty(H) \oplus c_0/H = c_0 \oplus S$ thus making c_0/H a complemented subspace of $c_0 \oplus S$. Since S is Schur, it is totally incomparable with c_0 and thus by the Edelstein-Wojtasczyk theorem (see [18, Theorem 2.c.13]) c_0/H decomposes as $A \oplus B$ where A is a complemented subspace of c_0 and B a complemented subspace of S. Since c_0/H is a subspace of c_0 then B has to be finite dimensional, and therefore $c_0/H = c_0$, which we assumed not to happen. □

The Johnson-Zippin theorem

We reformulate Proposition 6 via almost splitting sequences; of course the extension result is also valid for $\mathcal{L}_\infty$-valued operators and not only for $C(K)$-valued ones.

Theorem 2. Let H be a subspace of c_0. Every exact sequence $0 \longrightarrow H^\perp \longrightarrow \ell_1 \longrightarrow H^* \longrightarrow 0$ almost splits.

Proof. First of all, a fundamental difference from other extension results is our access to the tool which we mentioned earlier: a sequence $0 \longrightarrow W \longrightarrow \ell_1 \longrightarrow X \longrightarrow 0$ almost splits if and only if $\operatorname{Ext}(X, C(K)) = 0$ for all $C(K)$-spaces. Thus the result asserts that whenever H is a subspace of c_0 then $\operatorname{Ext}(H^*, C(K)) = 0$ for all $C(K)$-spaces.

Now, entering into the proof, since every subspace H of c_0 admits a decomposition $0 \longrightarrow c_0(A_n) \longrightarrow H \longrightarrow c_0(B_n) \longrightarrow 0$, every dual H^* of a subspace of c_0 admits a decomposition $0 \longrightarrow \ell_1(B_n^*) \longrightarrow H^* \longrightarrow \ell_1(A_n^*) \longrightarrow 0$. The property $\operatorname{Ext}(\cdot, \spadesuit) = 0$ is a 3-space property (e.g. [7, Corollary 1.2]), so it is enough to prove that $\operatorname{Ext}(\ell_1(F_n), \mathcal{L}_\infty) = 0$.

Every exact sequence $0 \longrightarrow \mathcal{L}_\infty \longrightarrow X \longrightarrow Z \longrightarrow 0$ locally splits, so there is a constant C such that every exact sequence $0 \longrightarrow \mathcal{L}_\infty \longrightarrow X_F \longrightarrow F \longrightarrow 0$ with F finite dimensional splits and there exists a projection $X_F \longrightarrow \mathcal{L}_\infty$ with norm at most C (see [6]); i.e., $\operatorname{Ext}(F, \mathcal{L}_\infty) = 0$ uniformly on F. So, $\operatorname{Ext}(\ell_1(F_n), \mathcal{L}_\infty)) = \prod \operatorname{Ext}(F_n, \mathcal{L}_\infty) = 0$. □

An estimate of the norm of the extension can be obtained through a non-linear argument:

Lemma 2. Let $0 \longrightarrow Y \longrightarrow X \xrightarrow{\pi} F \longrightarrow 0$ be an exact sequence with F finite-dimensional. For every $\varepsilon > 0$ there exists a homogeneous lifting $\varrho : F \longrightarrow X$ for π with finite-dimensional range, and $\|\varrho(x)\| \leq (1+\varepsilon)\|x\|$ for all $x \in F$. If furthermore Y is an $\mathcal{L}_{\infty,\lambda}$ space then there is a projection $P : X \longrightarrow Y$ with norm $\|P\| \leq \lambda + 2 + \varepsilon$.

Proof. Let A be a finite (symmetric) subset of the unit sphere of F whose convex hull contains the ball of radius $1-\varepsilon$. For each $a \in A$ choose a point $\varrho(a) \in a+Y$ with $\|\varrho(a)\| \leq 1+\varepsilon$, and put $\varrho(0)=0$. Any $x \in F$ with norm $1-\varepsilon$ is a finite convex combination of elements of A; define $\varrho(x)$ as the corresponding convex combination from $\varrho(A)$. Extending ϱ to all of F by homogeneity gives us a selection for the quotient mapping with finite dimensional range. It also satisfies the inequality $\|\varrho(x)\| \leq (1+\varepsilon)(1-\varepsilon)^{-1}\|x\|$ for all $x \in F$, which is near enough.

Now suppose that Y is an $\mathcal{L}_{\infty,\lambda}$ space. Let $\ell : F \longrightarrow X$ be any linear lifting for the quotient mapping, and define $\Omega : F \longrightarrow Y$ by $\Omega(z) = \ell(z) - \varrho(z)$. This is a zero-linear map with constant $\Delta_\Omega \leq 1+\varepsilon$. Since $\Omega(F)$ is finite dimensional, it is contained in some finite-dimensional subspace E of Y which is within Banach-Mazur distance λ of a finite dimensional ℓ_∞^n space. Now we apply the so-called nonlinear Hahn-Banach Theorem [5, Lemma 1] coordinatewise to obtain a linear mapping $L : F \longrightarrow Y$ satisfying the inequality $\|L(z) - \Omega(z)\| \leq \lambda\|z\|$.

The operator $(\ell - L)\circ\pi : X \longrightarrow X$ is a projection with kernel Y (one has $((\ell - L)\circ\pi)^2 = (\ell - L)\circ\pi$ since $\pi\circ L = 0$ and $\pi\circ\ell = \mathrm{id}$) and verifies, for any x,

$$\|(\ell - L)\pi x\| \leq \|\ell\pi x - \Omega\pi x\| + \|\Omega\pi x - L\pi x\| \leq (1+\varepsilon+\lambda)\|\pi x\|.$$

Hence $\mathrm{id} - (\ell - L)\pi$ is the desired projection onto Y with norm $\leq 2+\lambda+\varepsilon$. □

Since $C(K)$-spaces are $\mathcal{L}_{\infty,\lambda}$-spaces for every $\lambda > 1$, the decomposition technique of the previous section gives us an estimate of $3+\varepsilon$ for the norm of a projection in an exact sequence $0 \longrightarrow C(K) \longrightarrow X \longrightarrow H^* \longrightarrow 0$ when H is a subspace of $c_0(F_n^*)$. This is the same estimate that Johnson and Zippin obtain in [12], although our proof is considerably simpler. However, under the additional assumption that E has the approximation property, they were able to obtain the stronger conclusion that the extension operator $\tilde{T}$ could be chosen so that $\|\tilde{T}\| \leq (1+\varepsilon)\|T\|$. They ask if this estimate could be true without the approximation property; this remains unknown.

Further results and open problems

Let's begin this section with another especially interesting example of almost splitting obtained by Zippin in [20,21]:

Proposition 9. Every sequence $0 \longrightarrow W \stackrel{\imath}{\longrightarrow} \ell_p \longrightarrow \ell_p/W \longrightarrow 0$ (with $1 < p < \infty$) almost-splits.

Since the proof does not explicitly appear in the literature, let us present here a sketch: the dual space $\ell_p^* = \ell_{p^*}$ is strictly convex, so there exists a unique Hahn-Banach extension operator $\omega : W^* \longrightarrow \ell_{p^*}$. We show that its restriction $\omega : \mathrm{Ball}\, W^* \longrightarrow \mathrm{Ball}\, \ell_{p^*}$ is weak*-to-weak* continuous. The map ω is the composition of three maps

$$\begin{array}{ccc} W^* & \xrightarrow{\omega} & l_{p^*} \\ J_W^{-1}\downarrow & & \uparrow J_p \\ W & \xrightarrow{\imath} & l_p \end{array}$$

where J_p and J_W are the support mappings (also called the duality maps). Since $J_p(x) = \|x\|^{2-p}|x|^{p-1}\text{sgn } x$, it is clear that J_p is weakly sequentially continuous, as is the inclusion $\imath$. The support mapping J_W is just $J_p \circ \imath$. Its restriction to Ball W endowed with the weak topology is a continuous map from a compact space to a Hausdorff space, hence a homeomorphism; this makes its inverse weakly continuous.

In general, however, the extension of operators on a reflexive subspace is not to be expected. Suppose for example that a sequence $0 \longrightarrow Y \longrightarrow X \longrightarrow Z \longrightarrow 0$ does not locally split (as will be the case when Y is complemented in its bidual but not in X). Then there exists, by Proposition 4 and the well known factorization of weakly compact operators, an operator $T : Y \longrightarrow R$ into some reflexive space R that cannot be extended to X. This gives rise to the question of what happens when R is embedded into some $C(K)$ space. Does the resulting composition $Y \longrightarrow R \longrightarrow C(K)$ extend to X? In particular

Question 1. Does every sequence $0 \longrightarrow \ell_2 \longrightarrow C[0,1] \longrightarrow C[0,1]/\ell_2 \longrightarrow 0$ almost split?

We remark that, in general, $\mathcal{L}_\infty$-valued operators cannot be extended: consider the diagram

$$\begin{array}{ccccccccc} 0 & \longrightarrow & \ell_2 & \xrightarrow{\imath} & C[0,1] & \longrightarrow & C[0,1]/\ell_2 & \longrightarrow & 0 \\ & & \| & & & & & & \\ 0 & \longrightarrow & \ell_2 & \xrightarrow{\jmath} & \mathcal{L}_\infty(\ell_2) & \longrightarrow & S & \longrightarrow & 0 \end{array}$$

in which $\mathcal{L}_\infty(\ell_2)$ is the Bourgain-Pisier space associated to ℓ_2 (Proposition 8). Since S is Schur, $\mathcal{L}_\infty(\ell_2)$ does not contain c_0 and therefore any possible extension $J : C(K) \longrightarrow \mathcal{L}_\infty(\ell_2)$ of $\jmath$ would be weakly compact, hence completely continuous (by the Dunford-Pettis property of $C(K)$-spaces) and therefore $J \circ \imath$ could not be the identity on ℓ_2.

A simpler argument (using the fact that every operator $\ell_\infty \longrightarrow C[0,1]$ is weakly compact) shows that if X is a separable non-Schur space, then no sequence $0 \longrightarrow X \longrightarrow \ell_\infty \longrightarrow \ell_\infty/X \longrightarrow 0$ almost-splits.

The notion of almost-complementation puts all $C(K)$-spaces at the same level. We could, however, try to distinguish between different $C(K)$-spaces regarding extension properties of operators into them. As a token that such finer classification is possible, we consider the sequence $0 \longrightarrow W_T \longrightarrow \ell_1 \longrightarrow T \longrightarrow 0$ where T denotes Tsirelson's space. Since Tsirelson's space emphatically fails the Schur property, it follows from [7,14] that this sequence admits operators $W_T \longrightarrow C[0,1]$ which do not extend to ℓ_1, yet every operator $W_T \longrightarrow C(\omega^\omega)$ does extend to ℓ_1. This suggests

Question 2. Given a fixed compact space K, characterize the sequences $0 \longrightarrow Y \longrightarrow \ell_1 \longrightarrow Z \longrightarrow 0$ such that all operators $Y \longrightarrow C(K)$ extend to ℓ_1.

The problem of distinguishing between $C(K)$ and $\mathcal{L}_\infty$-spaces pervades the paper, so another natural question is:

Question 3. If $\mathrm{Ext}(Z, C(K)) = 0$ for every compact K, does $\mathrm{Ext}(Z, \mathcal{L}_\infty) = 0$ for all $\mathcal{L}_\infty$-spaces?

And, more generally and not only for subspaces of ℓ_1,

Question 4. Let $\mathcal{H}$ be the class of all subspaces of c_0, and let $\mathcal{L}_\infty(\mathcal{H})$ be the class of all $\mathcal{L}_\infty$ spaces with the property that every operator $T : H \longrightarrow L$ from $H \in \mathcal{H}$ into $L \in \mathcal{L}_\infty(\mathcal{H})$ can be extended to c_0. Characterize the class $\mathcal{L}_\infty(\mathcal{H})$.

The same question for the class of subspaces of ℓ_p, of subspaces of L_p, and for Hilbert subspaces of $C[0,1]$ does not lack interest. Or difficulty: although many Banach spaces (e.g. complemented subspaces of $C(K)$ spaces) belong to these classes, any characterization seems to be quite elusive. Now let $0 \longrightarrow \ell_2 \longrightarrow \mathcal{L}_\infty(\ell_2) \longrightarrow S \longrightarrow 0$ be a Bourgain-Pisier sequence. If H is a subspace of c_0 then $\mathfrak{L}(H, \ell_2) = \mathfrak{K}(H, \ell_2)$ and also $\mathfrak{L}(H, S) = \mathfrak{K}(H, S)$ since S is a Schur space. It is not hard to verify that $\mathfrak{L}(H, \spadesuit) = \mathfrak{K}(H, \spadesuit)$ is a 3-space property (see [8, §6.1 and §6.7]). In particular $\mathfrak{L}(H, \mathcal{L}_\infty(\ell_2)) = \mathfrak{K}(H, \mathcal{L}_\infty(\ell_2))$ and thus, by Proposition 1, operators $H \longrightarrow \mathcal{L}_\infty(\ell_2)$ extend to c_0. Since $\mathcal{L}_\infty(\ell_2)$ does not contain c_0 (another 3-space property, see [8]) it cannot be a complemented subspace of a $C(K)$-space.

References

1. Y. Benyamini and J. Lindenstrauss, Geometric nonlinear functional analysis, Vol. 1, Amer. Math. Soc., Providence (1999).
2. J. Bourgain, New classes of $\mathcal{L}_p$–spaces. *Lecture Notes in Math.* **889** Springer, Berlin-New York, 1981.
3. J. Bourgain and F. Delbaen, A class of special $\mathcal{L}_\infty$-spaces, *Acta Math.* **145** (1980) 155–176.
4. J. Bourgain and G. Pisier, A construction of $\mathcal{L}_\infty$-spaces and related Banach spaces, *Bol. Soc. Bras. Mat.* **14** (1983) 109–123.
5. F. Cabello Sánchez and J.M.F. Castillo, Duality and twisted sums of Banach spaces, *J. Funct. Anal.* **175** (2000) 1–16.
6. F. Cabello Sánchez and J.M.F. Castillo, Uniform boundedness and twisted sums of Banach spaces, *Houston J. Math.* to appear.
7. F. Cabello Sánchez, J.M.F. Castillo, N.J. Kalton and D. Yost, Twisted sums with $C(K)$-spaces, *Trans. Amer. Math. Soc.* to appear.
8. J.M.F. Castillo and M. González, Three Space Problems in Banach Space Theory, *Lecture Notes in Math.* **1667** Springer, New York-Berlin, 1977.
9. J.M.F. Castillo and Y. Moreno, On isomorphically equivalent extensions of quasi-Banach spaces, in *Recent Progress in Functional Analysis*, (K.D. Bierstedt, J. Bonet, M. Maestre, J. Schmets (eds.)), North-Holland Math. Studies **187** (2000) 263–272.
10. J. Diestel, H. Jarchow and A. Tonge, Absolutely summing operators. *Cambridge Studies in Advanced Mathematics*, **43** Cambridge University Press, 1995.

11. H. Fakhoury, Sélections linéaires associées au théorème de Hahn-Banach, *J. Functional Analysis* **11** (1972) 436–452.
12. W.B. Johnson and M. Zippin, Extension of operators from weak*-closed subspaces of ℓ_1 into $C(K)$ spaces, *Studia Math.* **117** (1995) 43–55.
13. W.B. Johnson and M. Zippin, Extension of operators from subspaces of $c_0(\Gamma)$ into $C(K)$ spaces, *Proc. Amer. Math. Soc.* **107** (1989) 751–754.
14. N.J. Kalton, On subspaces of c_0 and extension of operators into $C(K)$-spaces, Quart. *J. Math,* (Oxford) **52** (2001) 313–328.
15. J. Lindenstrauss, On the extension of compact operators, *Memoirs of the AMS* **48** (1964).
16. J. Lindenstrauss and A. Pełczyński, Contributions to the theory of the classical Banach spaces, *J. Funct. Anal.* **8** (1971) 225–249.
17. J. Lindenstrauss and H.P. Rosenthal, Automorphisms in c_0, ℓ_1 and m, *Israel J. Math.* **7** (1967) 227–239.
18. J. Lindenstrauss and L. Tzafriri, *Classical Banach spaces, Springer-Verlag.* Berlin-Heidelberg-New York, 1977.
19. E. Straeuli, On Hahn-Banach extension for certain operator ideals, *Arch. Math.* (Basel) **47** (1986) 49–54.
20. M. Zippin, The embedding of Banach spaces into spaces with structure, *Illinois J. Math.* **34** (1990) 586–606.
21. M. Zippin, A global approach to certain operator extension problems, in *Springer Lecture Notes in Math.* **1470** (1990) 78–84.
22. M. Zippin, Applications of Michael's continuous selection theorem to operator extension problems, *Proc. Amer. Math. Soc.* **127** (1999) 1371–1378.

Functional Analysis and its Applications
V. Kadets and W. Żelazko

Un théorème de point fixe pour les fonctions multivoques acycliques

Robert Cauty

Université Paris 6, UFR 920, Boîte courrier 172, 4, place Jussieu, 75252 Paris Cedex 05

Abstract

We prove that equiconnected spaces have the fixed point property for multivalued u.s.c. acyclic compact mappings.

Key words: points fixes, fonctions multivoques acycliques
2000 MSC: 54H25, 54C60, 55M20

1. Introduction

Soit $I = [0,1]$. Un espace topologique X est dit uniformément contractile, ou UC, s'il existe une fonction continue $\lambda : X \times X \times I \to X$ vérifiant, quels que soient x, y dans X et t dans I,

$$\lambda(x,y,0) = x, \quad \lambda(x,y,1) = y$$
$$\lambda(x,x,t) = x$$

S'il est nécessaire de fixer la fonction λ vérifiant ces conditions, nous parlerons de l'espace UC (X, λ).

La classe des espaces UC est vaste. Elle contient les sous-ensembles convexes des espaces topologiques, les groupes contractiles et les rétractes de tels espaces. En particulier, tout rétracte absolu est UC.

Email address: cauty@math.jussieu.fr (Robert Cauty).

Pour tout compact K et tout entier $n \geq 0$, nous notons $H^n(K, \mathbb{Q})$ le n-ème groupe de cohomologie de Čech de K a coefficients rationnels (cohomologie réduite si $n = 0$). Le compact K est dit acyclique si $H^n(K, \mathbb{Q}) = 0$ pour tout $n \geq 0$.

Soient X, Y des espaces séparés. Une fonction multivoque compacte de X dans Y est une fonction faisant correspondre à tout élément x de X un sous-ensemble compact non vide $F(x)$ de Y et telle que $\bigcup_{x \in X} F(x)$ soit contenu dans un compact de Y. Une telle fonction est dite semi-continue supérieurement, ou s.c.s., si, pour tout point x de X et tout voisinage U de $F(x)$ dans Y, il existe un voisinage V de x tel que $F(y) \subset U$ pour tout $y \in V$. Elle est dite acyclique si le compact $F(x)$ est acyclique pour tout $x \in X$.

Si F est une fonction multivoque de X dans lui-même, un point x_0 de X est un point fixe de F si $x_0 \in F(x_0)$. Il est connu que si X est un rétracte absolu, alors toute fonction multivoque compacte s.c.s. et acyclique F de X dans lui-même a un point fixe. Lorsque X est compact, c'est un cas particulier d'un théorème d'Eilenberg et Montgomery [4], et le cas général se ramène au cas compact comme suit. Soit K un compact de X contenant $\bigcup_{x \in X} F(x)$, et soit φ un plongement de K dans le cube de Hilbert I^ω. Puisque X est un rétracte absolu, il existe une fonction continue $\psi : I^\omega \to X$ telle que $\psi|\varphi(K) = \varphi^{-1}$. La formule $G(y) = \varphi(F(\psi(y)))$ définit une fonction multivoque compacte s.c.s. et acyclique de I^ω dans lui-même. D'après le résultat cité d'Eilenberg et Montgomery, G a un point fixe y_0, et il est facile de voir que $\varphi^{-1}(y_0)$ est un point fixe de F. Nous nous proposons ici de généraliser ce résultat à tous les espaces UC séparés.

Théorème. *Toute fonction multivoque compacte s.c.s. et acyclique d'un espace UC séparé dans lui-même a un point fixe.*

La démonstration de ce théorème suit le schéma introduit dans [1] pour résoudre le problème de Schauder, et qui en fait fournit un canevas général pour étendre aux espaces ULC les théorèmes de point fixe connus pour les applications compactes des rétractes absolus de voisinage. Nous utiliserons le foncteur construit dans [3], qui associe à tout compact X un espace UC $(U(X), \lambda_X)$ contenant X et ayant la propriété universelle suivante: pour tout espace UC (Y, λ) et toute fonction continue f de X dans Y, il existe une unique fonction continue g de $U(X)$ dans Y prolongeant f et vérifiant $g(\lambda_X(x, y, t)) = \lambda(g(x), g(y), t)$ quels que soient x, y et t.

Soit Y un espace UC, et soit F une fonction multivoque compacte s.c.s. et acyclique de Y dans lui-même. Fixons un sous-ensemble compact X de Y contenant $\bigcup_{y \in Y} F(y)$. La propriété universelle de notre foncteur nous permet de trouver une fonction continue r de $U(X)$ dans Y telle que $r(x) = x$ pour tout $x \in X$. La formule $G(x) = F(r(x))$ définit alors une fonction multivoque compacte s.c.s. et acyclique de $U(X)$ dans X. Si x_0 est un point fixe de G, il appartient à X et, comme $r(x_0) = x_0$, c'est aussi un point fixe de F. La démonstration est donc ramenée à la vérification du cas particulier suivant:

Proposition. *Pour tout compact X, toute fonction multivoque compacte s.c.s. et acyclique de $U(X)$ dans X a un point fixe.*

La démonstration de cette proposition se fera en deux étapes. Nous la prouverons d'abord pour les compacts métrisables en utilisant l'existence d'un compact Z et d'une fonction continue φ de Z sur X ayant des propriétés particulières qui permettent de se ramener au résultat connu pour les rétractes absolus. Nous passerons ensuite au cas général par un argument de limite projective.

Rappelons que, pour tout compact X, $U(X)$ est la limite inductive d'une suite croissante $\{U_n(X)\}_{n=1}^{\infty}$ de compacts. Les $U_n(X)$ sont des foncteurs qui commutent aux limites projectives et préservent la surjectivité des fonctions. Si X est métrisable, il en est de même des $U_n(X)$, et $U(X)$ est donc alors parfaitement normal.

2. Le cas métrisable

Si le compact X est métrisable, nous notons $\mathcal{T}(X)$ l'ensemble des topologies métrisables τ sur $U(X)$ qui sont moins fines que la topologie libre et telles que $\lambda_X : (U(X), \tau) \times (U(X), \tau) \times I \to (U(X), \tau)$ soit continue.

Lemme 1. Soit X un compact métrisable, et soit F une fonction multivoque compacte de $U(X)$ dans X. Si F est s.c.s., il existe $\tau \in \mathcal{T}(X)$ telle que F soit τ-s.c.s..

Démonstration. Soit $\mathcal{B} = \{B_i\}_{i=0}^{\infty}$ une base dénombrable de X stable par réunion finie. Pour $i \geq 0$, soit $V_i = \{x \in U(X) \mid F(x) \subset B_i\}$. Pour tout compact K de X et tout voisinage U de K dans X, il existe i tel que $K \subset B_i \subset U$, donc F sera τ-s.c.s. si tous les V_i sont τ-ouverts. Puisque F est s.c.s., V_i est ouvert. Puisque $U(X)$ est parfaitement normal, il existe une fonction continue $\alpha_i : U(X) \to I$ telle que $\alpha_i^{-1}(]0,1]) = V_i$. Définissons $\alpha : U(X) \to I^{\omega}$ par $\alpha(x) = (\alpha_0(x), \alpha_1(x), \dots)$. Puisque α est continue, le lemme 2 de [3] montre l'existence d'une topologie $\tau \in \mathcal{T}(X)$ telle que α soit τ-continue. Cette topologie a la propriété souhaitée.

Soit $\mathcal{A}$ la classe des compacts X tels que $(U(X), \tau)$ soit un rétracte absolu quelle que soit $\tau \in \mathcal{T}(X)$. Un compact K est de forme triviale s'il est non vide et si toute fonction continue de K dans un rétracte absolu de voisinage est homotope à une fonction constante; en particulier, un compact de forme triviale est acyclique.

Le lemme suivant est prouvé dans [2].

Lemme 2. Soit X un compact métrisable. Il existe un compact métrisable Z et une fonction continue $\varphi : Z \to X$ vérifiant

(i) Z appartient á $\mathcal{A}$,

(ii) pour tout $x \in X$, $\varphi^{-1}(x)$ est de forme triviale.

Démonstration de la proposition pour X métrisable. Soit $F : U(X) \to X$ une fonction multivoque compacte s.c.s. et acyclique. Le lemme 1 nous fournit une topologie $\tau \in \mathcal{T}(X)$ telle que F soit τ-s.c.s. Fixons un compact Z et une fonction continue $\varphi : Z \to X$ vérifiant les conditions du lemme 2. Soit $\hat{\varphi} : U(Z) \to U(X)$ le prolongement naturel de φ. Le lemme 2 de [3] nous permet de trouver $\tau' \in \mathcal{T}(Z)$ telle que $\hat{\varphi} : (U(Z), \tau') \to (U(X), \tau)$ soit continue. La formule

$$G(z) = \varphi^{-1}(F(\hat{\varphi}(z)))$$

définit une fonction multivoque compacte de $(U(Z), \tau')$ dans Z, qui est s.c.s. comme composée des fonctions s.c.s. $\hat{\varphi}$, F et φ^{-1}. Puisque $\varphi^{-1}(x)$ est acyclique pour tout $x \in X$, le théorème de Vietoris-Begle ([7], théorème 6.9.15) entraîne que $H^n(\varphi^{-1}(F(\hat{\varphi}(z))), \mathbb{Q}) = H^n(F(\hat{\varphi}(z)), \mathbb{Q}) = 0$ pour tout $n \geq 0$, donc G est aussi acyclique. Puisque Z appartient à $\mathcal{A}$, $(U(Z), \tau')$ est un rétracte absolu. Le théorème de point fixe mentionné dans l'introduction montre que G a un point fixe z_0. Alors z_0 appartient à Z, donc $\hat{\varphi}(z_0) = \varphi(z_0) \in \varphi(\varphi^{-1}(F(\hat{\varphi}(z_0)))) = F(\varphi(z_0))$.

3. Résultats auxiliaires

Un système projectif d'espaces topologiques sur un ensemble ordonné filtrant A sera noté $\mathbb{S} = (X_\alpha, p_\alpha^\beta, A)$ (les X_α sont des espaces topologiques et, pour $\alpha \leq \beta$, $p_\alpha^\beta : X_\beta \to X_\alpha$ est continue). Nous notons $\lim \mathbb{S}$ la limite projective de ce système et p_α la projection de $\lim \mathbb{S}$ dans X_α. Si B est un sous-ensemble filtrant de A, nous notons $\mathbb{S}|B = (X_\alpha, p_\alpha^\beta, B)$ le système obtenu en restreignant l'ensemble des indices à B. Si B est cofinal dans A, nous identifions naturellement $\lim(\mathbb{S}|B)$ à $\lim \mathbb{S}$.

Un système projectif $\mathbb{S} = (X_\alpha, p_\alpha^\beta, A)$ est appelé un ω-système projectif s'il vérifie les conditions suivantes:

(a) Tout sous-ensemble dénombrable totalement ordonné de A a une borne supérieure.

(b) Pour tout sous-ensemble totalement ordonné B de A admettant une borne supérieure β, la fonction $\lim_{\alpha \in B} p_\alpha^\beta : X_\beta \to \lim(\mathbb{S}|B)$ est un homéomorphisme.

(c) Chaque X_α a une base dénombrable.

Il est connu que tout compact X est la limite d'un ω-système projectif $(X_\alpha, p_\alpha^\beta, A)$ tel que les projections $p_\alpha : X \to X_\alpha$ soient surjectives (voir la démonstration de la proposition 3.2.17 de [6]).

Un sous-ensemble B d'un ensemble filtrant A est dit fermé si, pour tout sous-ensemble totalement ordonné C de B admettant une borne supérieure dans A, cette borne supérieure appartient à B. Si $\mathbb{S} = (X_\alpha, p_\alpha^\beta, A)$ est un ω-système projectif et si B est un sous-ensemble cofinal et fermé de A, alors $\mathbb{S}|B$ est aussi un ω-système projectif.

Soient $(X_\alpha, p_\alpha^\beta, A)$ et $(Y_\alpha, q_\alpha^\beta, A)$ deux ω-systèmes projectifs de compacts de limites X et Y respectivement, et soit F une fonction multivoque compacte s.c.s. et acyclique de X dans Y. Supposons les projections p_α surjectives. Nous pouvons alors, pour tout $\alpha \in A$, définir une fonction multivoque compacte $F_\alpha : X_\alpha \to Y_\alpha$ par la formule

$$F_\alpha(x) = q_\alpha(F(p_\alpha^{-1}(x))).$$

Étant composée des fonctions s.c.s. q_α, F et p_α^{-1}, F_α est s.c.s.. Pour ce qui est de son acyclicité, nous avons le résultat suivant:

Lemme 3. L'ensemble B des $\alpha \in A$ tels que F_α soit acyclique est cofinal et fermé dans A.

Démonstration. Soit M un sous-ensemble totalement ordonné de A admettant une borne supérieure μ. Pour $\alpha \leq \beta$, nous avons $p_\beta^{-1}(x) \subset p_\alpha^{-1}(p_\alpha^\beta(x))$, d'où $q_\alpha^\beta(F_\beta(x)) \subset F_\alpha(p_\alpha^\beta(x))$ pour tout $x \in X_\beta$. Par conséquent, pour tout $x \in X_\mu$, nous avons un système projectif $(F_\alpha(p_\alpha^\mu(x)), q_\alpha^\beta|F_\beta(p_\beta^\mu(x)), M)$ et $q_\alpha^\mu(F_\mu(x)) \subset F_\alpha(p_\alpha^\mu(x))$ pour tout $\alpha \in M$.

Affirmation 1. Pour tout $x \in X_\mu$, les fonctions $q_\alpha^\mu|F_\mu(x) : F_\mu(x) \to F_\alpha(p_\alpha^\mu(x))$ représentent $F_\mu(x)$ comme limite du système projectif $(F_\alpha(p_\alpha^\mu(x)), q_\alpha^\beta|F_\beta(p_\beta^\mu(x)), M)$.

En effet, d'après (b), la limite de ce système projectif est un sous-ensemble de Y_μ. Puisque $q_\alpha^\mu(F_\mu(x)) \subset F_\alpha(p_\alpha^\mu(x))$ pour tout $\alpha \in M$, $F_\mu(x)$ est contenu dans cette limite. Si $y \in Y_\mu$ n'appartient pas à $F_\mu(x)$, il y a, d'après (b), un $\alpha \in M$ tel que $q_\alpha^\mu(y) \notin q_\alpha^\mu(F_\mu(x))$. Puisque F_μ est s.c.s., l'ensemble V des $z \in X_\mu$ tels que $F_\mu(z) \subset Y_\mu \setminus (q_\alpha^\mu)^{-1}(q_\alpha^\mu(y))$ est un voisinage de x. Il existe donc $\beta \in M$ avec $\alpha \leq \beta$ et un voisinage W de $p_\beta^\mu(x)$ dans X_β tels que $(p_\beta^\mu)^{-1}(W) \subset V$. Alors $(p_\beta^\mu)^{-1}(p_\beta^\mu(x)) \subset V$, d'où

$$\begin{aligned} F_\beta(p_\beta^\mu(x)) &= q_\beta(F(p_\beta^{-1}(p_\beta^\mu(x)))) = q_\beta(F(p_\mu^{-1}((p_\beta^\mu)^{-1}(p_\beta^\mu(x))))) \\ &\subset q_\beta(F(p_\mu^{-1}(V))) = q_\beta^\mu q_\mu(F(p_\mu^{-1}(V))) = q_\beta^\mu(F_\mu(V)) \\ &\subset q_\beta^\mu(Y_\mu \setminus (q_\alpha^\mu)^{-1}(q_\alpha^\mu(y))) \subset Y_\beta \setminus (q_\alpha^\beta)^{-1}(q_\alpha^\mu(y)), \end{aligned}$$

ce qui entraîne que $q_\beta^\mu(y)$ n'appartient pas à $F_\beta(p_\beta^\mu(x))$ puisqu'il est contenu dans $(q_\alpha^\beta)^{-1}(q_\alpha^\mu(y))$. Le point y n'appartient donc pas à la limite projective du système $(F_\alpha(p_\alpha^\mu(x)), q_\alpha^\beta|F_\beta(p_\beta^\mu(x)), M)$, d'où l'affirmation.

Le fait que B est fermé résulte immédiatement de l'affirmation 1 et de la continuité de la cohomologie de Čech ([5, chapitre 10]).

Affirmation 2. Pour tout $\alpha_0 \in A$, il existe $\alpha_1 \geq \alpha_0$ tel que, pour tout $x \in X$, l'homomorphisme de $H^n(F_{\alpha_0}(p_{\alpha_0}^{\alpha_1}(x)), \mathbb{Q})$ dans $H^n(F_{\alpha_1}(x), \mathbb{Q})$ induit par $q_{\alpha_0}^{\alpha_1}|F_{\alpha_1}(x)$ soit trivial pour tout n.

Avant de prouver cette affirmation, remarquons qu'elle entraîne la cofinalité de B. En effet, partant d'un α_0 arbitraire, elle nous fournit une suite croissante $\{\alpha_i\}_{i=0}^\infty$ d'éléments de A telle que, pour tout i et tout x dans $X_{\alpha_{i+1}}$, l'homomorphisme de $H^n(F_{\alpha_i}(p_{\alpha_i}^{\alpha_{i+1}}(x), \mathbb{Q})$ dans $H^n(F_{\alpha_{i+1}}(x), \mathbb{Q})$ induit par $q_{\alpha_i}^{\alpha_{i+1}}|F_{\alpha_i}(x)$ soit trivial pour tout n. D'après (a), l'ensemble $A_0 = \{\alpha_0, \alpha_1, \dots\}$ a une borne supérieure α. D'après l'affirmation 1, pour tout $x \in X_\alpha$, $F_\alpha(x)$ est la limite du système projectif $(F_{\alpha_i}(p^{\alpha}_{\alpha_i}(x)), q^{\alpha_j}_{\alpha_i}|F_{\alpha_j}(p^{\alpha}_{\alpha_j}(x)), A_0)$ et le choix des α_i garantit que les fonctions $q^{\alpha_{i+1}}_{\alpha_i}|F_{\alpha_{i+1}}(p^{\alpha}_{\alpha_{i+1}}(x))$ de ce système induisent des homomorphismes triviaux en cohomologie de Čech. La continuité de la cohomologie de Čech entraîne alors que $H^n(F_\alpha(x), \mathbb{Q}) = 0$ pour tout $n \geq 0$ et tout $x \in X_\alpha$, donc α appartient à B.

Pour simplifier les notations dans la démonstration de l'affirmation 2, nous ferons les conventions suivantes. Pour $\beta \leq \gamma$ dans A, $K \subset Y_\beta$ et $e \in H^n(K, \mathbb{Q})$, nous notons $(q^{\gamma}_{\beta})^{-1}(e)$ (resp. $q_\beta^{-1}(e)$) l'image de e dans $H^n((q^{\gamma}_{\beta})^{-1}(K), \mathbb{Q})$ (resp. $H^n(q_\beta^{-1}(K), \mathbb{Q})$) par l'homomorphisme induit par la fonction de K dans $(q^{\gamma}_{\beta})^{-1}(K)$ (resp. $q_\beta^{-1}(K)$) induite par q^{γ}_{β} (resp. q_β). Pour $L \subset K$, nous noterons $e|L$ l'image de e dans $H^n(L, \mathbb{Q})$ par l'homomorphisme induit par l'inclusion de L dans K.

Fixons une base dénombrable $\mathcal{O} = \{O_i\}_{i=1}^{\infty}$ de la topologie de Y_{α_0} stable par réunion finie. Pour $i \geq 1$, soit $U_i = \{x \in X_{\alpha_0} \mid F_{\alpha_0}(x) \subset O_i\}$. Puisque F_{α_0} est s.c.s., U_i est ouvert, donc est σ-compact puisque le compact X_{α_0} est métrisable. Soit $U_i = \bigcup_{j=1}^{\infty} K_i^j$, où K_i^j est compact. Pour tout compact métrisable Z, $H^n(Z, \mathbb{Q})$ est dénombrable pour tout $n \geq 0$, donc nous pouvons trouver une partition $\mathbb{N} = \bigcup_{i=1}^{\infty} N_i$ de l'ensemble des entiers et, pour tout i, une surjection $k \mapsto e_k$ de N_i sur $\bigcup_{n=0}^{\infty} H^n(\overline{O_i}, \mathbb{Q})$.

Soit (i, j, k) un triplet d'entiers tel que k appartienne à N_i, et soit n tel que $e_k \in H^n(\overline{O_i}, \mathbb{Q})$. Nous allons montrer qu'il existe $\beta = \beta(i, j, k) \geq \alpha_0$ tel que, pour tout $x \in (p^{\beta}_{\alpha_0})^{-1}(K_i^j)$, $(q^{\beta}_{\alpha_0})^{-1}(e_k)|F_\beta(x) = 0$. Pour cela, remarquons d'abord que l'argument utilisé pour prouver l'affirmation 1 montre aussi que, pour tout $x \in X$, $F(x)$ est la limite du système projectif $(F_\alpha(p_\alpha(x)), q^{\beta}_{\alpha}|F_\beta(p_\beta(x)), A)$. Soit $x \in p_{\alpha_0}^{-1}(K_i^j)$; nous avons donc $F_{\alpha_0}(p_{\alpha_0}(x)) \subset O_i$. Puisque $F(x)$ est acyclique, la continuité de la cohomologie de Čech entraîne l'existence d'un $\beta_x \geq \alpha_0$ tel que $(q^{\beta_x}_{\alpha_0})^{-1}(e_k)|F_{\beta_x}(p_{\beta_x}(x)) = 0$, puis celle d'un voisinage V_{β_x} de $F_{\beta_x}(p_{\beta_x}(x))$ dans Y_{β_x}, contenu dans $(q^{\beta_x}_{\alpha_0})^{-1}(O_i)$ et tel que $(q^{\beta_x}_{\alpha_0})^{-1}(e_k)|V_x = 0$. Puisque F_{β_x} est s.c.s., il y a un voisinage ouvert W_x de $p_{\beta_x}(x)$ dans X_{β_x} tel que $F_{\beta_x}(W_x) \subset V_x$, d'où

$$(q^{\beta_x}_{\alpha_0})^{-1}(e_k)|F_{\beta_x}(y) = 0 \quad \text{pour tout } y \in W_x \tag{1}$$

Les ouverts $p_{\beta_x}^{-1}(W_x)$ recouvrent le compact $p_{\alpha_0}^{-1}(K_i^j)$, donc il existe $x_1, \ldots, x_s$ tels que $p_{\alpha_0}^{-1}(K_i^j) \subset \bigcup_{r=1}^{s} p_{\beta_{x_r}}^{-1}(W_{x_r})$. Fixons $\beta = \beta(i, j, k)$ tel que $\beta \geq \beta_{x_r}$ pour $1 \leq r \leq s$. Soit $x \in (p^{\beta}_{\alpha_0})^{-1}(K_i^j)$. Il existe $x' \in X$ tel que $p_\beta(x') = x$ et $1 \leq r \leq s$ tel que $x' \in p_{\beta_{x_r}}^{-1}(W_{x_r})$. Alors $p^{\beta}_{\beta_{x_r}}(x) = p_{\beta_{x_r}}(x') \in W_{x_r}$ et $q^{\beta}_{\beta_{x_r}}(F_\beta(x)) \subset F_{\beta_{x_r}}(p^{\beta}_{\beta_{x_r}}(x))$, d'où, d'après (1),

$$(q^{\beta}_{\alpha_0})^{-1}(e_k)|F_\beta(x) = (q^{\beta}_{\beta_{x_r}})^{-1}((q^{\beta_{x_r}}_{\alpha_0})^{-1}(e_k)|F_{\beta_{x_r}}(p^{\beta}_{\beta_{x_r}}(x)))|F_\beta(x) = 0$$

Puisque A est filtrant, la condition (a) entraîne l'existence d'un $\alpha_1 \in A$ tel que $\beta(i, j, k) \leq \alpha_1$ pour tout triplet d'indices (i, j, k) tel que $k \in N_i$. Soit $x \in X_{\alpha_1}$ et soit $e \in H^n(F_{\alpha_0}(p^{\alpha_1}_{\alpha_0}(x)), \mathbb{Q})$. Puisque $\mathcal{O}$ est stable par réunion finie, tout voisinage de $F_{\alpha_0}(p^{\alpha_1}_{\alpha_0}(x))$ dans Y_{α_0} contient la fermeture d'un O_i contenant $F_{\alpha_0}(p^{\alpha_1}_{\alpha_0}(x))$. La continuité de la cohomologie de Čech entraîne donc l'existence d'un i tel que O_i contienne $F_{\alpha_0}(p^{\alpha_1}_{\alpha_0}(x))$ et d'un élément $e' \in H^n(\overline{O_i}, \mathbb{Q})$ tel que $e = e'|F_{\alpha_0}(p^{\alpha_1}_{\alpha_0}(x))$.

Il existe j tel que $p^{\alpha_1}_{\alpha_0}(x) \in K^j_i$ et un $k \in N_i$ tel que $e' = e_k$. L'image de e dans $H^n(F_{\alpha_1}(x), \mathbb{Q})$ par l'homomorphisme induit par $q^{\alpha_1}_{\alpha_0}|F_{\alpha_1}(x))$ est donc égale à $(q^{\alpha_1}_{\alpha_0})^{-1}(e_k)|F_{\alpha_1}(x)$. Soit $\beta = \beta(i, j, k)$. Nous avons $\alpha_0 \leq \beta \leq \alpha_1$ et $q^{\alpha_1}_{\beta}(F_{\alpha_1}(x)) \subset F_\beta(p^{\alpha_1}_\beta(x))$, d'où

$$\begin{aligned}(q^{\alpha_1}_{\alpha_0})^{-1}(e_k)|F_{\alpha_1}(x) &= (q^{\alpha_1}_{\beta})^{-1}[(q^{\beta}_{\alpha_0})^{-1}(e_k)|F_\beta(p^{\alpha_1}_\beta(x))]|F_{\alpha_1}(x)\\ &= (q^{\alpha_1}_{\beta})^{-1}(0)|F_{\alpha_1}(x) = 0,\end{aligned}$$

et l'affirmation 2 en résulte.

Pour $\alpha \leq \beta$ et $n \geq 1$, nous notons $p^\beta_{\alpha,n}$ (resp. $p_{\alpha,n}$) le prolongement naturel de p^β_α (resp. p_α) à $U_n(X_\beta)$ (resp. $U_n(X)$) ($p^\beta_{\alpha,1} = p^\beta_\alpha$ et $p_{\alpha,1} = p_\alpha$). Pour $m \leq n$, nous avons $p^\beta_{\alpha,m} = p^\beta_{\alpha,n}|U_m(X_\beta)$ (resp. $p_{\alpha,m} = p_{\alpha,n}|U_m(X_\beta)$). Puisque le foncteur U_n commute aux limites projectives et préserve la surjectivité des fonctions, $(U_n(X_\alpha), p^\beta_{\alpha,n}, A)$ est un ω-système projectif de limite $U_n(X)$ et les projections $p_{\alpha,n}$ sont surjectives si les p_α le sont.

Soit F une fonction multivoque compacte s.c.s. de $U(X)$ dans Y. Supposons les projections p_α surjectives. Alors, pour $\alpha \in A$ et $n \geq 1$, la formule

$$F^n_\alpha(x) = q_\alpha(F(p^{-1}_{\alpha,n}(x)))$$

définit une fonction multivoque compacte s.c.s. de $U_n(X_\alpha)$ dans Y_α. Notant $U_0(X_\alpha) = \emptyset$, nous définissons alors une fonction multivoque G_α de $U(X_\alpha)$ dans Y_α en posant

$$G_\alpha(x) = F^n_\alpha(x) \quad \text{si } x \in U_n(X_\alpha) \setminus U_{n-1}(X_\alpha),\ n \geq 1.$$

Lemme 4. L'ensemble des $\alpha \in A$ tels que G_α soit s.c.s. est cofinal dans A.

Démonstration. Étant donné $\alpha_0 \in A$, nous allons construire une suite croissante $\{\alpha_n\}^\infty_{n=0}$ d'éléments de A et, pour tout $n \geq 1$, une base dénombrable $\mathcal{B}_n = \{B^i_n\}^\infty_{i=1}$ de Y_{α_n} stable par réunion finie, de façon que la condition suivante soit vérifiée:

($\star$) Pour tout $m \leq n$, tout $x \in U_m(X_{\alpha_{n+1}})$ tel que $p^{\alpha_{n+1}}_{\alpha_n,n}(x) \in U_m(X_{\alpha_n}) \setminus U_{m-1}(X_{\alpha_n})$ et tout $i \geq 1$, si $G_{\alpha_n}(p^{\alpha_{n+1}}_{\alpha_n,n}(x)) \subset B^i_n$, alors il existe un voisinage V de x dans $U_n(X_{\alpha_{n+1}})$ tel que $q^{\alpha_{n+1}}_{\alpha_n}(F^n_{\alpha_{n+1}}(y)) \subset B^i_n$ pour tout $y \in V$.

Puisque $U_0(X_{\alpha_0}) = \emptyset$, nous pouvons prendre $\alpha_1 = \alpha_0$ et pour $\mathcal{B}_1$ n'importe quelle base dénombrable stable par réunion finie de Y_{α_1}. Soit $n \geq 1$ et supposons α_n et $\mathcal{B}_n$ construits. Pour $1 \leq m \leq n$ et $i \geq 1$, soit $W(m,i) = \{x \in U_m(X_{\alpha_n}) \setminus U_{m-1}(X_{\alpha_n}) | G_{\alpha_n}(x) \subset B^i_n\}$. Puisque $G_{\alpha_n}|U_m(X_{\alpha_n}) \setminus U_{m-1}(X_{\alpha_n}) = F^m_{\alpha_n}|U_m(X_{\alpha_n}) \setminus U_{m-1}(X_{\alpha_n})$ est s.c.s., $W(m,i)$ est ouvert dans $U_m(X_{\alpha_n}) \setminus U_{m-1}(X_{\alpha_n})$, donc est σ-compact. Soit $W(m,i) = \bigcup^\infty_{j=1} K^j_{m,i}$, où les $K^j_{m,i}$ sont compacts.

Soit $O_i = \{x \in U_n(X) \mid q_{\alpha_n}(F(x)) \subset B^i_n\}$. Puisque F est s.c.s., O_i est ouvert dans $U_n(X)$. Puisque $G_{\alpha_n}(x) = F^m_{\alpha_n}(x)$ pour $x \in U_m(X_{\alpha_n}) \setminus U_{m-1}(X_{\alpha_n})$, la définition de $F^m_{\alpha_n}$ entraîne que le compact $p^{-1}_{\alpha_n,n}(K^j_{m,i}) \setminus O_i$ est disjoint de $U_m(X)$, donc il existe $\beta = \beta(m,i,j) \geq \alpha$ tel que

$$p_{\beta,n}(p^{-1}_{\alpha_n,n}(K^j_{m,i}) \setminus O_i) \cap U_m(X_\beta) = \emptyset. \tag{2}$$

La condition (a) entraîne l'existence d'un $\alpha_{n+1} \in A$ tel que $\alpha_{n+1} \geq \beta(m,i,j)$ quels que soient m, i et j. Soient $m \leq n$ et $x \in U_m(X_{\alpha_{n+1}}) \setminus U_{m-1}(X_{\alpha_{n+1}})$ tel que $p_{\alpha_n}^{\alpha_{n+1}}(x) \in U_m(X_{\alpha_n}) \setminus U_{m-1}(X_{\alpha_n})$. Si $G_{\alpha_n}(p_{\alpha_n,n}^{\alpha_{n+1}}(x)) \subset B_i$, il existe j tel que $p_{\alpha_n,n}^{\alpha_{n+1}}(x) \in K_{m,i}^j$. Soit $\beta = \beta(m,i,j)$. Puisque x appartient à $U_m(X_{\alpha_{n+1}})$, $p_{\beta,n}^{\alpha_{n+1}}(x)$ appartient à $U_m(X_\beta)$ et (2) entraîne

$$p_{\beta,n}^{-1}(x) \subset p_{\beta,n}^{-1}((p_{\alpha_n,n}^{\beta})^{-1}(K_{m,i}^j) \cap U_m(X_\beta)) \subset O_i.$$

Il existe donc un voisinage V de $p_{\beta,n}^{\alpha_{n+1}}(x)$ dans $U_n(X_\beta)$ tel que $p_{\beta,n}^{-1}(V) \subset O_i$. Pour $y \in (p_{\beta,n}^{\alpha_{n+1}})^{-1}(V)$, nous avons

$$\begin{aligned} q_{\alpha_n}^{\alpha_{n+1}}(F_{\alpha_{n+1}}^n(y)) = q_{\alpha_n}^{\alpha_{n+1}} \circ q_{\alpha_{n+1}}(F(p_{\alpha_{n+1},n}^{-1}(y))) &\subset q_{\alpha_n}(F(p_{\beta,n}^{-1}(p_{\beta,n}^{\alpha_{n+1}}(y))) \\ &\subset q_{\alpha_n}(F(p_{\beta,n}^{-1}(V))) \subset q_{\alpha_n}(F(O_i)) \subset B_n^i, \end{aligned}$$

donc la condition ($\star$) est vérifiée. Prenons pour $\mathcal{B}_{n+1}$ n'importe quelle base dénombrable de $Y_{\alpha_{n+1}}$ stable par réunion finie.

Pour prouver le lemme, il suffit, puisque α_0 était arbitraire, de montrer que si α est la borne supérieure de la suite $\{\alpha_n\}_{n=0}^{\infty}$, alors G_α est s.c.s. et, comme $U(X_\alpha)$ est la limite inductive de la suite croissante de compacts $\{U_k(X_\alpha)\}$, il suffit pour cela de montrer que $G_\alpha | U_k(X_\alpha)$ est s.c.s. pour tout $k \geq 1$.

Fixons un entier $k \geq 1$, et soit $x \in U_k(X_\alpha)$. Soit $m \leq k$ tel que $x \in U_m(X_\alpha) \setminus U_{m-1}(X_\alpha)$. Puisque $U_{m-1}(X_\alpha)$ est fermé dans $U_k(X_\alpha)$, (b) entraîne l'existence d'un $n_0 \geq k$ tel que $p_{\alpha_{n_0},k}^{\alpha}(x) \in U_m(X_{\alpha_{n_0}}) \setminus U_{m-1}(X_{\alpha_{n_0}})$. Alors, pour $n_0 \leq n$, nous avons $p_{\alpha_n,k}^{\alpha}(x) \in U_m(X_{\alpha_n}) \setminus U_{m-1}(X_{\alpha_n})$, donc $G_{\alpha_n}(p_{\alpha_n,k}^{\alpha}(x)) = F_{\alpha_n}^m(x)$ et $G_\alpha(x) = F_\alpha^m(x)$. L'affirmation 1 s'applique à la fonction F_α^m et montre que $G_\alpha(x)$ est la limite du système projectif $(G_{\alpha_n}(p_{\alpha_n,k}^{\alpha}(x)), q_{\alpha_n}^{\alpha_r} | G_{\alpha_r}(p_{\alpha_r,k}^{\alpha}(x)), n \geq n_0)$, donc si M est un voisinage de $G_\alpha(x)$ dans Y_α, il existe $n \geq n_0$ et un voisinage N de $G_{\alpha_n}(p_{\alpha_n,k}^{\alpha}(x))$ dans Y_{α_n} tel que $G_\alpha(x) \subset (q_{\alpha_n,k}^{\alpha})^{-1}(N) \subset M$. Puisque la base $\mathcal{B}_n$ de Y_{α_n} est stable par réunion finie, il existe i tel que $G_{\alpha_n}(p_{\alpha_n,k}^{\alpha}(x)) \subset B_n^i \subset N$. Puisque $m \leq k \leq n$ et que $p_{\alpha_{n+1},k}^{\alpha}(x)$ appartient à $U_m(X_{\alpha_{n+1}})$, il existe d'après ($\star$) un voisinage V de $p_{\alpha_{n+1},k}^{\alpha}(x)$ dans $U_n(X_{\alpha_{n+1}})$ tel que $q_{\alpha_n}^{\alpha_{n+1}}(F_{\alpha_{n+1}}^n(y)) \subset B_n^i$ pour tout y dans V. Alors $(p_{\alpha_{n+1},k}^{\alpha})^{-1}(V) \cap U_k(X_\alpha)$ est un voisinage de x dans $U_k(X_\alpha)$ et si y appartient à ce voisinage, alors

$$G_\alpha(y) \subset F_\alpha^n(y) \subset (q_{\alpha_{n+1}}^{\alpha})^{-1}(F_{\alpha_{n+1}}^n(p_{\alpha_{n+1},k}^{\alpha}(y))),$$

d'où

$$q_{\alpha_n}^{\alpha}(G_\alpha(y)) = q_{\alpha_n}^{\alpha_{n+1}} \circ q_{\alpha_{n+1}}^{\alpha}(G_\alpha(y)) \subset q_{\alpha_n}^{\alpha_{n+1}}(F_{\alpha_{n+1}}^n(p_{\alpha_{n+1},k}^{\alpha}(y))) \subset B_n^i$$

puisque $p_{\alpha_{n+1},k}^{\alpha} \in V$. Par conséquent

$$G_\alpha(y) \subset (q_{\alpha_n}^{\alpha})^{-1}(B_n^i) \subset (q_{\alpha_n}^{\alpha})^{-1}(N) \subset M,$$

ce qui montre que $G_\alpha | U_k(X_\alpha)$ est bien s.c.s. et achève de prouver le lemme.

4. Le cas général

Soit X un compact, et soit $F : U(X) \to X$ une fonction multivoque compacte s.c.s. et acyclique. Représentons X comme limite d'un ω-système projectif $(X_\alpha, p_\alpha^\beta, A)$ de compacts tels que les projections p_α soient surjectives. Alors $U_n(X)$ est la limite de l'ω-système projectif $(U_n(X), p_{\alpha,n}^\beta, A)$ et, pour $\alpha \in A$, nous définissons une fonction multivoque compacte s.c.s. $F_\alpha^n : U_n(X_\alpha) \to X_\alpha$ par

$$F_\alpha^n(x) = p_\alpha(F(p_{\alpha,n}^{-1}(x))),$$

et une fonction multivoque compacte $G_\alpha : U(X_\alpha) \to X_\alpha$ par

$$G_\alpha(x) = F_\alpha^n(x) \quad \text{si } x \in U_n(X_\alpha) \setminus U_{n-1}(X_\alpha),\ n \geq 1.$$

Le lemme 3 entraîne que l'ensemble A_n des $\alpha \in A$ tels que F_α^n soit acyclique est cofinal et fermé dans A. D'après la proposition 3.1.1 de [6], $A_\infty = \bigcap_{n=1}^\infty A_n$ est cofinal et fermé dans A. D'après le lemme 4, l'ensemble B des $\alpha \in A_\infty$ tels que G_α soit s.c.s. est cofinal dans A_∞, donc dans A. Si $x \in B$, alors G_α est une fonction multivoque compacte s.c.s. et acyclique de $U(X_\alpha)$ dans X_α. Puisque X_α est métrisable, l'ensemble P_α des points fixes de G_α est non vide pour tout $\alpha \in B$ et, comme G_α est s.c.s., P_α est fermé dans $U(X_\alpha)$. Évidemment, P_α est contenu dans X_α.

Soient $\alpha \leq \beta$ dans B et $x \in P_\beta$. Puisque x appartient à X_β, nous avons $G_\beta(x) = F_\beta^1(x)$ donc, puique $p_{\alpha,1}^\beta = p_\alpha^\beta$ et $p_{\alpha,1} = p_\alpha$,

$$\begin{aligned} G_\alpha(p_\alpha^\beta(x)) &= F_\alpha^1(p_\alpha^\beta(x)) = p_\alpha(F(p_\alpha^{-1}(p_\alpha^\beta(x))) \supset p_\alpha(F(p_\beta^{-1}(x))) \\ &= p_\alpha^\beta p_\beta(F(p_\beta^{-1}(x))) = p_\alpha^\beta(G_\beta(x)) \ni p_\alpha^\beta(x), \end{aligned}$$

ce qui montre que $p_\alpha^\beta(P_\beta) \subset P_\alpha$. Nous obtenons donc un système projectif de compacts $(P_\alpha, p_\alpha^\beta | P_\beta, B)$. Comme ces compacts sont non vides, leur limite P n'est pas vide, et est contenue dans la limite du système projectif $(X_\alpha, p_\alpha^\beta, B)$, qui est égale à X puisque B est cofinal dans A. Si y appartient à P, alors, en raisonnant comme dans la démonstration de l'affirmation 1, on constate que $F(y)$ est la limite du système projectif $(F_\alpha^1(p_\alpha(y)), p_\alpha^\beta | F_\beta^1(p_\beta(y)), B)$, et comme $p_\alpha(y)$ appartient à $F_\alpha^1(p_\alpha(y)) = G_\alpha(p_\alpha(y))$ pour tout $\alpha \in B$, nous avons $y \in F(y)$, donc y est un point fixe de F et la proposition est prouvée.

References

1. R. Cauty, Solution du problème de point fixe de Schauder, *Fund. Math.* **170** (2001) 231-246.
2. R. Cauty, Le problème de Schauder; correction et complément, Prépublication.
3. R. Cauty, Le théorème de point fixe de Lefschetz-Hopf pour les compacts ULC, Prépublication.
4. S. Eilenberg and D. Montgomery, Fixed point theorems for multivalued transformations, *Amer. J. Math.* **68** (1946) 214-222.

5. S. Eilenberg and N. Steenrod, *Foundations of algebraic topology*. Princeton University Press, Princeton, 1952.
6. V.V. Fedorchuk et A. Chigogidze, *Rétractes absolus et variétés de dimension infinie*. Nauka, Moscou, 1992 (en Russe).
7. E.H. Spanier, *Algebraic topology*. Mc Graw Hill, New York, 1996.

Functional Analysis and its Applications
V. Kadets and W. Żelazko

Almost Periodic Mappings to Complex Manifolds

Sergej Favorov and Natalia Parfyonova

Karazin Kharkiv National University, Department of Mechanics and Mathematics, Kharkiv, Ukraine

Key words: Almost periodic holomorphic mappings, meromorphic functions
2000 MSC: 32A20

A continuous mapping F of a tube $T_K = \{z = x + iy : x \in \mathbb{R}^m, y \in K \subset \mathbb{R}^m\}$ to a metric space X is *almost periodic* if the family $\{F(z+t)\}_{t\in\mathbb{R}^m}$ of shifts along $\mathbb{R}^m$ is a relatively compact set with respect to the topology of the uniform convergence on T_K.

Further, let X be a complex manifold, and F be a holomorphic mapping of a tube $T_\Omega = \{z = x+iy : x \in \mathbb{R}^m, y \in \Omega\}$, with the convex open base $\Omega \subset \mathbb{R}^m$, to X. We will say that F is *almost periodic* if the restriction of F to each tube T_K, with the compact base $K \subset \Omega$, is almost periodic.

For $X = \mathbb{C}$ we obtain the well-known class of holomorphic almost periodic functions; for $X = \mathbb{C}^q$ the corresponding class was being studied in [3], [7], [8]; for $X = \mathbb{CP}$, we get the class of meromorphic almost periodic functions that was being studied in [4], [10], [5]; the class of holomorphic almost periodic curves, corresponding to the case $X = \mathbb{CP}^q$, was being studied in [6].

The following theorem is well known:

Theorem B (H. Bohr [2]). *If a holomorphic bounded function on a strip is almost periodic on some straight line in this strip, then this function is almost periodic on the whole strip.*

This theorem was extended to holomorphic functions on a tube domain in [11]; besides usual uniform metric, various integral metric were being studied there.

The direct generalization of Theorem B to complex manifold is not valid:

Email address: `Natalia.D.Parfyonova@univer.kharkov.ua` (Natalia Parfyonova).

Example. Let $f(z)$ be the quotient of two periodic functions $\sin\sqrt{2}z$ and $\sin z$. It is clear that the restriction of $f(z)$ to any straight line $\{z = x+iy_0 : x \in \mathbb{R}\}, y_0 \neq 0$, is almost periodic on this line. Besides, $f(z)$ is bounded as a mapping to compact manifold $\mathbb{CP}$. Nevertheless, zeros and poles of $f(z)$ are not separated on the axis $z = x \in \mathbb{R}$, therefore $f(z)$ is not almost periodic on any substrip containing the real axis (see [4]).

In order to give the right version of Theorem B, we need the following proposition.

Proposition. *If K is compact subset of $\mathbb{R}^m$, then an almost periodic mapping F of T_K to a metric space X is uniformly continuous and $F(T_K)$ is a relatively compact subset of X.*

Proof. Let

$$\varphi(t) = \sup_{y\in K} d(F(z+t), F(z)), \quad t \in \mathbb{R}^m.$$

It is easy to prove that $\varphi(t)$ is an almost periodic function on $\mathbb{R}^m$, therefore the function $\varphi(t)$ extends continuously to Bohr's compactification B of the set $\mathbb{R}^m$ (see, for example, [1]). Hence for any point $\tau \in B$ and any $\varepsilon > 0$ there exists a neighborhood $U \subset B$ of this point such that

$$|\varphi(t) - \varphi(t')| < \varepsilon$$

for each $t, t' \in U \cap \mathbb{R}^m$.

Using the obvious inequality

$$\varphi(t - t') \leq |\varphi(t) - \varphi(t')|$$

we get for all $z \in T_K, t, t' \in U \cap \mathbb{R}^m$

$$d(F(z+t'), F(z+t)) \leq \varepsilon.$$

Therefore the function

$$\overline{F}(\tau, y) = \lim_{x\to\tau,\ x\in\mathbb{R}^m} F(x+iy)$$

is well-defined and continuously maps the compact set $B \times K$ to X. Since $\overline{F}(x, y) = F(x+iy)$ for $x \in \mathbb{R}^m$, we obtain all the statements of our proposition.

Note that a bounded holomorphic function on a tube domain is uniformly continuous on every subtube with the compact base, but bounded holomorphic mappings, in general, have not this property. It motivates the following result is natural.

Theorem. *Let F be a holomorphic mapping of T_Ω to a complex manifold X such that for every compact subset $K \subset \Omega$ the mapping F is uniformly continuous on T_K and $F(T_K)$ is a relatively compact subset of X. If the restriction of $F(z)$ to some hyperplane $\mathbb{R}^m + iy'$ is almost periodic, then $F(z)$ is an almost periodic mapping of T_Ω to X.*

Corollary. *Let F be a holomorphic mapping form T_Ω to a compact complex manifold X such that F is uniformly continuous on T_K for every compact set $K \subset \Omega$. If* *the restriction of $F(z)$ to some hyperplane $\mathbb{R}^m + iy'$ is almost periodic, then $F(z)$ is an almost periodic mapping of T_Ω to X.*

Proof the Theorem. Take an arbitrary sequence $\{t_n\} \subset \mathbb{R}^m$. Since the function $F(z)$ is uniformly continuous, the family $\{F(z + t_n\}$ is equicontinuous on each compact set $S \subset T_\Omega$. Further, it follows from the condition of the Theorem that the union of all the images of S under mappings of this family is contained compact subset of X. Therefore, passing on to a subsequence if necessary, we may assume that the sequence $\{F(z+t_n)\}$ converges to a holomorphic mapping $G(z)$ uniformly on every compact subset of T_Ω. It easy to see that the mapping $G(z)$ is bounded and uniformly continuous on every tube T_K with the compact base $K \subset \Omega$. Let us prove that this convergence is uniform on every T_K. Assume the contrary. Then we get

$$d(F(z_n + t_n), G(z_n)) \geq \varepsilon_0 > 0 \qquad (1)$$

for some sequence $z_n = x_n + iy_n \in T_{K'}$, where K' is some compact subset of Ω.

Replacing sequence by a subsequence if necessary, we may assume that the mappings $G(x_n + z)$ converge to a holomorphic mapping $H(z)$, and the mappings $F(z+x_n+t_n)$ converge to a holomorphic mapping $\tilde{H}(z)$ uniformly on every compact subsets of T_Ω. We may also assume that $y_n \to y_0 \in K'$. Using (1) we get

$$|\tilde{H}(iy_0) - H(iy_0))| \geq \varepsilon_0.$$

Since the mapping $F(x+iy')$ of $\mathbb{R}^m$ to X is almost periodic, we may assume that a subsequence of the mappings $F(x+t_n+iy')$ converges to $G(x+iy')$ uniformly in $x \in \mathbb{R}^m$. Therefore the sequences of mappings $F(x+x_n+t_n+iy')$ and $G(x+x_n+iy')$ have the same limit, i. e., $\tilde{H}(x+iy') = H(x+iy')$ for all $x \in \mathbb{R}^m$. Since $\tilde{H}(z)$, $H(z)$ are holomorphic mappings, we get $\tilde{H}(x+iy') \equiv H(x+iy')$ on T_Ω. This contradiction proves the Theorem.

References

1. Ch. Berg, Introduction to the almost periodic functions of Bohr, Proceedings of a symposium held in Copenhagen, April 24-25 (1987), 15-24.
2. H. Bohr, Ueber analytische fastperiodische Funktionen, *Math. Ann.* **103** (1930) 1-14.
3. S.Ju. Favorov, A.Ju. Rashkovskii, L.I. Ronkin, Almost periodic divisors in a strip, *Journal D'analyse Mathematique* **74** (1998) 325-345.
4. N.D. Parfyonova, S.Yu. Favorov, Meromorphic almost periodic functions, *Mathematychni Studii* **13** (2000) 190-198.
5. N.D. Parfyonova, Meromorphic almost periodic functions and their continuous extension on the Bohr's compact, *Vestnik Kharkovskogo Universiteta* **542** (2002) 73-84. (In Russian).
6. N.D. Parfyonova, Holomorphic almost periodic mappings into complex manifolds, *Mat. Fiz. Anal. Geom.* **9** (2002) 294-305. (In Russian).

7. A.Yu. Rashkovskii, Monge-Ampere operators and Jessen functions of holomorphic almost periodic mappings. *Mat. Fiz. Anal. Geom.* **5** (1998) 274-296.
8. A.Yu. Rashkovskii, Flows associated with holomorphic almost periodic functions, *Mat. Fiz. Anal. Geom.* **2** (2) (1995) 150-169.
9. L.I. Ronkin, Jessens' theorems for holomorphic almost periodic functions in tube domains, *Sib. Math. J.* **28** (3) (1987) 199-204. (In Russian).
10. F. Sunyer i Balaguer, Una nova generalitzacio de les functions gairebe-periodiques, Inst.d'Estudis Catalans, Arxius de la seccio de ciencies XVII, 1949, 3-46.
11. O.I. Udodova, Holomorphic almost periodic functions in tube donains,. *Vestnik Kharkovskogo Universiteta* **542** (2002) 96-105.

Functional Analysis and its Applications 85
V. Kadets and W. Żelazko

Some new differential equations of the first-order in the spaces $M(1,3) \times R(u)$ and $M(1,4) \times R(u)$ with given symmetry groups

Vasyl Fedorchuk [a] **and Volodymyr Fedorchuk** [b]

[a]Pedagogical Academy, Institute of Mathematics, 2 Podchorążych Street, 30-084, Kraków, Poland; Pidstryhach Institute of Applied Problems of Mechanics and Mathematics of NAS of Ukraine, 3b Naukova Street, 79-601, Lviv, Ukraine;
[b]Ivan Franko Lviv National University, 1 Universytetska Street, Lviv, 79000, Ukraine

Abstract

We present some new first-order differential equations invariant under continuous subgroups of the generalized Poincaré group $P(1,4)$ *and defined in the spaces* $M(1,3) \times R(u)$ *and* $M(1,4) \times R(u)$.

Key words: Lie groups, Lie algebras, subgroup structure, differential invariants, differential equations
2000 MSC: 58J70, 35A30, 59D19

1. Introduction

The mathematical modeling of various natural processes very often requires that some symmetry properties of the processes considered be taken into account. This leads to mathematical models with non-trivial symmetry groups. In many cases this type of models can be expressed in the form of (linear or nonlinear) differential equations with non-trivial symmetry properties in the spaces of different dimensions.

It is well known that among equations, important for theoretical and mathematical physics, there are also ones which have nontrivial symmetry groups. For example, in the space $M(1,3) \times R(u)$ we have the Eikonal equation, the Euler–Lagrange–

Born–Infeld equation and homogeneous and inhomogeneous Monge–Ampere equations. In the space $M(1,4) \times R(u)$ we have the linear and nonlinear wave equations and Dirac equation. Here, and in what follows, $R(u)$ is the number axis of the dependent variable u. These equations are invariant under the generalized Poincaré group $P(1,4)$ (see, for example, [7, 11–16]). The group $P(1,4)$ is the group of rotations and translations of the five-dimensional Minkowski space $M(1,4)$. This group has many applications in theoretical and mathematical physics [7, 11, 16]. The group $P(1,4)$ has many subgroups used in theoretical physics [3–6, 9]. Among these subgroups there are the Poincaré group $P(1,3)$ and the extended Galilei group $\widetilde{G}(1,3)$ (see also [10]). Thus, the results obtained with the help of the subgroup structure of the group $P(1,4)$ will be useful in relativistic and non-relativistic physics. Therefore, it is important from the physical and mathematical points of view that we are able to construct, in the spaces $M(1,3) \times R(u)$ and $M(1,4) \times R(u)$, new differential equations invariant under continuous subgroups of the generalized Poincaré group $P(1,4)$.

The papers [1, 2] are devoted to the construction of the first-order differential equations invariant under splitting subgroups of the group $P(1,4)$, defined in the spaces $M(1,3) \times R(u)$ and $M(1,4) \times R(u)$.

In the present paper we continue to study this type of equations. We concentrate our attention on the first-order differential equations invariant under continuous subgroups of the group $P(1,4)$, defined in the spaces $M(1,3) \times R(u)$ and $M(1,4) \times R(u)$. We give the number of all first-order differential equations of a certain form (in these spaces) which are invariant under splitting subgroups of the group $P(1,4)$ as well as some new obtained equations.

Recently, we have constructed the first-order differential equations (in the spaces $M(1,3) \times R(u)$ and $M(1,4) \times R(u)$) which are invariant under non-splitting subgroups of the group $P(1,4)$ for the majority of these subgroups.

Our paper is organized as follows. In the first section we introduce some notation and results concerning the Lie algebra of the group $P(1,4)$ which we use in the following. Sections 3 and 4 present our main results.

2. The Lie algebra of the group $P(1,4)$ and its non-conjugate subalgebras

The Lie algebra of the group $P(1,4)$ is given by the 15 basis elements $M_{\mu\nu} = -M_{\nu\mu}$ $(\mu, \nu = 0,1,2,3,4)$ and P'_μ $(\mu = 0,1,2,3,4)$, which satisfy the commutation relations

$$[P'_\mu, P'_\nu] = 0, \qquad [M'_{\mu\nu}, P'_\sigma] = g_{\mu\sigma} P'_\nu - g_{\nu\sigma} P'_\mu,$$

$$[M'_{\mu\nu}, M'_{\rho\sigma}] = g_{\mu\rho} M'_{\nu\sigma} + g_{\nu\sigma} M'_{\mu\rho} - g_{\nu\rho} M'_{\mu\sigma} - g_{\mu\sigma} M'_{\nu\rho},$$

where $g_{00} = -g_{11} = -g_{22} = -g_{33} = -g_{44} = 1$, $g_{\mu\nu} = 0$, if $\mu \neq \nu$. Here, and in what follows, $M'_{\mu\nu} = iM_{\mu\nu}$.

Below, we will use the following basis elements:

$$G = M'_{40}, \quad L_1 = M'_{32}, \quad L_2 = -M'_{31}, \quad L_3 = M'_{21},$$

$$P_a = M'_{4a} - M'_{a0}, \quad C_a = M'_{4a} + M'_{a0}, \quad (a = 1, 2, 3),$$

$$X_0 = \frac{1}{2}\left(P'_0 - P'_4\right), \quad X_k = P'_k \quad (k = 1, 2, 3), \quad X_4 = \frac{1}{2}\left(P'_0 + P'_4\right).$$

For the study of the subgroup structure of the group $P(1,4)$ we used the method proposed in [20]. Continuous subgroups of the group $P(1,4)$ have been described in [3–6, 9]. From the results obtained (see also [10]) it follows, in particular, that the Lie algebra of the group $P(1,4)$ contains as subalgebras the Lie algebra of the Poincaré group $P(1,3)$ and the Lie algebra of the extended Galilei group $\widetilde{G}(1,3)$.

3. The first-order differential equations in the space $M(1,3) \times R(u)$

The group $P(1,4)$ acts on $M(1,3) \times R(u)$ (i.e. on the Cartesian product of the four–dimensional Minkowski space (of the independent variables x_0, x_1, x_2, x_3) and the number axis of the dependent variable u). The group $P(1,4)$ usually acts on $M(1,3) \times R(u)$ as a group generated by translations and rotations of this space.

Let

$$X = \sum_{i=0}^{3} \xi_i(x,u)\frac{\partial}{\partial x_i} + \eta(x,u)\frac{\partial}{\partial u}$$

be one of the basis infinitesimal operators. The first prolongation of X has the form

$$X^{(1)} = X + \sum_{i=0}^{3}\left(\frac{\partial \eta}{\partial x_i} + \frac{\partial \eta}{\partial u}u_i - \sum_{j=0}^{3} u_j\frac{\partial \xi_j}{\partial x_i} - \sum_{j=0}^{3} u_i u_j \frac{\partial \xi_j}{\partial u}\right)\frac{\partial}{\partial u_i},$$

where $u_i \equiv \frac{\partial u}{\partial x_i}$, $i = 0, 1, 2, 3$.

Now, a function $J(x, u^{(1)})$ is a first-order differential invariant if

$$X^{(1)} \cdot J(x, u^{(1)}) = 0.$$

Here $u^{(1)} = (u, u_0, u_1, u_2, u_3)$ is an element of the first prolongation $R(u)^{(1)}$.

The results obtained in [12–14] enable us to use the subgroup structure [3–6, 9] of the generalized Poincaré group $P(1,4)$ in order to construct the first-order differential equations invariant under continuous subgroups of the group $P(1,4)$, defined in the space $M(1,3) \times R(u)$.

Let us consider the following representation of the Lie algebra of the group $P(1,4)$:

$$P_0' = \frac{\partial}{\partial x_0}, \quad P_1' = -\frac{\partial}{\partial x_1}, \quad P_2' = -\frac{\partial}{\partial x_2}, \quad P_3' = -\frac{\partial}{\partial x_3},$$

$$P_4' = -\frac{\partial}{\partial u}, \quad M_{\mu\nu}' = -\left(x_\mu P_\nu' - x_\nu P_\mu'\right), \; x_4 \equiv u.$$

More details about this representation can be found in [12–15].

The considered equations can be written in the following form (see, for example [17–19, 21]):

$$F(J_1, J_2, ..., J_t) = 0, \tag{1}$$

where F is an arbitrary smooth function of its arguments, $\{J_1, J_2, ..., J_t\}$ is a functional basis of the first-order differential invariants of continuous subgroups of the group $P(1,4)$.

In the construction of the differential equations with non-trivial symmetry groups it has turned out that different non-conjugate subalgebras can have the same functional basis of the first-order differential invariants. Consequently, there is no one-to-one correspondence between non-conjugate subalgebras of the Lie algebra of the group $P(1,4)$ and corresponding differential equations.

Definition 1. *We call two subalgebras L^1 and L^2 of the Lie algebra of the group $P(1,4)$ equivalent if they have the same functional basis of the first-order differential invariants.*

It is possible to prove that the relation of equivalence of subalgebras L^1 and L^2 given by Definition 1 is the set-theoretical equivalence relation. With respect to this equivalence relation, all non-conjugate subalgebras of the Lie algebra of the group $P(1,4)$ split into classes of equivalent subalgebras. Each two different classes have different functional bases of the first-order differential invariants.

Definition 2. *We call two classes of the first-order differential equations of the form (1) equivalent if functional bases corresponding to them belong to the equivalent subalgebras.*

First, we give the results obtained with the help of the splitting subgroups of the group $P(1,4)$. One of the results in this section can be formulated as follows.

Proposition 1. *In the space $M(1,3)\times R(u)$, there exist 242 non-equivalent classes of the first-order differential equations of the form (1) which are invariant under the splitting subgroups of the group $P(1,4)$.*

This proposition has been proved with use of:
– the list of the splitting subalgebras of the Lie algebra of the group $P(1,4)$ [8];
– the general ranks of the matrices which contain coordinates of the basis elements of the subalgebras of the considered Lie algebra;
– theorem on the number of invariants of the Lie group of point transformations (see, for example, [18, 19]).

The differential first-order equations in the space $M(1,3) \times R(u)$ which are invariant under splitting subgroups of the group $P(1,4)$ are constructed. Some of the results obtained have been presented in [2].

Since we cannot present all the results obtained here, in this section we only give some of the new results which cannot be found in [2].

Below, for some splitting subgroups of the group $P(1,4)$, we write the basis elements of its Lie algebras and corresponding arguments $J_1, J_2, ..., J_t$ of the function F.

1. $\langle G \rangle$,

$$J_1 = x_1, \quad J_2 = x_2, \quad J_3 = x_3, \quad J_4 = (x_0^2 - u^2)^{1/2}, \quad J_5 = (x_0 + u)^2 \tfrac{u_0 - 1}{u_0 + 1},$$

$$J_6 = \tfrac{u_1}{u_2}, \quad J_7 = \tfrac{u_1}{u_3}, \quad J_8 = \tfrac{u_1^2}{u_0^2 - 1}, \qquad u_\mu \equiv \tfrac{\partial u}{\partial x_\mu}, \ \mu = 0,\ 1,\ 2,\ 3;$$

2. $\langle G, P_3 \rangle$,

$$J_1 = x_1, \quad J_2 = x_2, \quad J_3 = (x_0^2 - x_3^2 - u^2)^{1/2}, \quad J_4 = \tfrac{x_0 + u}{u_0 + 1} u_1,$$

$$J_5 = x_3 + \tfrac{x_0 + u}{u_0 + 1} u_3, \quad J_6 = \tfrac{u_1}{u_2}, \quad J_7 = \frac{u_0^2 - u_3^2 - 1}{u_1^2};$$

3. $\langle G, L_3, X_4 \rangle$,

$$J_1 = x_3, \quad J_2 = (x_1^2 + x_2^2)^{1/2}, \quad J_3 = \tfrac{x_1 u_2 - x_2 u_1}{x_1 u_1 + x_2 u_2}, \quad J_4 = (x_0 + u)\tfrac{(u_0 - 1)}{u_3},$$

$$J_5 = \tfrac{u_0^2 - 1}{u_3^2}, \quad J_6 = \tfrac{u_1^2 + u_2^2}{u_3^2};$$

4. $\langle G, P_1, P_2, X_4 \rangle$,

$$J_1 = x_3, \quad J_2 = \tfrac{x_0 + u}{u_0 + 1} u_3, \quad J_3 = x_1 + \tfrac{x_0 + u}{u_0 + 1} u_1, \quad J_4 = x_2 + \tfrac{x_0 + u}{u_0 + 1} u_2,$$

$$J_5 = \frac{u_0^2 - u_1^2 - u_2^2 - 1}{u_3^2};$$

5. $\langle L_3 + bG, P_1, P_2, P_3, X_1, X_2, X_4, b > 0 \rangle$,

$$J_1 = x_3 + \tfrac{x_0 + u}{u_0 + 1} u_3, \quad J_2 = (u_0^2 - u_1^2 - u_2^2 - u_3^2 - 1)\left(\tfrac{x_0 + u}{u_0 + 1}\right)^2;$$

6. $\langle G, P_1, P_2, P_3, X_1, X_2, X_3, X_4 \rangle$,

$$J_1 = (u_0^2 - u_1^2 - u_2^2 - u_3^2 - 1)\left(\tfrac{x_0 + u}{u_0 + 1}\right)^2;$$

7. $\langle L_1, L_2, L_3, P_1, P_2, P_3, X_1, X_2, X_3, X_4 \rangle$,

$$J_1 = x_0 + u, \quad J_2 = \frac{u_1^2 + u_2^2 + u_3^2 + 2(u_0 + 1)}{(u_0 + 1)^2};$$

8. $\langle G, L_1, L_2, L_3, P_1, P_2, P_3, X_1, X_2, X_3, X_4 \rangle$,

$$J_1 = (u_0^2 - u_1^2 - u_2^2 - u_3^2 - 1)\left(\tfrac{x_0 + u}{u_0 + 1}\right)^2.$$

Now, we present results obtained with the help of the non-splitting subgroups of the group $P(1,4)$.

We have constructed classes of the first-order differential equations (in the space $M(1,3) \times R(u)$) which are invariant under non-splitting subgroups of the group $P(1,4)$ for the majority of these subgroups. These classes of equations can be written in the form (1), where $\{J_1, J_2, ..., J_t\}$ is a functional basis of the first-order differential invariants of the non-splitting subgroups of the group $P(1,4)$.

Here, we only write some of the new results obtained.

Below, for some non–splitting subgroups of the group $P(1,4)$, we give the basis elements of their respective Lie algebras and corresponding arguments $J_1, J_2, ..., J_t$ of the function F.

1. $\langle L_3 - X_4,\ P_3\rangle$,

$$J_1 = x_0 + u, \quad J_2 = (x_1^2 + x_2^2)^{1/2}, \quad J_3 = x_0^2 - x_3^2 - u^2 + (x_0 + u)\arctan \tfrac{x_1}{x_2},$$

$$J_4 = \tfrac{x_3}{x_0+u} + \tfrac{u_3}{u_0+1}, \quad J_5 = \tfrac{x_1u_2 - x_2u_1}{x_1u_1 + x_2u_2}, \quad J_6 = \tfrac{u_1^2+u_2^2}{(u_0+1)^2},$$

$$J_7 = \tfrac{u_3^2}{(u_0+1)^2} + \tfrac{2}{u_0+1}, \qquad u_\mu \equiv \tfrac{\partial u}{\partial x_\mu},\ \mu = 0,1,2,3;$$

2. $\langle G + a_3X_3,\ P_1,\ P_2,\ a_3 < 0\rangle$,

$$J_1 = x_3 - a_3\ln(x_0+u), \quad J_2 = (x_0^2 - x_1^2 - x_2^2 - u^2)^{1/2},$$

$$J_3 = (x_0+u)\tfrac{u_3}{u_0+1}, \quad J_4 = x_1 + (x_0+u)\tfrac{u_1}{u_0+1},$$

$$J_5 = x_2 + (x_0+u)\tfrac{u_2}{u_0+1}, \quad J_6 = \tfrac{u_0^2 - u_1^2 - u_2^2 - 1}{u_3^2};$$

3. $\langle G + a_3X_3,\ L_3,\ P_1,\ P_2,\ a_3 < 0\rangle$,

$$J_1 = (x_0^2 - x_1^2 - x_2^2 - u^2)^{1/2}, \quad J_2 = (x_0+u)\tfrac{u_3}{u_0+1},$$

$$J_3 = x_3 - a_3\ln(x_0+u), \quad J_4 = \left(x_1 + \tfrac{x_0+u}{u_0+1}u_1\right)^2 + \left(x_2 + \tfrac{x_0+u}{u_0+1}u_2\right)^2,$$

$$J_5 = \tfrac{u_0^2 - u_1^2 - u_2^2 - 1}{u_3^2};$$

4. $\langle L_3,\ P_1 + X_2,\ P_2 - X_1,\ P_3 + \tilde{h}_3X_3,\ X_4\rangle$,

$$J_1 = x_0 + u, \quad J_2 = x_3 + (x_0+u)\tfrac{u_3}{u_0+1} - \tfrac{\tilde{h}_3u_3}{u_0+1},$$

$$J_3 = \left(x_1 + \tfrac{u_2}{u_0+1} + \tfrac{x_0+u}{u_0+1}u_1\right)^2 + \left(x_2 - \tfrac{u_1}{u_0+1} + \tfrac{x_0+u}{u_0+1}u_2\right)^2,$$

$$J_4 = \tfrac{u_1^2+u_2^2+u_3^2+2(u_0+1)}{(u_0+1)^2};$$

5. $\langle G + aX_3,\ L_3,\ P_1,\ P_2,\ P_3,\ X_4,\ a < 0\rangle$,

$$J_1 = (x_0+u)^2\left(\tfrac{u_1^2+u_2^2+u_3^2+2u_0+2}{(u_0+1)^2} - 1\right),$$

$$J_2 = \left(x_1 + \tfrac{x_0+u}{u_0+1}u_1\right)^2 + \left(x_2 + \tfrac{x_0+u}{u_0+1}u_2\right)^2,$$

$$J_3 = x_3 - a\ln(x_0+u) + (x_0+u)\tfrac{u_3}{u_0+1};$$

6. $\langle G + aX_3,\ P_1,\ P_2,\ P_3,\ X_1,\ X_2,\ X_4,\ a < 0\rangle$,

$$J_1 = a\ln(x_0+u) - (x_0+u)\tfrac{u_3}{u_0+1} - x_3,$$

$$J_2 = (u_0^2 - u_1^2 - u_2^2 - u_3^2 - 1)\tfrac{(x_0+u)^2}{(u_0+1)^2};$$

7. $\langle L_1 + \tfrac{1}{2}(P_1 + C_1),\ L_2 + \tfrac{1}{2}(P_2 + C_2),\ L_3 + \tfrac{1}{2}(P_3 + C_3),$

$L_3 - \frac{1}{2}(P_3 + C_3) + \alpha(X_0 + X_4),\ X_1,\ X_2,\ X_3,\ X_0 - X_4,\ b < 0\rangle,$

$J_1 = \frac{u_1^2+u_2^2+u_3^2+1}{u_0^2}$.

4. The first-order differential equations in the space $M(1,4) \times R(u)$

In this section we present some of the new results concerning the first-order differential equations in the space $M(1,4) \times R(u)$.

The group $P(1,4)$ acts on $M(1,4) \times R(u)$ (i.e. on the Cartesian product of the five–dimensional Minkowski space (of the independent variables x_0, x_1, x_2, x_3, x_4) and the number axis of the dependent variable u). The group $P(1,4)$ usually acts on $M(1,4)$ as a group generated by translations and rotations of this space, and it acts trivially on $R(u)$ in this specific case.

Let

$$X = \sum_{i=0}^{4} \xi_i(x)\frac{\partial}{\partial x_i}$$

be one of the basis infinitesimal operators. It generates the action

$$g_t(x, u(x)) = (g_t x, u(x)) = (y, u(g_{-t}y)),$$

where $g_t = \exp tX \in$ P(1,4), $x \in$ M(1,4), $y = g_t x$. From this, one obtains the first prolongation of X in the form

$$X^{(1)} = X - \sum_{i=0}^{4}(\sum_{j=0}^{4} \frac{\partial \xi_j}{\partial x_i} u_j)\frac{\partial}{\partial u_i}, \quad u_j \equiv \frac{\partial u}{\partial x_j}, \quad j = 0, 1, 2, 3, 4.$$

Now, a function $J(x, u^{(1)})$ is a first-order differential invariant if

$$X^{(1)} \cdot J(x, u^{(1)}) = 0.$$

Here $u^{(1)} = (u, u_0, u_1, u_2, u_3, u_4)$ is an element of the first prolongation $R(u)^{(1)}$.

Let us consider the following representation of the Lie algebra of the group $P(1,4)$:

$$P_0' = \frac{\partial}{\partial x_0}, \quad P_1' = -\frac{\partial}{\partial x_1}, \quad P_2' = -\frac{\partial}{\partial x_2}, \quad P_3' = -\frac{\partial}{\partial x_3},$$

$$P_4' = -\frac{\partial}{\partial x_4}, \quad M_{\mu\nu}' = -\left(x_\mu P_\nu' - x_\nu P_\mu'\right).$$

The equations in the space $M(1,4) \times R(u)$ which are invariant under continuous subgroups of the group $P(1,4)$ can also be written in the form (1).

First, we also give the results which are based on splitting subgroups of the group $P(1,4)$.

One of the results obtained can be formulated as follows.

Proposition 2. *In the space $M(1,4)\times R(u)$, there exist 243 non-equivalent first-order differential equations of the form (1) which are invariant under the splitting subgroups of the group $P(1,4)$.*

The proof of Proposition 2 has been obtained in analogous manner as the proof of Proposition 1.

The first-order differential equations in the space $M(1,4)\times R(u)$ which are invariant under splitting subgroups of the group $P(1,4)$ are constructed. Some of the results obtained have been presented in [1].

In this section we only give some of the new results which cannot be found in [1].

Below, for some splitting subgroups of the group $P(1,4)$, we also write the basis elements of their respective Lie algebras and corresponding arguments $J_1, J_2, \ldots, J_t$ of the function F.

1. $\langle P_3 + C_3 + 2L_3\rangle$,

$$J_1 = x_0,\quad J_2 = x_1x_4 - x_2x_3,\quad J_3 = (x_1^2 + x_2^2)^{1/2},\quad J_4 = (x_3^2 + x_4^2)^{1/2},$$
$$J_5 = u,\quad J_6 = x_1u_2 - x_2u_1,\quad J_7 = x_3u_4 - x_4u_3,$$
$$J_8 = u_0,\quad J_9 = u_1^2 + u_2^2,\quad J_{10} = u_3^2 + u_4^2,\qquad u_\mu \equiv \frac{\partial u}{\partial x_\mu},\ \mu = 0,\ 1,\ 2,\ 3,\ 4;$$

2. $\langle L_3,\ X_0 + X_4\rangle$,

$$J_1 = x_3,\quad J_2 = x_4,\quad J_3 = (x_1^2 + x_2^2)^{1/2},\quad J_4 = u,\quad J_5 = x_1u_2 - x_2u_1,$$
$$J_6 = u_0,\quad J_7 = u_3,\quad J_8 = u_4,\quad J_9 = u_1^2 + u_2^2\ ;$$

3. $\langle G,\ L_3,\ X_4\rangle$,

$$J_1 = x_3,\quad J_2 = (x_1^2 + x_2^2)^{1/2},\quad J_3 = u,\quad J_4 = x_1u_2 - x_2u_1,$$
$$J_5 = (x_0 + x_4)(u_0 + u_4),\quad J_6 = u_3,\quad J_7 = u_0^2 - u_4^2,\quad J_8 = u_1^2 + u_2^2\ ;$$

4. $\langle L_1,\ L_2,\ L_3,\ X_0 - X_4\rangle$,

$$J_1 = x_0,\quad J_2 = (x_1^2 + x_2^2 + x_3^2)^{1/2},\quad J_3 = u,$$
$$J_4 = x_1u_1 + x_2u_2 + x_3u_3,\quad J_5 = u_0,\quad J_6 = u_4,\quad J_7 = u_1^2 + u_2^2 + u_3^2\ ;$$

5. $\langle G,\ L_3,\ P_1,\ P_2,\ X_4\rangle$,

$$J_1 = x_3,\quad J_2 = u,\quad J_3 = \frac{x_0+x_4}{u_0-u_4},$$
$$J_4 = \left(x_1 + \frac{x_0+x_4}{u_0-u_4}u_1\right)^2 + \left(x_2 + \frac{x_0+x_4}{u_0-u_4}u_2\right)^2,\quad J_5 = u_3,$$
$$J_6 = u_0^2 - u_1^2 - u_2^2 - u_4^2\ ;$$

6. $\langle G,\ P_3,\ L_3,\ X_1,\ X_2,\ X_4\rangle$,

$$J_1 = u,\quad J_2 = \frac{x_0+x_4}{u_0-u_4},\quad J_3 = \frac{u_0-u_4}{x_0+x_4}x_3 + u_3,\quad J_4 = u_1^2 + u_2^2,$$
$$J_5 = u_0^2 - u_3^2 - u_4^2\ ;$$

7. $\langle P_1,\ P_2,\ P_3,\ X_1,\ X_2,\ X_3,\ X_4\rangle$,

$$J_1 = x_0 + x_4,\quad J_2 = u,\quad J_3 = u_0 - u_4,\quad J_4 = u_0^2 - u_1^2 - u_2^2 - u_3^2 - u_4^2\ ;$$

8. $\langle L_3,\ P_1,\ P_2,\ P_3,\ X_1,\ X_2,\ X_3,\ X_4\rangle$,

$J_1 = x_0 + x_4, \quad J_2 = u, \quad J_3 = u_0 - u_4, \quad J_4 = u_0^2 - u_1^2 - u_2^2 - u_3^2 - u_4^2$;

9. $\langle L_1 + \frac{1}{2}(P_1 + C_1),\ L_2 + \frac{1}{2}(P_2 + C_2),\ L_3 + \frac{1}{2}(P_3 + C_3),$

$L_3 - \frac{1}{2}(P_3 + C_3),\ X_0,\ X_1,\ X_2,\ X_3,\ X_4\rangle,$

$J_1 = u, \quad J_2 = u_0, \quad J_3 = u_1^2 + u_2^2 + u_3^2 + u_4^2$.

Here, we also present some of the new results concerning the non-splitting subgroups of the group $P(1,4)$.

For the majority of non-splitting subgroups, the first-order differential equations in the space $M(1,4) \times R(u)$ which are invariant under these subgroups are found.

These equations can also be written in the form (1), where $\{J_1, J_2, ..., J_t\}$ is a functional basis of the first-order differential invariants of the non-splitting subgroups of the group $P(1,4)$.

Now, we only write some of the new results obtained.

Below, for some non-splitting subgroups of the group $P(1,4)$, we give the basis elements of their respective Lie algebras and corresponding arguments $J_1, J_2, ..., J_t$ of the function F.

1. $\langle L_3,\ P_3 + X_0\rangle,$

$J_1 = (x_1^2 + x_2^2)^{1/2}, \quad J_2 = (x_0 + x_4)^2 - 2x_3,$

$J_3 = 2(x_0 + x_4)^3 - 6x_3(x_0 + x_4) + 3(x_0 - x_4), \quad J_4 = u,$

$J_5 = x_1 u_2 - x_2 u_1, \quad J_6 = x_0 + x_4 + \frac{u_3}{u_0 - u_4}, \quad J_7 = u_0 - u_4,$

$J_8 = u_1^2 + u_2^2, \quad J_9 = u_0^2 - u_3^2 - u_4^2, \qquad u_\mu \equiv \frac{\partial u}{\partial x_\mu}, \quad \mu = 0, 1, 2, 3, 4;$

2. $\langle G + a_3 X_3,\ P_1,\ P_2,\ a_3 < 0\rangle,$

$J_1 = x_3 - a_3 \ln(x_0 + x_4), \quad J_2 = (x_0^2 - x_1^2 - x_2^2 - x_4^2)^{1/2}, \quad J_3 = u,$

$J_4 = \frac{x_0 + x_4}{u_0 - u_4}, \quad J_5 = x_1 + u_1 \frac{x_0 + x_4}{u_0 - u_4}, \quad J_6 = x_2 + u_2 \frac{x_0 + x_4}{u_0 - u_4},$

$J_7 = u_3, \quad J_8 = u_0^2 - u_1^2 - u_2^2 - u_4^2$;

3. $\langle G + a_3 X_3,\ L_3,\ P_1,\ P_2,\ a_3 < 0\rangle,$

$J_1 = (x_0^2 - x_1^2 - x_2^2 - x_4^2)^{1/2}, \quad J_2 = u, \quad J_3 = \frac{u_0 - u_4}{x_0 + x_4},$

$J_4 = x_3 - a_3 \ln(x_0 + x_4), \quad J_5 = \left(u_1 + x_1 \frac{u_0 - u_4}{x_0 + x_4}\right)^2 + \left(u_2 + x_2 \frac{u_0 - u_4}{x_0 + x_4}\right)^2,$

$J_6 = u_3, \quad J_7 = u_0^2 - u_1^2 - u_2^2 - u_4^2$;

4. $\langle L_3,\ P_1,\ P_2,\ P_3 + X_3,\ X_4\rangle,$

$J_1 = x_0 + x_4, \quad J_2 = u, \quad J_3 = x_3 + \frac{x_0 + x_4 - 1}{u_0 - u_4} u_3,$

$J_4 = \left(\frac{u_0 - u_4}{x_0 + x_4} x_1 + u_1\right)^2 + \left(\frac{u_0 - u_4}{x_0 + x_4} x_2 + u_2\right)^2,$

$J_5 = u_0 - u_4, \quad J_6 = u_0^2 - u_1^2 - u_2^2 - u_3^2 - u_4^2$;

5. $\langle G + a X_3,\ L_3,\ P_1,\ P_2,\ P_3,\ X_4,\ a < 0\rangle,$

$J_1 = u, \quad J_2 = \frac{x_0 + x_4}{u_0 - u_4}, \quad J_3 = \left(\frac{u_0 - u_4}{x_0 + x_4} x_1 + u_1\right)^2 + \left(\frac{u_0 - u_4}{x_0 + x_4} x_2 + u_2\right)^2,$

$J_4 = x_3 - a\ln(u_0 - u_4) + \frac{x_0+x_4}{u_0-u_4}u_3, \quad J_5 = u_0^2 - u_1^2 - u_2^2 - u_3^2 - u_4^2$;

6. $\langle L_3 + bG + kX_3,\ P_1,\ P_2,\ P_3,\ X_1,\ X_2,\ X_4,\ b > 0,\ k < 0\rangle$,

$J_1 = u, \quad J_2 = \frac{x_0+x_4}{u_0-u_4}, \quad J_3 = bx_3 - k\ln(x_0 + x_4) + b\frac{x_0+x_4}{u_0-u_4}u_3,$

$J_4 = u_0^2 - u_1^2 - u_2^2 - u_3^2 - u_4^2$;

7. $\langle L_1 + \frac{1}{2}(P_1 + C_1),\ L_2 + \frac{1}{2}(P_2 + C_2),\ L_3 + \frac{1}{2}(P_3 + C_3),$

$L_3 - \frac{1}{2}(P_3 + C_3) + \alpha(X_0 + X_4),\ X_1,\ X_2,\ X_3,\ X_0 - X_4,\ \alpha < 0\rangle,$

$J_1 = u, \quad J_2 = u_0, \quad J_3 = u_1^2 + u_2^2 + u_3^2 + u_4^2$.

Since the Lie algebra of the group $P(1,4)$ contains, as subalgebras, the Lie algebra of the Poincaré group $P(1,3)$ and the Lie algebra of the extended Galilei group $\widetilde{G}(1,3)$ [10], the results obtained can be used in relativistic and non-relativistic physics.

References

1. Fedorchuk V.I., Differential equations of the first-order in the space $M(1,4) \times R(u)$ with nontrivial symmetry groups, *Proc. of the Institute of Mathematics of NAS of Ukraine* **36** (2001) 283–292.
2. Fedorchuk V.I., On Differential Equations of First- and Second-Order in the Space $M(1,3) \times R(u)$ with Nontrivial Symmetry Groups, *Proc. of the Fourth Internat. Conf. Symmetry in Nonlinear Mathematical Physics (9–15 July 2001, Kyiv, Ukraine), Proceedings of Institute of Mathematics of NAS of Ukraine*, Kyiv, **43**, Part 1, 145–148.
3. Fedorchuk V.M., Continuous subgroups of the inhomogeneous de Sitter group $P(1,4)$, *Preprint, Inst. Matemat. Acad. Nauk Ukr. SSR* N 78.18, 1978.
4. Fedorchuk V.M., Nonsplitting subalgebras of the Lie algebra of the generalized Poincaré group $P(1,4)$, *Ukr. Mat. Zh.* **33** (5) (1981) 696–700.
5. Fedorchuk V.M., Splitting subalgebras of the Lie algebra of the generalized Poincaré group $P(1,4)$, *Ukr. Mat. Zh.* **31** (6) (1979) 717–722.
6. Fedorchuk V.M. and Fushchych W.I., On subgroups of the generalized Poincaré group, in *Proceedings of the International Seminar on Group Theoretical Methods in Physics*, Moscow, Nauka, V. 1, 1980, 61–66.
7. Fushchych W.I., Representations of full inhomogeneous de Sitter group and equations in five-dimensional approach. I, *Teoret. i mat. fizika* **4** (3) (1970) 360–367.
8. Fushchych W., Barannyk L. and Barannyk A., *Subgroup analysis of the Galilei and Poincaré groups and reductions of nonlinear equations*. Kiev, Naukova Dumka, 1991.
9. Fushchich W.I., Barannik A.F., Barannik L.F. and Fedorchuk V.M., Continuous subgroups of the Poincaré group $P(1,4)$, *J. Phys. A: Math. Gen.* **18** (4) (1985) 2893–2899.

10. Fushchich W.I. and Nikitin A.G., Reduction of the representations of the generalized Poincaré algebra by the Galilei algebra, *J. Phys.A: Math. and Gen.* **13** (7) (1980) 2319–2330.
11. Fushchych W.I. and Nikitin A.G., *Symmetries of Equations of Quantum Mechanics*. Allerton Press Inc., New York, 1994.
12. Fushchych W.I. and Serov N.I., On some exact solutions of the multidimensional nonlinear Euler-Lagrange equation, *Dokl. Akad. Nauk SSSR* **278** (4) (1984) 847–851.
13. Fushchych W.I. and Serov N.I., The symmetry and some exact solutions of the multidimensional Monge-Ampere equation, *Dokl. Akad. Nauk SSSR* **283** (3) (1983) 543–546.
14. Fushchych W.I. and Shtelen W.M., The symmetry and some exact solutions of the relativistic Eikonal equation, *Lett. Nuovo Cim.* **34** (16) (1982) 498–502.
15. Fushchych W.I., Shtelen W.M. and Serov N.I., *Symmetry Analysis and Exact Solutions of Equations of Nonlinear Mathematical Physics*. Dordrecht, Kluver Academic Publishers, 1993.
16. Kadyshevsky V.G., New approach to theory electromagnetic interactions, *Fizika elementar. chastitz. i atomn. yadra* **11** (1) (1980) 5–39.
17. Lie S., Über Differentialinvarianten, *Math. Ann.* **24** (1) (1884) 52–89.
18. Olver P.J., *Applications of Lie Groups to Differential Equations*. Springer-Verlag, New York, 1986.
19. Ovsiannikov L.V., *Group Analysis of Differential Equations*. Academic Press, New York, 1982.
20. Patera J., Winternitz P. and Zassenhaus H., Continuous subgroups of the fundamental groups of physics. I. General method and the Poincaré group, *J. Math. Phys.* **16** (8) (1975) 1597–1614.
21. Tresse A., Sur les invariants differentiels des groupes continus de transformations, *Acta math.* **18** (1894) 1–88.

V. Kadets and W. Żelazko

Inverse Spectral Problems for Sturm-Liouville Operators with Singular Potentials, II. Reconstruction by Two Spectra ⋆

Rostyslav O. Hryniv [a,b] **and Yaroslav V. Mykytyuk** [b]

[a]Institute for Applied Problems of Mechanics and Mathematics, 3b Naukova st., 79601 Lviv, Ukraine
[b]Lviv National University, 1 Universytetska st., 79602 Lviv, Ukraine

Abstract

We solve the inverse spectral problem of recovering the singular potentials $q \in W_2^{-1}(0,1)$ *of Sturm-Liouville operators by two spectra. The reconstruction algorithm is presented and necessary and sufficient conditions on two sequences to be spectral data for Sturm-Liouville operators under consideration are given.*

Key words: Inverse spectral problems, Sturm-Liouville operators, singular potentials
2000 MSC: 34A55, 34B24, 34L05, 34L20

1. Introduction

Suppose that q is a real-valued distribution from the space $W_2^{-1}(0,1)$. A Sturm-Liouville operator T in a Hilbert space $\mathcal{H} := L_2(0,1)$ corresponding formally to the differential expression

⋆ The work was partially supported by Ukrainian Foundation for Basic Research DFFD under grant No. 01.07/00172. R. H. acknowledges support of the Alexander von Humboldt Foundation.
Email addresses: rhryniv@iapmm.lviv.ua (Rostyslav O. Hryniv), yamykytyuk@yahoo.com (Yaroslav V. Mykytyuk).

$$-\frac{\mathrm{d}^2}{\mathrm{d}x^2} + q \tag{1.1}$$

can be defined as follows [18]. We take a distributional primitive $\sigma \in \mathcal{H}$ of q, put

$$l_\sigma(u) := -(u' - \sigma u)' - \sigma u'$$

for absolutely continuous functions u whose *quasi-derivative* $u^{[1]} := u' - \sigma u$ is again absolutely continuous, and then set

$$Tu := l_\sigma(u)$$

for suitable u (see Section 2 for rigorous definitions). When considered on this natural domain subject to the boundary conditions

$$u^{[1]}(0) - Hu(0) = 0, \quad u^{[1]}(1) + hu(1) = 0, \qquad H, h \in \mathbb{R} \cup \{\infty\}, \tag{1.2}$$

the operator $T = T(\sigma, H, h)$ becomes [17,18] selfadjoint, bounded below, and possesses a simple discrete spectrum accumulating at $+\infty$. (If $H = \infty$ or $h = \infty$, then the corresponding boundary condition is understood as a Dirichlet one.)

We observe that among the singular potentials included in this scheme there are, e.g., the Dirac δ-potentials and Coulomb $1/x$-like potentials that have been widely used in quantum mechanics to model particle interactions of various types (see, e.g., [1,2]). It is also well known that singular potentials of this kind usually do not produce a single Sturm-Liouville operator. Still for many reasons $T(\sigma, H, h)$ can be regarded as a natural operator associated with differential expression (1.1) and boundary conditions (1.2). For instance, $l_\sigma(u) = -u'' + qu$ in the sense of distributions and hence for regular (i.e., locally integrable) potentials q the above definition of $T(\sigma, H, h)$ coincides with the classical one. Also $T(\sigma, H, h)$ depends continuously in the uniform resolvent sense on $\sigma \in \mathcal{H}$, which allows us to regard $T(\sigma, H, h)$ for singular $q = \sigma'$ as a limit of regular Sturm-Liouville operators.

Sturm-Liouville and Schrödinger operators with distributional potentials from the space $W_2^{-1}(0,1)$ and $W_{2,unif}^{-1}(\mathbb{R})$ respectively have been shown to possess many properties similar to those for operators with regular potentials, see, e.g., [9,10,17,18]. In particular, just as in the regular case, the potential $q = \sigma'$ and boundary conditions (1.2) of the Sturm-Liouville operator $T(\sigma, H, h)$ can be recovered via the corresponding spectral data, the sequences of eigenvalues and so-called norming constants [11].

The main aim of the present paper is to treat the following inverse spectral problem. Suppose that $\{\lambda_n^2\}$ is the spectrum of the operator $T(\sigma, H, h_1)$ and $\{\mu_n^2\}$ is the spectrum of the operator $T(\sigma, H, h_2)$ with $h_2 \neq h_1$. Is it possible to recover σ, H, h_1, and h_2 from the two given spectra? More generally, the task is to find necessary and sufficient conditions for two sequences (λ_n^2) and (μ_n^2) so that they could serve as spectra of Sturm-Liouville operators on $(0,1)$ with some potential $q = \sigma' \in W_2^{-1}(0,1)$ and boundary conditions (1.2) for two different values of h and, then, to present an algorithm recovering these two operators (i.e., an algorithm determining σ and the boundary conditions).

For the case of a regular (i.e., locally integrable) potential q the above problem was treated by Levitan and Gasymov [14,15] when h_1 and h_2 are finite and by

Marchenko [16, Ch. 3.4] when one of h_1, h_2 is infinite. Earlier, Borg [4] proved that two such spectra determine the regular potential uniquely.

Observe that $T(\sigma + h', H - h', h + h') = T(\sigma, H, h)$ for any $h' \in \mathbb{C}$, so that it is impossible to recover σ, H, h_1, and h_2 uniquely; however, we shall show that the potential $q = \sigma'$ and the very operators $T(\sigma, H, h_1)$ and $T(\sigma, H, h_2)$ are determined uniquely by the two spectra. The reconstruction procedure consists in reduction of the problem to recovering the potential $q = \sigma'$ and the boundary conditions based on the spectrum and the sequence of so-called norming constants. The latter problem has been completely solved in our paper [11], and this allows us to give a complete description of the set of spectral data and to develop a reconstruction algorithm for the inverse spectral problem under consideration.

The organization of the paper is as follows. In the next two sections we restrict ourselves to the case where $H = \infty$ and $h_1, h_2 \in \mathbb{R}$. In Section 2 refined eigenvalue asymptotics is found and some other necessary conditions on spectral data are established. These results are then used in Section 3 to determine the set of norming constants and completely solve the inverse spectral problem. The case where one of h_1 and h_2 is infinite is treated in Section 4, and in Section 5 we comment on the changes to be made if H is finite. Finally, Appendix A contains some facts about Riesz bases of sines and cosines in $L_2(0,1)$ that are frequently used throughout the paper.

2. Spectral asymptotics

In this section (and until Section 4) we shall consider the case of the Dirichlet boundary condition at the point $x = 0$ and the boundary conditions of the third type at the point $x = 1$ (i. e., the case where $H = \infty$ and $h_1, h_2 \in \mathbb{R}$). We start with the precise definition of the Sturm-Liouville operators under consideration.

Suppose that q is a real-valued distribution from the class $W_2^{-1}(0,1)$ and $\sigma \in \mathcal{H}$ is any of its real-valued distributional primitives. For $h \in \mathbb{R}$ we denote by $T_{\sigma,h} = T(\sigma, \infty, h)$ the operator in $\mathcal{H}$ given by

$$T_{\sigma,h} u = l_\sigma(u) := -(u^{[1]})' - \sigma u'$$

and the boundary conditions

$$u(0) = 0, \qquad u^{[1]}(1) + hu(1) = 0, \qquad h \in \mathbb{R}$$

(we recall that $u^{[1]}$ stands for the quasi-derivative $u' - \sigma u$ of u). More precisely, the domain of $T_{\sigma,h}$ equals

$$\mathfrak{D}(T_{\sigma,h}) = \{u \in W_1^1[0,1] \mid u^{[1]} \in W_1^1[0,1],\ l_\sigma(u) \in \mathcal{H},\\ u(0) = 0,\ u^{[1]}(1) + hu(1) = 0\}.$$

It follows from [18] that so defined $T_{\sigma,h}$ is a selfadjoint operator with simple discrete spectrum accumulating at $+\infty$. The results of [19] also imply that the eigenvalues λ_n^2 of $T_{\sigma,h}$, when ordered increasingly, obey the asymptotic relation $\lambda_n = \pi(n - 1/2) + \widetilde{\lambda}_n$

with an ℓ_2-sequence $(\widetilde{\lambda}_n)$. However, we shall need a more precise form of $\widetilde{\lambda}_n$, and shall derive it next.

To start with, we introduce an operator T_σ defined by $T_\sigma u = l_\sigma(u)$ on the domain

$$\mathfrak{D}(T_\sigma) = \{u \in W_1^1[0,1] \mid u^{[1]} \in W_1^1[0,1],\ l_\sigma(u) \in \mathcal{H},\ f(0) = 0\}.$$

In other words, T_σ is a one-dimensional extension of $T_{\sigma,h}$ discarding the boundary condition at the point $x = 1$. We invoke now the following results of [12] on the operator T_σ. Just as in the classical situation with regular potential q the operators $T_{\pm\sigma}$ turn out to be similar to the potential-free operator T_0; the similarity is performed by the *transformation operators* $I + K_\sigma^\pm$, where $K_\sigma^\pm$ are integral Volterra operators of Hilbert-Schmidt class given by

$$K_\sigma^\pm u(x) = \int_0^x k_\sigma^\pm(x,t)u(t)\ dt.$$

It follows that any function u from $\mathfrak{D}(T_\sigma)$ has the form $u = (I + K_\sigma^+)v$ for some function $v \in \mathfrak{D}(T_0)$ (i.e., for some $v \in W_2^2[0,1]$ satisfying the boundary condition $v(0) = 0$) and that $T_\sigma(I + K_\sigma^+)v = -(I + K_\sigma^+)v''$. Moreover, there exists a Hilbert-Schmidt kernel $r_\sigma(x,t)$ such that

$$[(I + K_\sigma^+)v]^{[1]}(x) = [(I + K_\sigma^-)v'](x) + \int_0^x r_\sigma(x,t)v(t)\ dt$$

for every $v \in \mathfrak{D}(T_0)$. Also, for every fixed $x \in [0,1]$ the functions $k_\sigma^\pm(x,\cdot)$ and $r_\sigma(x,\cdot)$ belong to $\mathcal{H}$.

Due to the above similarity of T_σ and T_0, for any nonzero $\lambda \in \mathbb{C}$ the function

$$u(x,\lambda) := \sin\lambda x + \int_0^x k_\sigma^+(x,t)\sin\lambda t\ dt$$

is an eigenfunction of the operator T_σ corresponding to the eigenvalue $\lambda^2 \in \mathbb{C}$. If in addition $u(x,\lambda)$ satisfies the boundary condition $u^{[1]}(1) + hu(1) = 0$, then $u(x,\lambda)$ is also an eigenfunction of the operator $T_{\sigma,h}$ corresponding to the eigenvalue λ^2. Therefore the nonzero spectrum of $T_{\sigma,h}$ consists of the squared nonzero λ-solutions of the equation

$$\begin{aligned}\cos\lambda + \int_0^1 k_\sigma^-(1,t)\cos\lambda t\ dt + \frac{1}{\lambda}\int_0^1 r_\sigma(1,t)\sin\lambda t\ dt& \\ + \frac{h}{\lambda}\sin\lambda + \frac{h}{\lambda}\int_0^1 k_\sigma^+(1,t)\sin\lambda t\ dt = 0.&\end{aligned} \tag{2.1}$$

Recall that the operator $T_{\sigma,h}$ is bounded below and hence after addition to q a suitable constant $T_{\sigma,h}$ becomes positive. Therefore without loss of generality we may assume that all solutions of equation (2.1) are real and nonzero. Observe also that due to the symmetry we may consider only the positive zeros. The asymptotics of these zeros is given in the following theorem.

Theorem 2.1. Suppose that $\sigma \in \mathcal{H}$ and $h \in \mathbb{R}$ are such that the operator $T_{\sigma,h}$ is positive and denote by $\lambda_1^2 < \lambda_2^2 < \cdots$ the eigenvalues of $T_{\sigma,h}$. Then the numbers λ_n satisfy the relation

$$\lambda_n = \pi(n - 1/2) + \frac{h}{\pi n} + \delta_n(h) + \gamma_n,$$

in which the sequences $\big(\delta_n(h)\big)_{n=1}^{\infty}$ and $(\gamma_n)_{n=1}^{\infty}$ belong to ℓ_1 and ℓ_2 respectively with γ_n independent of h.

Proof. Recall that $\lambda_n \to \infty$ as $n \to \infty$ and that the functions $k_\sigma^{\pm}(1,t)$ and $r_\sigma(1,t)$ belong to $\mathcal{H}$. Therefore (2.1) and the Riemann lemma imply that

$$\cos\lambda_n = -\int_0^1 k_\sigma^-(1,t)\cos\lambda_n t\;dt + O(1/\lambda_n) = o(1)$$

as $n \to \infty$. By the standard Rouché-type arguments (see details, e.g., in [16, Ch. 1.3]) we conclude that $\lambda_n = \pi(n-1/2) + \widetilde{\lambda}_n$ with $\widetilde{\lambda}_n \to 0$ as $n \to \infty$. As a result (see Appendix A), the system $\{\sin\lambda_n x\}$ turns out to be a Riesz basis of $\mathcal{H}$ and the sequences $(c_n)_{n=1}^{\infty}$ and $(c_n')_{n=1}^{\infty}$ with

$$c_n := \int_0^1 k_\sigma^+(1,t)\sin\lambda_n t\;dt, \quad c_n' := \int_0^1 r_\sigma(1,t)\sin\lambda_n t\;dt$$

belong to ℓ_2. Relation (2.1) now implies that

$$\sin\widetilde{\lambda}_n = (-1)^{n+1}\int_0^1 k_\sigma^-(1,t)\cos\lambda_n t\;dt + h\frac{\cos\widetilde{\lambda}_n}{\lambda_n} + (-1)^{n+1}\frac{hc_n + c_n'}{\lambda_n} =$$
$$= \gamma_n + \frac{h}{\pi n} + \delta_n' + \delta_n'',$$

where

$$\begin{aligned}
\gamma_n &:= (-1)^{n+1}\int_0^1 k_\sigma^-(1,t)\cos\pi(n-1/2)t\;dt,\\
\delta_n' &:= (-1)^{n+1}\int_0^1 k_\sigma^-(1,t)[\cos\lambda_n t - \cos\pi(n-1/2)t]\;dt, \qquad (2.2)\\
\delta_n'' &:= \frac{h\cos\widetilde{\lambda}_n}{\lambda_n} - \frac{h}{\pi n} + (-1)^{n+1}\frac{hc_n + c_n'}{\lambda_n}.
\end{aligned}$$

Next we observe that the systems $\{\cos\pi(n-1/2)t\}_{n=1}^{\infty}$ and $\{\cos\lambda_n t\}_{n=1}^{\infty}$ form Riesz bases of $\mathcal{H}$, so that (γ_n) and (δ_n') belong to ℓ_2. It follows that $(\sin\widetilde{\lambda}_n) \in \ell_2$ and therefore $(\widetilde{\lambda}_n) \in \ell_2$. Lemma A.2 now implies that the numbers δ_n' form an ℓ_1-sequence, and simple arguments justify the same statement about δ_n''.

Thus we have shown that

$$\sin\widetilde{\lambda}_n = \gamma_n + \frac{h}{\pi n} + \delta_n' + \delta_n''$$

with the sequences (γ_n) from ℓ_2 and (δ_n') and (δ_n'') from ℓ_1. Applying arcsin to both parts of this equality, we arrive at

$$\widetilde{\lambda}_n = \gamma_n + \frac{h}{\pi n} + \delta_n$$

with some ℓ_1-sequence (δ_n), and the theorem is proved. □

It was shown in [11] that any sequence (λ_n^2) of pairwise distinct positive numbers satisfying the relation $\lambda_n = \pi(n-1/2)+\widetilde{\lambda}_n$ with an ℓ_2-sequence $(\widetilde{\lambda}_n)$ is the spectrum of some Sturm-Liouville operator $T_{\sigma,h}$ with potential $q = \sigma'$ from $W_2^{-1}(0,1)$, the Dirichlet boundary condition at $x = 0$, and third type boundary condition (1.2) at $x = 1$ with some $h \in \mathbb{R}$. Suppose that (μ_n^2) is the spectrum of a Sturm-Liouville operator with the same potential q, the Dirichlet boundary condition at $x = 0$, and third type boundary condition (1.2) at $x = 1$ with a different $h \in \mathbb{R}$. We ask whether there exists any relation between the two spectra.

The first restriction, just as in the regular case, is that the spectra should interlace.

Lemma 2.2. Suppose that $\sigma \in \mathcal{H}$ is real valued and that sequences $\{\lambda_n^2\}$ and $\{\mu_n^2\}$ are the spectra of Sturm-Liouville operators T_{σ,h_1} and T_{σ,h_2} with distinct $h_1, h_2 \in \overline{\mathbb{R}}$. Then these sequences interlace.

Proof. Denote by $u = u(x,\lambda)$ a solution of equation $l_\sigma(u) = \lambda u$ satisfying the boundary condition $u(0) = 0$. We recall that u being a solution of $l_\sigma(u) = \lambda u$ means that

$$\frac{\mathrm{d}}{\mathrm{d}x}\begin{pmatrix} u^{[1]} \\ u \end{pmatrix} = \begin{pmatrix} -\sigma & -\sigma^2 - \lambda \\ 1 & \sigma \end{pmatrix}\begin{pmatrix} u^{[1]} \\ u \end{pmatrix} \tag{2.3}$$

and hence u enjoys the standard uniqueness properties of solutions to first order differential systems with regular (i.e. locally integrable) coefficients. The numbers λ_n and μ_n are then solutions of the equations $u^{[1]}(1,\lambda) + h_1 u(1,\lambda) = 0$ and $u^{[1]}(1,\lambda) + h_2 u(1,\lambda) = 0$ respectively.

We introduce a continuous function $\theta(x,\lambda)$ through $\cot\theta(x,\lambda) \equiv u^{[1]}/u$. After differentiating both sides of this identity in x and using (2.3) we get

$$-\frac{\theta'}{\sin^2\theta} = \frac{-\sigma u' u - u^{[1]}u'}{u^2} - \lambda = -(\cot\theta + \sigma)^2 - \lambda,$$

or

$$\theta' = \lambda\sin^2\theta + (\cos\theta + \sigma\sin\theta)^2.$$

It follows from [3, Ch. 8.4] or [7, proof of Theorem XI.3.1] that the function $\theta(1,\lambda)$ is strictly increasing in λ; hence solutions of the equations $\cot\theta(1,\lambda) = -h_1$ and $\cot\theta(1,\lambda) = -h_2$ interlace, and the proof is complete. □

The next restriction concerns asymptotics of λ_n and μ_n.

Lemma 2.3. Suppose that (λ_n^2) and (μ_n^2) are spectra of Sturm-Liouville operators T_{σ,h_1} and T_{σ,h_2} with real-valued $\sigma \in \mathcal{H}$ and $h_1, h_2 \in \mathbb{R}$. Then there exists an ℓ_2-sequence (ν_n) such that

$$\lambda_n - \mu_n = \frac{h_1 - h_2}{\pi n} + \frac{\nu_n}{n}. \tag{2.4}$$

Proof. Put $\widetilde{\lambda}_n := \lambda_n - \pi(n - 1/2)$ and $\widetilde{\mu}_n := \mu_n - \pi(n - 1/2)$ so that $\lambda_n - \mu_n = \widetilde{\lambda}_n - \widetilde{\mu}_n$. Recall that

$$\sin \widetilde{\lambda}_n - \sin \widetilde{\mu}_n = \frac{h_1 - h_2}{\pi n} + \delta_n'(h_1) - \delta_n'(h_2) + \delta_n''(h_1) - \delta_n''(h_2)$$

with the quantities $\delta_n'(h), \delta_n''(h)$ of (2.2). Relation

$$\delta_n'(h_1) - \delta_n'(h_2) = (-1)^{n+1} \int_0^1 k_\sigma^-(1,t)[\cos \lambda_n t - \cos \mu_n t]\, dt$$

and Lemma A.2 show that $\delta_n'(h_1) - \delta_n'(h_2) = (\widetilde{\lambda}_n - \widetilde{\mu}_n)\nu_n'$ for some ℓ_2-sequence (ν_n'). Also there exists an ℓ_2-sequence (ν_n'') such that $\delta_n''(h_1) - \delta_n''(h_2) = \nu_n''/n$. Finally, using the relations

$$\sin \widetilde{\lambda}_n - \sin \widetilde{\mu}_n = (\widetilde{\lambda}_n - \widetilde{\mu}_n) \cos \widetilde{\nu}_n,$$

where $\widetilde{\nu}_n$ are points between $\widetilde{\lambda}_n$ and $\widetilde{\mu}_n$ (so that $(1 - \cos \widetilde{\nu}_n)$ is an ℓ_2-sequence), and combining the above relations we arrive at

$$(\widetilde{\lambda}_n - \widetilde{\mu}_n)(\cos \widetilde{\nu}_n - \nu_n') = \frac{h_1 - h_2}{\pi n} + \frac{\nu_n''}{n},$$

which implies (2.4). □

Corollary 2.4. Under the above assumptions, $\lim_{n\to\infty}(\lambda_n^2 - \mu_n^2) = 2(h_1 - h_2)$.

3. Reduction to the inverse spectral problem by one spectrum and norming constants

Suppose that $\sigma \in \mathcal{H}$ is real-valued and that $\{\lambda_n^2\}$ and $\{\mu_n^2\}$ are eigenvalues of the operators T_{σ,h_1} and T_{σ,h_2} respectively introduced in the previous section. We shall show how the problem of recovering σ, h_1, and h_2 based on these spectra can be reduced to the problem of recovering σ and h_1 based on the spectrum $\{\lambda_n^2\}$ of T_{σ,h_1} and the so-called norming constants $\{\alpha_n\}$ defined below. This latter problem for the class of Sturm-Liouville operators with singular potentials from $W_2^{-1}(0,1)$ is completely solved in [11]; see also [13,16] for the regular case of integrable potentials.

Denote by $u_1(x,\lambda)$ and $u_2(x,\lambda)$ solutions of the equation $l_\sigma(u) = \lambda^2 u$ satisfying the terminal conditions $u_1(1,\lambda) = u_2(1,\lambda) = 1$ and $u_j^{[1]}(1,\lambda) = -h_j$, $j = 1,2$.

Then $u_1(x,\lambda_n)$ are the eigenfunctions of the operator T_{σ,h_1} corresponding to the eigenvalues λ_n^2, and we put

$$\alpha_n := 2\int_0^1 |u_1(x,\lambda_n)|^2\,\mathrm{d}x.$$

Our next aim is to show that α_n can be expressed in terms of the spectra $\{\lambda_k^2,\mu_k^2\}$ only.

Set $\Phi_j(\lambda) := u_j(0,\lambda)$, $j=1,2$; then zeros of Φ_1 and Φ_2 are precisely the numbers $\pm\lambda_n$ and $\pm\mu_n$ respectively. We show that Φ_1 and Φ_2 uniquely determine α_n and are uniquely determined by their zeros.

Lemma 3.1. The norming constants α_n satisfy the following equality:

$$\alpha_n = \frac{h_1-h_2}{\lambda_n}\frac{\Phi_1'(\lambda_n)}{\Phi_2(\lambda_n)}. \tag{3.1}$$

Proof. The proof is rather standard and the details can be found, e.g., in [13], so we shall only sketch its main points here.

Put $m(\lambda) := -\Phi_2(\lambda)/\Phi_1(\lambda)$; then the function

$$f(x,\lambda) := u_2(x,\lambda) + m(\lambda)u_1(x,\lambda)$$

for all $\lambda\in\mathbb{C}$ such that λ^2 is not in the spectrum of T_{σ,h_1} satisfies the equation $l_\sigma(f)=\lambda^2 f$ and the boundary condition $f(0,\lambda)=0$. In other words, $f(x,\lambda)$ is an eigenfunction of the operator T_σ corresponding to the eigenvalue λ^2. Since $u_1(x,\lambda_n)$ is an eigenfunction of the operator T_σ corresponding to the eigenvalue λ_n^2, we have

$$\begin{aligned}(\lambda^2-\lambda_n^2)\int_0^1 f(x,\lambda)u_1(x,\lambda_n)\,\mathrm{d}x &= (T_\sigma f, u_1) - (f, T_\sigma u_1)\\ &= -f^{[1]}(1,\lambda)\overline{u_1(1,\lambda)} + f(1,\lambda)\overline{u^{[1]}(1,\lambda_n)} = h_2 - h_1.\end{aligned}$$

On the other hand,

$$\begin{aligned}(\lambda^2-\lambda_n^2)\int_0^1 f(x,\lambda)u_1(x,\lambda_n)\,\mathrm{d}x &= (\lambda^2-\lambda_n^2)\int_0^1 u_2(x,\lambda)u_1(x,\lambda_n)\,\mathrm{d}x\\ &\quad - (\lambda^2-\lambda_n^2)\frac{\Phi_2(\lambda)}{\Phi_1(\lambda)}\int_0^1 u_1(x,\lambda)u_1(x,\lambda_n)\,\mathrm{d}x,\end{aligned}$$

and after combining the two relations and letting $\lambda\to\lambda_n$ we arrive at

$$\alpha_n = \frac{h_1-h_2}{\lambda_n}\frac{\Phi_1'(\lambda_n)}{\Phi_2(\lambda_n)}.$$

The proof is complete. □

The following statement has appeared in many variants in numerous sources, but we include its proof here for the sake of completeness.

Lemma 3.2. In order that a function $f(\lambda)$ admit the representation

$$f(\lambda) = \cos\lambda + \int_0^1 g(t)\cos\lambda t \, dt \tag{3.2}$$

with an $\mathcal{H}$-function g, it is necessary and sufficient that

$$f(\lambda) = \prod_{k=1}^{\infty} \frac{(f_k^2 - \lambda^2)}{\pi^2(k-1/2)^2}, \tag{3.3}$$

where $f_k := \pi(k - 1/2) + \widetilde{f}_k$ and $(\widetilde{f}_k)$ is an ℓ_2-sequence.

Proof. *Necessity.* The function f of (3.2) is an even entire function of order 1, and the standard Rouché-type arguments (see the analysis of Section 2) show that zeros $\pm f_k$ of f have the asymptotics $f_k = \pi(k - 1/2) + \widetilde{f}_k$ with $(\widetilde{f}_k) \in \ell_2$. On the other hand, up to a scalar factor C, the function f is recovered from its zeros as

$$f(\lambda) = C\prod_{k=1}^{\infty}\Bigl(1 - \frac{\lambda^2}{f_k^2}\Bigr) = C_1 \prod_{k=1}^{\infty} \frac{(f_k^2 - \lambda^2)}{\pi^2(k-1/2)^2}, \tag{3.4}$$

where

$$C_1 = C\prod_{k=1}^{\infty}\frac{\pi^2(k-1/2)^2}{f_k^2}.$$

To determine C_1, we observe that, in view of (3.2), $\lim_{\nu\to\infty} f(i\nu)/\cos i\nu = 1$ and that

$$\cos\lambda = \prod_{k=1}^{\infty}\frac{(\pi^2(k-1/2)^2 - \lambda^2)}{\pi^2(k-1/2)^2}; \tag{3.5}$$

comparing now (3.4) and (3.5), we conclude that $C_1 = 1$ and the necessity is justified.

Sufficiency. Suppose now that f has a representation of the form (3.3), in which $f_k = \pi(k-1/2) + \widetilde{f}_k$ with $(\widetilde{f}_k) \in \ell_2$. We assume first that the zeros f_k are pairwise distinct. Then due to the asymptotics of f_k the system $(\cos f_k x)_{k=1}^{\infty}$ is a Riesz basis of $\mathcal{H}$ (see Lemma A.1) and the numbers $c_k := -\cos f_k = (-1)^k \sin\widetilde{f}_k$ define an ℓ_2-sequence. Therefore there exists a unique function $g \in \mathcal{H}$ with Fourier coefficients c_k in the Riesz basis $(\cos f_k x)_{k=1}^{\infty}$. Now the function

$$\widetilde{f}(\lambda) := \cos\lambda + \int_0^1 g(x)\cos\lambda x \, dx \tag{3.6}$$

is an even entire function of order 1 and has zeros at the points $\pm f_k$. It follows that $\widetilde{f}$ and f differ by a scalar factor, which as before is shown to equal 1.

Due to the asymptotics of the zeros f_k only finitely many of them can repeat. If, e.g., $f_n = f_{n+1} = \cdots = f_{n+p-1}$ is a zero of the function f of multiplicity $p = p(n)$, then we include to the above system the functions $\cos f_n x$, $x\sin f_n x, \cdots$, $x^{p-1}\sin(\pi p/2 - f_n x)$. After this modification has been done for every multiple root, we again get a Riesz basis of $\mathcal{H}$ and can find a function $g \in \mathcal{H}$ with Fourier coefficients $c_n := -\cos f_n, c_{n+1} = -\sin f_n, \cdots, c_{n+p(n)} := -\sin(\pi p(n)/2 - f_n x)$. Then $\pm f_n$ becomes a zero of the function $\widetilde{f}$ in (3.6) of multiplicity $p(n)$, and the proof is completed as above. □

Lemma 3.3. The following equalities hold:

$$\begin{aligned}\Phi_1(\lambda) &= \prod \pi^{-2}(k-1/2)^{-2}(\lambda_k^2 - \lambda^2),\\ \Phi_2(\lambda) &= \prod \pi^{-2}(k-1/2)^{-2}(\mu_k^2 - \lambda^2).\end{aligned} \tag{3.7}$$

Proof. We recall [12] that the function u_1 has a representation via a transformation operator connected with the point $x = 1$ of the form

$$u_1(x,\lambda) = \cos\lambda(1-x) + \int_x^1 k(x,t)\cos\lambda(1-t)\,\mathrm{d}t,$$

which implies that

$$\Phi_1(\lambda) = \cos\lambda + \int_0^1 k_1(t)\cos\lambda t\,\mathrm{d}t,$$

where $k_1(t) := k(0, 1-t)$ is a function from $\mathcal{H}$. The claim now follows from Lemma 3.2. The representation for Φ_2 is derived analogously, and the lemma is proved. □

Now, given two spectra (λ_n^2) and (μ_n^2) of two Sturm-Liouville operators T_{σ,h_1} and T_{σ,h_2} respectively, with unknown $\sigma \in \mathcal{H}$ and $h_1, h_2 \in \mathbb{R}$, we proceed as follows. First, we identify $h_1 - h_2$ as $\lim_{n\to\infty}(\lambda_n^2 - \mu_n^2)/2$ (see Corollary 2.4), then construct the functions Φ_1 and Φ_2 of (3.7) and determine the norming coefficients (α_n) via (3.1). The spectral data $\{(\lambda_n^2), (\alpha_n)\}$ determine σ and h_1 up to an additive constant by means of the algorithm of [11]. This gives the required function σ and two numbers h_1 and h_2 up to an additive constant, and the reconstruction is complete.

The second part of the inverse spectral problem is to identify those pairs of sequences (λ_n^2) and (μ_n^2) that are spectra of Sturm-Liouville operators T_{σ,h_1} and T_{σ,h_2} for some real-valued $\sigma \in \mathcal{H}$ and some real h_1, h_2, $h_1 \neq h_2$. We established in Theorem 2.1 and Lemmata 2.2 and 2.3 the necessary conditions on the two spectra; it turns out that these conditions are sufficient as well. Namely, the following statement holds true.

Theorem 3.4. Suppose that sequences (λ_n^2) and (μ_n^2) of positive pairwise distinct numbers satisfy the following assumptions:

(i) the sequences (λ_n^2) and (μ_n^2) interlace;

(ii) $\lambda_n = \pi(n-1/2) + \widetilde{\lambda}_n$ and $\mu_n = \pi(n-1/2) + \widetilde{\mu}_n$ with some ℓ_2-sequences $(\widetilde{\lambda}_n)$ and $(\widetilde{\mu}_n)$;

(iii) there exist a real number h and an ℓ_2-sequence (ν_n) such that $\lambda_n - \mu_n = \frac{h}{\pi n} + \frac{\nu_n}{n}$.

Then there exist a function $\sigma \in \mathcal{H}$ and two real numbers h_1 and h_2 such that (λ_n^2) and (μ_n^2) are the spectra of the Sturm-Liouville operators T_{σ,h_1} and T_{σ,h_2} respectively. Moreover, the operators T_{σ,h_1} and T_{σ,h_2} are recovered uniquely.

In order to prove the theorem we have to show first that the numbers α_n constructed for the two sequences through formula (3.1), with the functions Φ_1, Φ_2 of (3.7) and $h_1 - h_2 := h$ with h of item (3), are positive and obey the asymptotics $\alpha_n = 1 + \widetilde{\alpha}_n$ for some ℓ_2-sequence $(\widetilde{\alpha}_n)$. Then the algorithm of [11] uses (λ_n^2) and (α_n) to determine a (unique up to an additive constant) function $\sigma \in \mathcal{H}$ and $h_1 \in \mathbb{R}$ such that $\{\lambda_n^2\}$ is the spectrum of the Sturm-Liouville operator T_{σ,h_1} and α_n are the corresponding norming constants. The second and final step is to verify that $\{\mu_n^2\}$ is the spectrum of the Sturm-Liouville operator T_{σ,h_2} with $h_2 := h_1 - h$, where h is the number of item (3). These two steps are performed in the following two lemmata.

Lemma 3.5. Assume (1)–(3) of Theorem 3.4. Then the numbers α_n constructed via relation (3.1) with Φ_1, Φ_2 of (3.7) and $h_1 - h_2 := h$ with h of item (3) are all positive and obey the asymptotics $\alpha_n = 1 + \widetilde{\alpha}_n$, where $(\widetilde{\alpha}_n)$ is an ℓ_2-sequence.

Proof. By Lemma 3.2 there exist $\mathcal{H}$-functions f_1, f_2 such that the functions Φ_j, $j = 1, 2$, of (3.7) admit the representation

$$\Phi_j(\lambda) = \cos \lambda + \int_0^1 f_j(t) \cos \lambda t \, dt.$$

It follows that

$$\begin{aligned} \Phi_1'(\lambda_n) &= -\sin \lambda_n - \int_0^1 t f_1(t) \sin \lambda_n t \, dt \\ &= (-1)^n \cos \widetilde{\lambda}_n - \int_0^1 t f_1(t) \sin \lambda_n t \, dt = (-1)^n + \hat{\lambda}_n, \end{aligned}$$

where $(\hat{\lambda}_n)_{n=1}^{\infty} \in \ell_2$, and by similar arguments

$$\begin{aligned} \Phi_2(\lambda_n) &= (\lambda_n - \mu_n)\Phi_2'(\mu_n) + O(|\lambda_n - \mu_n|^2) \\ &= (\lambda_n - \mu_n)[(-1)^n + \hat{\mu}_n] \end{aligned}$$

for some ℓ_2-sequence $(\hat{\mu}_n)$. Next, assumptions (2) and (3) easily imply the relation

$$\frac{h}{\lambda_n(\lambda_n - \mu_n)} = 1 + \hat{\nu}_n,$$

for some ℓ_2-sequence $(\hat{\nu}_n)$. Therefore the numbers

$$\alpha_n := \frac{h}{\lambda_n} \frac{\Phi_1'(\lambda_n)}{\Phi_2(\lambda_n)}$$

obey the required asymptotics.

Finally, the interlacing property of the two sequences implies that all α_n are of the same sign, and thus are all positive in view of the asymptotics established. The proof is complete. □

With the above-stated properties of the sequences (λ_n^2) and (α_n) we can employ the result of [11] that guarantees existence of a unique Sturm-Liouville operator T_{σ,h_1} with a real-valued function $\sigma \in \mathcal{H}$ and a real number h_1 such that $\{\lambda_n^2\}$ coincides with the spectrum of T_{σ,h_1} and α_n are the corresponding norming constants. Now we put $h_2 := h_1 - h$ with h of assumption (3) and expect $\{\mu_n^2\}$ to be the spectrum of T_{σ,h_2}.

Lemma 3.6. The spectrum of the Sturm-Liouville operator T_{σ,h_2} with the above $\sigma \in \mathcal{H}$ and $h_2 \in \mathbb{R}$ coincides with (μ_n^2).

Proof. Denote by $v_1(x,\lambda)$ a solution of the equation $l_\sigma(v) = \lambda^2 v$ satisfying the boundary conditions $v(1) = 1, v^{[1]}(1) = -h_1$. Then $v_1(x,\lambda_n)$ is an eigenfunction of the operator T_{σ,h_1} corresponding to the eigenvalue λ_n^2, and by construction

$$2\int_0^1 |v_1(x,\lambda_n)|^2 \,\mathrm{d}x = \alpha_n := \frac{h}{\lambda_n}\frac{\Phi_1'(\lambda_n)}{\Phi_2(\lambda_n)}$$

with the functions Φ_1, Φ_2 of (3.7).

Denote by (ν_n^2) the spectrum of T_{σ,h_2} and put

$$\widetilde{\Phi}_2(\lambda) := \prod_{k=1}^{\infty} \pi^{-2}(k-1/2)^{-2}(\nu_n^2 - \lambda^2).$$

By Lemma 3.1 we have

$$\alpha_n = \frac{h}{\lambda_n}\frac{\Phi_1'(\lambda_n)}{\widetilde{\Phi}_2(\lambda_n)},$$

whence $\Phi_2(\lambda_n) = \widetilde{\Phi}_2(\lambda_n)$ for all $n \in \mathbb{N}$. Lemma 3.2 implies that

$$\Phi_2(\lambda) - \widetilde{\Phi}_2(\lambda) = \int_0^1 f(t)\cos\lambda t \,\mathrm{d}t$$

for some $f \in \mathcal{H}$. Equality $\Phi_2(\lambda_n) = \widetilde{\Phi}_2(\lambda_n)$ means that the function f is orthogonal to $\cos\lambda_n t$, $n \in \mathbb{N}$. Since the system $\{\cos\lambda_n t\}$ forms a Riesz basis of $\mathcal{H}$, we get $f \equiv 0$ and $\Psi_2 \equiv \widetilde{\Psi}_2$. Thus $\nu_n = \mu_n$, and the lemma is proved. □

4. Reconstruction by Dirichlet and Dirichlet-Neumann spectra

The analysis of the previous two sections does not cover the case where one of h_1, h_2 is infinite. In this case the other number may be taken 0 without loss of generality (recall that $T_{\sigma,h} = T_{\sigma+h,0}$), i.e., the boundary conditions under considerations become Dirichlet and Dirichlet-Neumann ones.

Suppose therefore that $\sigma \in \mathcal{H}$ is real valued and that (λ_n^2) and (μ_n^2) are spectra of the operators $T_{\sigma,0}$ and $T_{\sigma,\infty}$ respectively; without loss of generality we assume that λ_n and μ_n are positive and strictly increase with n. The reconstruction procedure remains the same as before; namely, we use the two spectra to determine a sequence of norming constants (α_n) and then recover σ by the spectral data $\{(\lambda_n^2), (\alpha_n)\}$.

Denote by $u_-(\cdot, \lambda)$ and $u_+(\cdot, \lambda)$ solutions of the equation $l_\sigma u = \lambda u$ satisfying the initial conditions $u_-(0, \lambda) = 0, u_-^{[1]}(0, \lambda) = \lambda$ and the terminal conditions $u_+(1, \lambda) = 1, u_+^{[1]}(1, \lambda) = 0$ respectively. Then $u_+(\cdot, \lambda_n)$ is an eigenfunction of the operator $T_{\sigma,0}$ corresponding to the eigenvalue λ_n^2 and

$$\alpha_n := 2 \int_0^1 |u_+(x, \lambda_n)|^2 \, dx$$

is the corresponding norming constant. We also put $\Psi_1(\lambda) := u_+(0, \lambda)$ and $\Psi_2(\lambda) := u_-(1, \lambda)$; then zeros of the functions Ψ_1 and Ψ_2 are numbers $\pm\lambda_n$ and $\pm\mu_n$ respectively. As earlier, the function Ψ_1 is uniquely determined by its zeros through formula (3.7). Observe that the Dirichlet eigenvalues μ_n^2 have asymptotics different from that of λ_n^2, so that Ψ_2 requires slight modification of formula (3.7). We first investigate the asymptotics of μ_n.

Theorem 4.1 ([12,17]). Suppose that $\sigma \in \mathcal{H}$ and $\{\mu_n^2\}$, $n \in \mathbb{N}$, is the spectrum of the operator $T_{\sigma,\infty}$. Then the numbers μ_n satisfy the relation

$$\mu_n = \pi n + \widetilde{\mu}_n,$$

in which the sequence $(\widetilde{\mu}_n)$ belongs to ℓ_2.

Proof. The solution u_- can be represented by means of the transformation operator [12] as $u_-(x, \lambda) = \sin \lambda x + \int_0^x k_\sigma^+(x, t) \sin \lambda t \, dt$, where k_σ^+ is the kernel of the transformation operator. Therefore the numbers $\pm\mu_n$ are zeros of the function

$$\Psi_2(\lambda) := \sin \lambda + \int_0^1 k_\sigma^+(1, t) \sin \lambda t \, dt, \tag{4.1}$$

which is entire of order 1. Since the function $k_\sigma^+(1, \cdot)$ belongs to $\mathcal{H}$ [12], the required asymptotics of μ_n is derived in a standard way (cf. [16] and Section 2). □

The function Ψ_2 is determined by its its zeros in the following way.

Lemma 4.2. The following equality holds:

$$\Psi_2(\lambda) = \lambda \prod \pi^{-2} k^{-2} (\mu_k^2 - \lambda^2). \tag{4.2}$$

The proof is completely analogous to that of Lemma 3.2 and is left to the reader. Now we show how the norming constants α_n are expressed via Ψ_1 and Ψ_2.

Lemma 4.3. The norming constants satisfy the following equality:

$$\alpha_n = -\frac{\Psi_1'(\lambda_n)}{\Psi_2(\lambda_n)}. \tag{4.3}$$

Proof. (cf. [5]) The Green's function $G(x, y, \lambda^2)$ of the operator $T_{\sigma,0}$ (i. e., the kernel of the resovlent $(T_{\sigma,0} - \lambda^2)^{-1}$) equals

$$G(x, y, \lambda^2) = \sum_{n=1}^{\infty} \frac{2}{\lambda_n^2 - \lambda^2} \frac{u_+(x, \lambda_n) u_+(y, \lambda_n)}{\alpha_n}.$$

On the other hand, we have

$$G(x, y, \lambda^2) = \frac{1}{W(\lambda)} \cdot \begin{cases} u_-(x, \lambda) u_+(y, \lambda), & 0 \le x \le y \le 1, \\ u_+(x, \lambda) u_-(y, \lambda), & 0 \le y < x \le 1, \end{cases}$$

where $W(\lambda) := u_+(x, \lambda) u_-^{[1]}(x, \lambda) - u_-(x, \lambda) u_+^{[1]}(x, \lambda)$ is the Wronskian of u_+ and u_-. The value of $W(\lambda)$ is independent of $x \in [0, 1]$; in particular, taking $x = 0$ we find that $W(\lambda) \equiv \lambda u_+(0, \lambda) = \lambda \Psi_1(\lambda)$.

Now we take $x = y = 1$ in the above expressions and find that

$$\sum_{n=1}^{\infty} \frac{2}{\lambda_n^2 - \lambda^2} \frac{1}{\alpha_n} \equiv \frac{\Psi_2(\lambda)}{\lambda \Psi_1(\lambda)}.$$

Comparing the residues at the poles $\lambda = \lambda_n$, we derive formula (4.3), and the lemma is proved. □

Now the reconstruction procedure is completed as follows: we determine the sequence of norming constants (α_n) as explained above and find a unique operator $T_{\sigma,0}$ with spectrum (λ_n^2) and norming constants (α_n), see [11]. This gives the function $\sigma \in \mathcal{H}$ and thus the operator $T_{\sigma,\infty}$.

Now we would like to give an explicit description of the set of all possible spectra of the operators $T_{\sigma,0}$ and $T_{\sigma,\infty}$ when a real-valued function σ runs through $\mathcal{H}$, i. e., to give the necessary and sufficient conditions on two sequences (λ_n^2) and (μ_n^2) to be the spectra of $T_{\sigma,0}$ and $T_{\sigma,\infty}$ with a real-valued $\sigma \in \mathcal{H}$. Necessary conditions are that the eigenvalues λ_n^2 and μ_n^2 should interlace and obey the asymptotics described in Theorems 2.1 and 4.1; we shall show that these conditions are in fact sufficient as well.

Theorem 4.4. Suppose that sequences (λ_n^2) and (μ_n^2) of positive pairwise distinct numbers satisfy the following assumptions:

(i) the sequences (λ_n^2) and (μ_n^2) interlace: $0 < \lambda_1^2 < \mu_1^2 < \lambda_2^2 < \dots$;

(ii) $\lambda_n = \pi(n - 1/2) + \widetilde{\lambda}_n$ and $\mu_n = \pi n + \widetilde{\mu}_n$ with some ℓ_2-sequences $(\widetilde{\lambda}_n)$ and $(\widetilde{\mu}_n)$.

Then there exist a unique function $\sigma \in \mathcal{H}$ such that (λ_n^2) and (μ_n^2) are the spectra of the Sturm-Liouville operators $T_{\sigma,0}$ and $T_{\sigma,\infty}$ respectively.

The main ingredients of the proof of this theorem are the same as for Theorem 3.4: first, we construct functions Ψ_1 and Ψ_2 by their zeros and define the sequence (α_n) as explained in Lemma 4.3; then we prove that α_n are all positive and have the required asymptotics $\alpha_n = 1 + \widetilde{\alpha}_n$ for some ℓ_2-sequence $(\widetilde{\alpha}_n)$. Using now the reconstruction algorithm of [11], we find a unique function $\sigma \in \mathcal{H}$ such that the corresponding Sturm-Liouville operator $T_{\sigma,0}$ with potential $q = \sigma' \in \mathcal{H}$ possesses the spectral data $\{(\lambda_n^2), (\alpha_n)\}$. Finally, we prove that (μ_n^2) is the spectrum of the operator $T_{\sigma,\infty}$ with σ just found, and the reconstruction procedure is complete.

5. The case of the Neumann boundary condition at $x = 0$

The analysis of the previous sections can easily be modified to cover the case $H = 0$, i.e., the Neumann boundary condition $u^{[1]}(0) = 0$. As before, we use the two spectra to determine the sequence of norming constants and then apply the reconstruction procedure of [11] to find the corresponding Sturm-Liouville operators. Also the necessary and sufficient conditions on the two spectra can be established. We formulate the corresponding results in the following two theorems.

Theorem 5.1. Suppose that sequences (λ_n^2) and (μ_n^2) of positive pairwise distinct numbers satisfy the following assumptions:

(i) the sequences (λ_n^2) and (μ_n^2) interlace;

(ii) $\lambda_n = \pi(n-1) + \widetilde{\lambda}_n$ and $\mu_n = \pi(n-1) + \widetilde{\mu}_n$ with some ℓ_2-sequences $(\widetilde{\lambda}_n)$ and $(\widetilde{\mu}_n)$;

(iii) there exist a real number h and an ℓ_2-sequence (ν_n) such that $\lambda_n - \mu_n = \frac{h}{\pi n} + \frac{\nu_n}{n}$.

Then there exist a unique function $\sigma \in \mathcal{H}$ and real constants h_1, h_2 such that (λ_n^2) and (μ_n^2) are the spectra of the Sturm-Liouville operators $T_{\sigma,0,h_1}$ and $T_{\sigma,0,h_2}$ respectively.

Conversely, the spectra (λ_n^2) and (μ_n^2) of Sturm-Liouville operators $T_{\sigma,0,h_1}$ and $T_{\sigma,0,h_2}$ with $\sigma \in \mathcal{H}$ and $h_1, h_2 \in \mathbb{R}$ satisfy assumptions (1)–(3).

For the case where one of h_1, h_2 is infinite (say, $h_2 = \infty$) the asymptotics of the corresponding spectrum is different; also assumption (3) becomes meaningless and should be omitted.

Theorem 5.2. Suppose that sequences (λ_n^2) and (μ_n^2) of positive pairwise distinct numbers satisfy the following assumptions:

(i) the sequences (λ_n^2) and (μ_n^2) interlace;

(ii) $\lambda_n = \pi(n-1) + \widetilde{\lambda}_n$ and $\mu_n = \pi(n-1/2) + \widetilde{\mu}_n$ with some ℓ_2-sequences $(\widetilde{\lambda}_n)$ and $(\widetilde{\mu}_n)$.

Then there exist a unique function $\sigma \in \mathcal{H}$ and a real constant h_1 such that (λ_n^2) and (μ_n^2) are the spectra of the Sturm-Liouville operators $T_{\sigma,0,h_1}$ and $T_{\sigma,0,\infty}$ respectively.

Conversely, the spectra (λ_n^2) and (μ_n^2) of Sturm-Liouville operators $T_{\sigma,0,h_1}$ and $T_{\sigma,0,\infty}$ with $\sigma \in \mathcal{H}$ and $h_1 \in \mathbb{R}$ satisfy assumptions (1) and (2).

Appendix Appendix A. Riesz bases

In this appendix we gather some well known facts about Riesz bases of sines and cosines (see, e.g., [6,8] and the references therein for a detailed exposition of this topic).

Recall that a sequence $(e_n)_1^\infty$ in a Hilbert space $\mathcal{H}$ is a Riesz basis if and only if any element $e \in \mathcal{H}$ has a unique expansion $e = \sum_{n=1}^\infty c_n e_n$ with $(c_n) \in \ell_2$. If (e_n) is a Riesz basis, then in the above expansion the *Fourier coefficients* c_n are given by $c_n = (e, e_n')$, where $(e_n')_1^\infty$ is a system biorthogonal to (e_n), i.e., a system which satisfies the equalities $(e_k, e_n') = \delta_{kn}$ for all $k, n \in \mathbb{N}$. Moreover, the biorthogonal system (e_n') is a Riesz basis of $\mathcal{H}$ as long as (e_n) is, in which case for any $e \in \mathcal{H}$ also the expansion $e = \sum (e, e_n) e_n'$ takes place. In particular, if (e_n) is a Riesz basis, then for any $e \in \mathcal{H}$ the sequence (c_n') with $c_n' := (e, e_n)$ belongs to ℓ_2.

Proposition A.1 ([8]). Suppose that $\mu_k \to 0$ as $k \to \infty$ and that the sequence $\pi k + \mu_k$ is strictly increasing. Then each of the following systems forms a Riesz basis of $L_2(0,1)$:
(a) $\{\sin(\pi k x + \mu_k x)\}_{k=1}^\infty$;
(b) $\{\sin(\pi[k - 1/2]x + \mu_k x)\}_{k=1}^\infty$;
(c) $\{\cos(\pi k x + \mu_k x)\}_{k=0}^\infty$;
(d) $\{\cos(\pi[k + 1/2]x + \mu_k x)\}_{k=0}^\infty$.

Lemma A.2. Suppose that (λ_n) and (μ_n) are two sequences of real numbers such that

$$\lim_{n\to\infty} \big[\lambda_n - \pi(n - 1/2)\big] = \lim_{n\to\infty} \big[\mu_n - \pi(n - 1/2)\big] = 0.$$

and assume that $f \in \mathcal{H}$. Then there exists an ℓ_2-sequence (ν_n) such that

$$\int_0^1 f(t)[\cos \lambda_n t - \cos \mu_n t] \, \mathrm{d}t = (\lambda_n - \mu_n)\nu_n + O(|\lambda_n - \mu_n|^3)$$

as $n \to \infty$.

Proof. Using the relation

$$\cos \lambda_n t - \cos \mu_n t = -2 \sin[(\lambda_n - \mu_n)t/2] \sin[(\lambda_n + \mu_n)t/2],$$

we find that

$$\int_0^1 f(t)[\cos\lambda_n t - \cos\mu_n t]\,\mathrm{d}t$$
$$= -2\int_0^1 f(t)\sin[(\lambda_n-\mu_n)t/2]\sin[(\lambda_n+\mu_n)t/2]\,\mathrm{d}t$$
$$= -(\lambda_n-\mu_n)\int_0^1 tf(t)\sin[(\lambda_n+\mu_n)t/2]\,\mathrm{d}t + O(|\lambda_n-\mu_n|^3).$$

We put $\nu_n := -\int_0^1 tf(t)\sin[(\lambda_n+\mu_n)t/2]\,\mathrm{d}t$ and observe that the sequence (ν_n) belongs to ℓ_2 since the system $\{\sin[(\lambda_n+\mu_n)t/2]\}$ is Riesz basic in $\mathcal{H}$ (at least for all n large enough) and the function $tf(t)$ belongs to $\mathcal{H}$. □

References

1. S. Albeverio, F. Gesztesy, R. Høegh-Krohn, and H. Holden, *Solvable Models in Quantum Mechanics*. Springer-Verlag, New York-Berlin-Heidelberg-London-Paris-Tokyo, 1988.
2. S. Albeverio and P. Kurasov, *Singular Perturbations of Differential Operators. Solvable Schrödinger Type Operators*. Cambridge University Press, Cambridge, 2000.
3. F. V. Atkinson, *Discrete and Continuous Boundary Problems*. Academic Press, New York-London, 1964.
4. G. Borg, Eine Umkehrung der Sturm-Liouvilleschen Eigenwertaufgabe, *Acta Math.* **78** (1) (1946) 1–96.
5. F. Gesztesy and B. Simon, On the determination of a potential from three spectra, in *Differential operators and spectral theory*, 85–92, Amer. Math. Soc. Transl. Ser. 2, 189, Amer. Math. Soc., Providence, RI, 1999.
6. I. Gohberg and M. Krein, *Introduction to the Theory of Linear Non-selfadjoint Operators in Hilbert Space*. Nauka Publ., Moscow, 1965 (in Russian); *Engl. transl.:* Amer. Math. Soc. Transl. Math. Monographs, vol. 18, Amer. Math. Soc., Providence, RI, 1969.
7. P. Hartman, *Ordinary Differential Equations*. John Wiley&Sons, New York, 1964.
8. X. He and H. Volkmer, Riesz bases of solutions of Sturm-Liouville equations, *J. Fourier Anal. Appl.* **7** (3) (2001) 297–307.
9. R. O. Hryniv and Ya. V. Mykytyuk, 1D Schrödinger operators with singular periodic potentials, *Meth. Funct. Anal. Topol.* **7** (4) (2001) 31–42.
10. R. O. Hryniv and Ya. V. Mykytyuk, 1D Schrödinger operators with singular Gordon potentials, *Meth. Funct. Anal. Topol.* **8** (1) (2002) 36–48.
11. R. O. Hryniv and Ya. V. Mykytyuk, Inverse spectral problems for Sturm-Liouville operators with singular potentials, *Inverse Probl.* **19** (2003) 665–684.
12. R. O. Hryniv and Ya. V. Mykytyuk, Transformation operators for Sturm-Liouville operators with singular potentials, *Math. Phys. Anal. Geom.* (2003), *to appear.*

13. B. M. Levitan, *Inverse Sturm-Liouville Problems*. Nauka Publ., Moscow, 1984 (in Russian); *Engl. transl.:* VNU Science Press, Utrecht, 1987.
14. B. M. Levitan, On determination of a Sturm-Liouville differential equation by two spectra, *Izv. AN SSSR, Ser. Math.* **28** (1) (1964) 63–78.
15. B. M. Levitan and M. G. Gasymov, Determination of a differential equation by two spectra, *Uspekhi Mat. Nauk* **19** (2) (1964) 3–63.
16. V. A. Marchenko, *Sturm-Liouville Operators and Their Applications*. Naukova Dumka Publ., Kiev, 1977 (in Russian); *Engl. transl.:* Birkhäuser Verlag, Basel, 1986.
17. A. M. Savchuk, On eigenvalues and eigenfunctions of Sturm-Liouville operators with singular potentials, *Matem. Zametki (Math. Notes)* **69** (2) (2001) 277–285.
18. A. M. Savchuk and A. A. Shkalikov, Sturm-Liouville operators with singular potentials, *Matem. Zametki (Math. Notes)* **66** (6) (1999) 897–912.
19. A. M. Savchuk and A. A. Shkalikov, The Sturm-Liouville operators with distributional potential, *Trudy Mosk. Matem Ob-va (Trans. Moscow Math. Soc.)* **64** (2003) 159–212.

V. Kadets and W. Żelazko

Incorrect Nonlocal Boundary Value Problem for Partial Differential Equations

Volodymyr Il'kiv

Lviv Polytechnic National University, Bandery str. 12, Lviv, Ukraine

Abstract

The existence and uniqueness conditions of the pseudosolutions of a nonlocal two-point boundary value problem for a system of PDE with continuous coefficients in Sobolev spaces are studying.

Key words: Nonlocal boundary value problem, PDE, pseudosolutions, lagrangian function, small denominators
2000 MSC: 35G05, 35G15

The nonlocal boundary value problems for PDE, generally, are incorrect in the Hadamar sense. For example, the problem with two-point nonlocal conditions with respect to one selected variable in the Sobolev space scale has no solution or has a non unique solution [1–3]. Incorrectness of this problem can be explain by a null (nonuniqueness of solution) or arbitrary closely to null (nonuniqueness of solution) quantities which in the last case enter as multipliers into denominators of terms of Fourier series which represent a formal solution of the problem. For this reason these type of quantities will be called the small denominators, and the problem of their lower estimation was known since the XVIII century in the problem of celestial mechanics as the problem of small denominators [4].

By using the metrical approach to the problem of small denominators the nonlocal problem solvability in the Sobolev space scales [2, 5] for almost all coefficients in differential quantity of this problem has been established. This is true in particular for a such problem.

Email address: ilkiv@polynet.lviv.ua (Volodymyr Il'kiv).

Let $Q = [0,T] \times \Omega_p$ be a cylindrical domain, where Ω_p a p-dimensional torus, $(t,x) \in Q$, $x = (x_1, \dots, x_p)$. In the domain Q we consider the following problem with nonlocal boundary conditions in time variable t for the canonical system of m partial n order differential equations

$$\frac{\partial^n u}{\partial t^n} = \sum_{j=1}^{n} L_j(t,D)\frac{\partial^{n-j} u}{\partial t^{n-j}} + f, \tag{1}$$

$$lu \equiv \sum_{j=1}^{n} \left(B_{0j}(D)\frac{\partial^{n-j} u}{\partial t^{n-j}}\bigg|_{t=0} + B_{Tj}(D)\frac{\partial^{n-j} u}{\partial t^{n-j}}\bigg|_{t=T} \right) = \varphi, \tag{2}$$

where $L_j(t,D) = \sum\limits_{|s|\le j} L_{js}(t)D^s$, $L_{js}(t)$ are complex m-matrix coefficients continuous on $[0,T]$, $l_{0j}(D) = \sum\limits_{|s|\le j} l_{0js}D^s$, $l_{Tj}(D) = \sum\limits_{|s|\le j} l_{Tjs}D^s$, l_{0js}, l_{Tjs} are $nm \times m$ matrix complex coefficients.

We denote by H_q, $q \in \mathbb{R}$, the Sobolev space of periodical in space variable x vector-function $h(x) = \sum\limits_{k\in\mathbb{Z}^p} \widehat{h}(k)e^{i(k,x)}$ with the inner product

$$(h,g)_q = (2\pi)^p \sum_{k\in\mathbb{Z}^p} \widetilde{k}^{2q}\widehat{g}^*(k)\widehat{h}(k),$$

where $g(x) = \sum\limits_{k\in\mathbb{Z}^p} \widehat{g}(k)e^{i(k,x)}$, $k = (k_1, \dots, k_p)$, $(k,x) = k_1x_1 + \cdots + k_px_p$, $\widetilde{k} = \sqrt{1+k_1^2+\cdots+k_p^2}$ and * denotes the Hermitian conjugate of matrix. We use the notation H_q for vector-function spaces of arbitrary dimension so that $(h,g)_q = (h_1,g_1)_q + (h_2,g_2)_q$ for vector-functions $h = \operatorname{col}(h_1,h_2)$ and $g = \operatorname{col}(g_1,g_2)$ of the same dimension.

We denote by H_q^n, $q \in \mathbb{R}$, $n \in \mathbb{N}$, the completion of the space, of infinitely differentiable finite sums $v(t,x) = \sum\limits_{k} \widehat{v}(t,k)e^{i(k,x)}$ by the norm, generated by the inner product

$$(v,w)_{q,n} = \frac{1}{T}\int_0^T \sum_{j=0}^{n} \left(\frac{\partial^j v}{\partial t^j}, \frac{\partial^j w}{\partial t^j}\right)_{q-j} dt.$$

The spaces H_q and H_q^n form the scales of spaces with respect to index $q \in \mathbb{R}$.

If a vector-function φ belongs to the space scale H_q and a vector-function f belongs to the space scale H_q^n then for almost all vectors of the elements of matrices formed by l_{0js} and l_{Tjs} there exists a unique solution of the problem (1), (2) in the space scale H_q^n [6]. But, generally, for an arbitrary vector the problem (1), (2) have no solution in the space scale H_q^n or it is non unique. So that we consider the existence and the construction of approximate solutions (pseudosolutions) of this problem and study their uniqueness.

We can utilize the following pseudodifferential operators [3]. Let $F(k)$ is a complex matrix elements which are functions of variable $k \in \mathbb{Z}^p$, then by operator $F(D) =$

$F(D_1, \ldots, D_p)$ we shall mean the operator acting on the vector-function $h(x) = \sum\limits_{k \in \mathbb{Z}^p} \widehat{h}(k) e^{i(k,x)} \in H_q$ by the rule

$$F(D)h(x) = \sum_{k \in \mathbb{Z}^p} F(k)\widehat{h}(k)e^{i(k,x)}.$$

By operations on these operators $F(D)$ we mean the same operations on $F(k)$ for $k \in \mathbb{Z}^p$, in particular, the conjugate operator $F^*(D)$ is defined by the Hermitian conjugate matrix $F^*(k)$ and the scalar operator $\tilde{D}$ is defined by the number sequence $\tilde{k}$, $k \in \mathbb{Z}^p$.

A pseudoresolution of the problem (1), (2) will be found in the space H^n_{q+n}, where $q \in \mathbb{R}$ is the some fixed number. The operators $L : H^n_{q+n} \to H^0_q$ and $l : H^n_{q+n} \to H_{q-1/2}$ are continuous, so that we may take the right-hand sides f and φ of the problem to be elements of the spaces H^0_q and $H_{q-1/2}$ respectively. Let $u_0 \in H^n_{q+n}$ be some fixed function, $\varepsilon > 0$ be an arbitrary number, $K_\varepsilon(u_0)$ be the ball of radius ε around u_0 in the space H^n_{q+n}.

A pseudosolution of the problem (1), (2) is defined to be a the function $u^\varepsilon \in H^n_{q+n}$ minimizing the estimation error

$$Ru \equiv \|Lu - f\|^2_{q,0} + \|lu - \varphi\|_{q-1/2}$$

on the set $K_\varepsilon(u_0)$, namely $K_\varepsilon(u_0) = \min\limits_{u \in K_\varepsilon(u_0)} R(u)$. If $u_0 = 0$ then a pseudosolution with the minimal norm will be called the normal pseudosolution; if $u_0 \neq 0$ then the u_0-normal pseudosolution be called the pseudosolution minimizing $\|v - u_0\|_{q+n,n}$ on the set of pseudosolutions of the problem (1), (2).

Let $u \in H^n_{q+n}$. Then $g = Lu \in H^0_q$, $C_j = \partial^{j-1}u(0,x)/\partial t^{j-1} \in H_{q+n-j+1/2}$, $j = 1, \ldots, n$, and $C = C(x) = \operatorname{col}(C_1, \ldots, C_n) \in H_{q+n-1/2} \times \cdots \times H_{q+1/2}$; the function u may be written as the solution of the Cauchy problem $Lu = g$, $\partial^{j-1}u/\partial t^{j-1}|_{t=0} = C_j$, $j = 1, \ldots, n$, in the form $u = F_u U = F_{u1}C + F_{u2}g$, where $F_{u1} = E(t,D)$, $(F_{u2}g)(t,x) = E(t,D)\int\limits_0^t \mathcal{E}_n^{-1}(\tau, D)g(\tau, x)\,d\tau$, $E(t,D)$ is the normal (on $t = 0$) fundamental system of solutions of the system (1), $\mathcal{E}(t,D) = \operatorname{col}\left(E(t,D), \partial E(t,D)/\partial t, \ldots, \partial^{n-1}E(t,D)/\partial t^{n-1}\right)$, $\mathcal{E}_n^{-1}(\tau, D)$ is a matrix formed by the last m column of the matrix $\mathcal{E}^{-1}(\tau, D)$. Similarly we obtain the following relations

$$H^0_q \ni Lu = F_L U = F_{L1}C + F_{L2}\,g,$$

$$H_q \ni \widetilde{D}^{-1/2} lu = F_l U = F_{l1}C + F_{l2}\,g,$$

$$H^0_q \ni Vu \equiv \operatorname{col}(\tilde{D}^n u, \tilde{D}^{n-1}\partial u/\partial t, \ldots, \partial^n u/\partial t^n) = F_V U = F_{V1}C + F_{V2}\,g,$$

where $F_{L1} = 0$, $F_{L2} = I_m$, $F_{l1} = l_0(D) + l_T(D)\mathcal{E}(T,D)$,

$$F_{l2}\,g = l_T(D)\mathcal{E}(T,D)\int_0^T \mathcal{E}_n^{-1}(\tau, D)g(\tau, \cdot)\,d\tau,\ \ F_{V1} = \operatorname{col}(Z, A)\mathcal{E}(t,D),$$

$$(F_{V2}\, g)(t,x) = \mathrm{col}\,(Z,A)\mathcal{E}(t,D)\int_0^t \mathcal{E}_n^{-1}(\tau,D)g(\tau,x)\,d\tau + \mathrm{col}\,(0,g(t,x)),$$

$$Z = \mathrm{diag}\,(\widetilde{D}^n,\dots,\widetilde{D}^2,\widetilde{D}), \qquad A \equiv A(t,D) = \big(L_n(t,D),\dots,L_1(t,D)\big),$$
$$l_0(D) = \widetilde{D}^{-1/2}\big(l_{0n}(D),\dots,l_{01}(D)\big), l_T(D) = \widetilde{D}^{-1/2}\big(l_{Tn}(D),\dots,l_{T1}(D)\big).$$

The estimate error $R(u) \equiv \mathcal{R}(U)$ can be represented in the form

$$\mathcal{R}(U) \equiv \|F_L U - f\|_{q,0}^2 + \|F_l U - \varphi\|_{q,0}^2 = \|\mathrm{col}\,(F_L,F_l)U - \xi\|_{q,0}^2,$$

where $\xi = \mathrm{col}\,(f,\tilde{D}^{-1/2}\varphi)$ and the pseudosolution $u^\varepsilon = F_u U^\varepsilon$ of the problem (1), (2) may be found by solving the following minimization problem with respect to the vector U^ε:

$$\mathcal{R}(U^\varepsilon) = \min\{\mathcal{R}(U) : F_u U \in H_{q+n}^n, \|F_V U - V u_0\|_{q,0} \le \varepsilon\}. \tag{3}$$

The problem (3) is a convex conditional minimization problem [7] in the space H_q^0 norm on the linear set of elements $\mathrm{col}\,(F_L,F_l)U - \xi$. For solving this problem we, can use the Lagrange multiplier method. Consider now the Lagrange function

$$\mathcal{L}_\omega(U) = R(u) + \omega(\|u-u_0\|_{q+n,n}^2 - \varepsilon^2) =$$
$$= \mathcal{R}(U) + \omega(\|F_V U - V u_0\|_{q,0}^2 - \varepsilon^2) = \|FU - \eta\|_{q,0}^2 - \omega\varepsilon^2,$$

where $F = \mathrm{col}\,(F_L, F_l, \sqrt{\omega}F_V)$, $\eta = \mathrm{col}\,(\xi, \sqrt{\omega}V u_0)$, $\omega \ge 0$ is a Lagrange multiplier, saddle points of which $(U^\varepsilon,\omega_\varepsilon)$ [7] will determine the solutions U_ε of the problem (3). By the Kuhn–Tucker theorem [7] the saddle point $(U^\varepsilon,\omega_\varepsilon)$ satisfies the following conditions:

$$\mathcal{L}_{\omega_\varepsilon}(U^\varepsilon) = \min\{\mathcal{L}_{\omega_\varepsilon}(U) : L_u U \in H_{q+n}^n\}, \tag{4}$$
$$\omega_\varepsilon(\|F_V U^\varepsilon - V u_0\|_{q,0}^2 - \varepsilon^2) = 0, \quad \|F_V U^\varepsilon - V u_0\|_{q,0}^2 \le \varepsilon^2. \tag{5}$$

Let F^* denote the conjugate operator to operator F, then

$$F^* = (F_L^*, F_l^*, \sqrt{\omega}F_V^*) = \begin{pmatrix} F_{L1}^* & F_{l1}^* & \sqrt{\omega}F_{V1}^* \\ F_{L2}^* & F_{l2}^* & \sqrt{\omega}F_{V2}^* \end{pmatrix}$$

and the operator $F^*F = F_L^*F_L + F_l^*F_l + \omega F_V^*F_V$ is the positive definite operator since $\|FU\|_{q,0}^2 = (F^*FU,U)_{q,0} \ge \omega\|F_V U\|_{q,0}^2 = \omega\|u\|_{q+n,n}^2 > 0$ for an arbitrary function $0 \ne u = L_u U \in H_{q+n}^n$. We construct the projector $P_F = F(F^*F)^{-1}F^*$ onto the range of F to make the lower estimation of Lagrange function $\mathcal{L}_\omega(U)$ and find its minimal value for an arbitrary $\omega > 0$. We obtain

$$\mathcal{L}_\omega(U) = \|FU - P_F\eta\|_{q,0}^2 + \|(I-P_F)\eta\|_{q,0}^2 - \omega\varepsilon^2. \tag{6}$$

The components of operator F^* have the following form

$$F_{L1}^* = 0, \quad F_{L2}^* = I_m, \quad F_{l1}^* = l_0^*(D) + \mathcal{E}^*(T,D)l_T^*(D),$$
$$F_{l2}^* = T\mathcal{E}^{-1*}(t,D)\mathcal{E}^*(T,D)l_T^*(D),$$
$$F_{V1}^* v = \frac{1}{T}\int_0^T \mathcal{E}^*(\tau,D)(Z,A^*(\tau,D))v(\tau,\cdot)\,d\tau,$$

$$(F_{V2}^{*}v)(t,x) = \mathcal{E}^{-1*}(t,D)\int_{t}^{T}\mathcal{E}^{*}(\tau,D)(Z,A^{*}(\tau,D))v(\tau,x)\,d\tau + (0,I_m)v(t,x).$$

Since $FU - P_F\eta = F(U - (F^*F)^{-1}F^*\eta)$, from the formula (16) we obtain $\min \mathcal{L}_\omega(U) = \mathcal{L}_\omega(U_\omega)$ for

$$u_\omega = F_u U_\omega, \qquad U_\omega = (F^*F)^{-1}F^*\eta. \tag{7}$$

The function u_ω is a unique function that minimizes the Lagrangion function $\mathcal{L}_\omega(U)$, in particular

$$\mathcal{L}_\omega(U_\omega) < \mathcal{L}_\omega(F_V^{-1}Vu_0) = R(u_0) < \infty. \tag{8}$$

Using formula (7) we also have

$$u_\omega = u_0 + F_uF_V^{-1}(G^*G)^{-1}G^*\xi_0 = u^+ + \omega u_1, \tag{9}$$

where $G = \operatorname{col}(F_L, F_l)F_V^{-1}$, $\xi_0 = \xi - \operatorname{col}(Lu_0, \tilde{D}^{-1/2}lu_0) = \xi - GVu_0$,

$$\xi = \operatorname{col}(f, \tilde{D}^{-1/2}\varphi),\ u^+ = F_uF_V^{-1}G^+\xi,\ u_1 = F_uF_V^{-1}(G^*G+\omega)^{-1}(Vu_0 - G^+\xi),$$

$G^+ = \lim_{\omega\to 0}(G^*G+\omega)^{-1}G^* = \lim_{\omega\to 0}G(G^*G+\omega)^{-1}$ is a pseudoinverse operator [7, 8] of G.

Theorem 1. *If $\varepsilon \le \varepsilon_0 = \|g_0\|^2_{q+n,n}$, $g_0 = F_uF_V^{-1}G^+\xi_0 = u^+ - F_uF_V^{-1}G^+GVu_0$, then a pseudosolution of the problem (1), (2) exists, is unique, has the form*

$$u^\varepsilon = u_{\omega_\varepsilon} = u_0 + F_uF_V^{-1}(G^*G+\omega_\varepsilon)^{-1}G^*\xi_0 = u^+ + \omega_\varepsilon u_{1\varepsilon}, \tag{10}$$

where ω_ε is a solution of the algebraic equation $\|u_\omega - u_0\| = \varepsilon$,

$$u_{1\varepsilon} = F_uF_V^{-1}(G^*G+\omega_\varepsilon)^{-1}(Vu_0 - G^+\xi),$$

and is not the solution of the problem (1), (2), $R(u^\varepsilon) > 0$. If $\varepsilon > \varepsilon_0$ then a pseudoresolution exists, generally, is not unique and has the form

$$u^\varepsilon = u^+ + F_uF_V^{-1}(I - G^+G)Q = u_0 + g_0 + F_uF_V^{-1}(I - G^+G)(Q - Vu_0), \tag{11}$$

where Q is an arbitrary vector-function in space H_q^0 that satisfies the inequality $\|(I - G^+G)(Q - Vu_0)\|^2_{q,0} \le \varepsilon^2 - \varepsilon_0^2$; there exists a unique u_0-normal pseudosolution $u^\varepsilon = u_0 + g_0$ and a unique normal pseudosolution $u^\varepsilon = u^+ + \beta F_uF_V^{-1}(I - G^+G)Vu_0$, where $\beta \ge 0$ is some constant. If $\xi = GG^+\xi$, then pseudosolutions (11) are the solutions of the problem (1), (2); if the inverse operator G^{-1} exists, then the pseudoresolution (11) is unique and is the solution of the problem (1), (2).

Proof. From (4) and (5) we find the saddle point of the Lagrangian function. The function $U^\varepsilon = U_{\omega_\varepsilon}$ determined in (7) satisfies condition (4) for all value $\omega_\varepsilon > 0$. Choose ω_ε so that condition (5) is valid. Consider the function $\|u_\omega - u_0\|_{q,0} \equiv \|F_VU_\omega - Vu_0\|_{q,0} = \|(G^*G+\omega)^{-1}G^*\xi_0\|_{q,0}$ which is continuous and strictly decreasing to zero as $\omega \to +\infty$; moreover, $\lim_{\omega\to+\infty} u_\omega = u_0$, $\lim_{\omega\to+0} u_\omega = u_0 + g_0$, $\varepsilon_0 = \|g_0\|_{q+n,n}$ may be a number if $g_0 \in H^n_{q+n}$ or $\varepsilon_0 = +\infty$ if $g_0 \notin H^n_{q+n}$. For this reason when $\varepsilon \leq \varepsilon_0$ the equation $\|u_\omega - u_0\| = \varepsilon$ has a unique solution $\omega = \omega_\varepsilon$ and $U^\varepsilon = U_{\omega_\varepsilon}$. Relation (10) follows from (9) at the change ω by ω_ε. The estimation error $R(u)$ at the pseudosolution u^ε is

$$R(u^\varepsilon) = \|(I - GG^+)\xi\|^2_{q,0} + \omega_\varepsilon^2\|(G^*G + \omega_\varepsilon)^{-1}(\xi_0 - (I - GG^+)\xi)\|^2_{q,0}.$$

If $R(u^\varepsilon) = 0$, then $(I - GG^+)\xi = 0$ and $\xi_0 = (I - GG^+)\xi$, namely $\xi_0 = 0$ and $0 = \|G^+\xi_0\|_{q,0} = \varepsilon_0 < \varepsilon$ that contradicts the condition $0 < \varepsilon \leq \varepsilon_0$. Consequently, $R(u^\varepsilon) > 0$ namely any pseudosolution of the problem (1), (2) is its solution. When $\varepsilon > \varepsilon_0$, the equation $\|u_\omega - u_0\|_{q+n,n} = \varepsilon$ has no solution and the condition (5) satisfies only at $\omega_\varepsilon = 0$. Then $\mathcal{L}_0(U) \equiv R(u) = \|GF_VU - \xi\|^2_{q,0}$ and

$$\mathcal{L}_0(U) \equiv R(u) = \|G(F_VU - G^+\xi)\|^2_{q,0} + \|(I - GG^+)\xi\|^2_{q,0} \geq \|(I - GG^+)\xi\|^2_{q,0}. \quad (12)$$

If $U^\varepsilon = F_V^{-1}G^+\xi) + F_V^{-1}(I - G^+G)Q$ then for all vector-functions $Q \in H_q^0$ the minimum $\mathcal{L}_0(U^\varepsilon) = \|(I - GG^+)\xi\|^2_{q,0}$ vanishes. Choose among vector-functions Q satisfying condition (5) which means that for functions

$$u^\varepsilon = F_uU^\varepsilon = u_0 + g_0 + F_uF_V^{-1}(I - G^+G)(Q - Vu_0) \quad (13)$$

the inequality $\|u^\varepsilon - u_0\|^2_{q+n,n} \equiv \|g_0\|^2_{q+n,n} + \|(I - G^+G)(Q - Vu_0)\|^2_{q,0} \leq \varepsilon^2$ is satisfied, this implies the condition on vector-functions Q.

For any pseudosolution (13) $u^\varepsilon \neq u_0 + g_0$ we have the inequality $\varepsilon^2 > \|u^\varepsilon - u_0\|^2_{q+n,n} > \|g_0\|^2_{q+n,n} = \varepsilon_0^2 = \|(u_0 + g_0) - u_0\|^2_{q+n,n}$ hence $u^\varepsilon = u_0 + g_0$ is the u_0-normal pseudosolution.

Compute the minimal value of the norm $\|u^\varepsilon\|_{q+n,n}$ for the set of the functions (13) satisfying condition $\|u^\varepsilon - u_0\|_{q+n,n} \leq \varepsilon$ to construct a normal pseudosolution. The Lagrangian function of this conditional minimization problem $\|u^\varepsilon\|^2_{q+n,n} + \Omega(\|u^\varepsilon - u_0\|^2_{q+n,n} - \varepsilon^2)$ satisfies inequality

$$\|u^\varepsilon\|^2_{q+n,n} + \Omega(\|u^\varepsilon - u_0\|^2_{q+n,n} - \varepsilon^2) \geq$$
$$\geq \frac{\Omega}{(1+\Omega)^2}\|(I - G^+G)Vu_0\|^2_{q,0} + \|G^+\xi\|^2_{q,0} + \Omega(\varepsilon_0^2 - \varepsilon^2),$$

so that when $(I - G^+G)Q = \Omega\|(I - G^+G)Vu_0\|^2_{q,0}/(1 + \Omega)$, namely $u^\varepsilon = u_0 + g_0 - F_uF_V^{-1}(I - G^+G)Vu_0/(1 + \Omega)$ it transformed into equality. If $\alpha \equiv \|(I - G^+G)Vu_0\|_{q,0} - \sqrt{\varepsilon^2 - \varepsilon_0^2} > 0$, then $\Omega = \Omega_\varepsilon = \alpha/\sqrt{\varepsilon^2 - \varepsilon_0^2}$ determines the Lagrangion multiplier with the help of which me construct the desired normal pseudosolution

$$u^\varepsilon = u^+ + \frac{\alpha}{\alpha + \sqrt{\varepsilon^2 - \varepsilon_0^2}}F_uF_V^{-1}(I - G^+G)Vu_0.$$

If $\alpha \leq 0$ then normal pseudosolution $u^\varepsilon = u^+$ corresponds to zero Lagrangian multiplier. Putting $\beta = \max(0, 1 - \sqrt{\varepsilon^2 - \varepsilon_0^2})/\|(I - G^+G)Vu_0\|_{q,0}$ we obtain the desired normal pseudosolution.

An equality $R(u^\varepsilon) = 0$ results from (12), when $\xi = GG^+\xi$, namely pseudosolutions (13) have been the solutions of the problem (1), (2). If operator G^{-1} exists, then $G^+G = I$, $G^{-1} = F_V(\mathrm{col}\,(F_L, F_l))^{-1}$ and formula (12) and (13) imply that $u^\varepsilon = u_0 + g_0 = u^+ = F_u(\mathrm{col}\,(F_L, F_l))^{-1}\xi$ is an unique solution of the problem (1), (2). The theorem is proved.

The theorem results that the condition $\varepsilon < \infty$ is a necessary condition of the existence of the solution of the problem (1), (2) in the space H^n_{q+n}. The case $\varepsilon = \infty$ indicates that this problem has small denominators and shows that the solution of this problem can exist only in the spaces of smaller smoothness and does not exist in the space H^n_{q+n}.

We note that other pseudosolutions may be obtained by using the minimization the weighted estimation error of the problem (1), (2), in particular one may examine the minimization problem of functional $\|Lu - f\|^2_{q+n,n}/\gamma_1 + \|lu - \varphi\|^2_{q-1/2}/\gamma_2$ on the ball $K_\varepsilon(u_0) \subset H^n_{q+n}$ where $\gamma_1, \gamma_2 \geq 0$ is a weighted coefficients, $\gamma_1\gamma_2 \neq 0$. If γ_1 or γ_2 is equal to zero then we erase respectively addend and equate respectively norm to zero. For instance, γ_1 denotes that we minimize the functional $\|lu - \varphi\|^2_{q-1/2}$ on the set of solutions of system $Lu = f$ in the ball $K_\varepsilon(u_0)$.

This problem is studied in the paper [9] for the PDE with constant coefficients at the case $f = 0$. In contrast to the problem (1), (2) this problem not always has pseudosolutions.

The other nonlocal problem, in particular, multipoint and also the nonlocal problem in other scales of functional spaces can be studied in the same way.

References

1. A.A. Dezin, *General questions of the boundary value problems theory.* Nauka, Moscow, 1980 (in Russian).
2. B.Yo. Ptashnyk, *Several ill-posed boundary value problems for PDE.* Naukova dumka, Kyiv, 1984 (in Russian).
3. V.S. Il'kiv, Multipoint nonlocal boundary value problem for PDE, *Diff. uravneniya (Diff. equations)* **23** (3) (1987) 487-492.
4. E.A. Grebennikov, Yu.A. Rjabov, *Resonances and small denominators in celestial mechanics.* Nauka, Moscow, 1978 (in Russian).
5. V.S. Il'kiv, Nonlocal boundary value problem for normal anizotropic systems of partial equations with constant coefficients, *Visnyk of the Lviv University. Series Mechanics and Mathematics* (1999) (54) 84–95 (in Ukrainian).
6. V.S. Il'kiv, Nonlocal boundary value problem for systems of partial equations in anizotropic spaces, *Nonlinear boundary value problems* **11** (2001) 57–64 (in Ukrainian).
7. Yu.P. Pytiev, *Mathematical methods of experiment interpretation.* Vysshaja shcola, Moscow, 1989 (in Russian).
8. V.V. Vojevodin, Yu.A. Kuznjecov, *Matrices and calculations.* Nauka, Moscow, 1984 (in Russian).
9. V.S. Il'kiv, Investigation of nonlocal boundary value problem for normal partial differential equation by means of minimization method in Sobolev spaces, *Mathematychni studii* **11** (1999) 167–176 (in Ukrainian).

Functional Analysis and its Applications
V. Kadets and W. Żelazko

Singular Integral Operators with Flip and Unbounded Coefficients on Rearrangement-Invariant Spaces

Alexei Karlovich [a,1]

[a] Departamento de Matemática, Instituto Superior Técnico,
Av. Rovisco Pais 1, 1049–001 Lisboa, Portugal

Abstract

We We obtain Fredholm criteria for singular integral operators of the form

$$P_+ + M_b P_- + M_u U P_-,$$

where $P_\pm$ are the Riesz projections, U is the flip operator, and M_b, M_u are operators of multiplication by functions b, u, respectively, on a reflexive rearrangement-invariant space with nontrivial Boyd indices over the unit circle. We assume a priori that M_b is bounded, but M_u may be unbounded. The function u belongs to a class of, in general, unbounded functions that relates to the Douglas algebra $H^\infty + C$.

Key words: Rearrangement-invariant space, Boyd indices, Douglas algebra $H^\infty + C$, outer function, Fredholmness, singular integral operator with flip
2000 MSC: 47A53, 46J15, 47B38, 48B60, 46B42

1. Introduction

Let $\mathbb{T}$ be the unit circle equipped with the normalized Lebesgue measure m and let $X := X(\mathbb{T}, m)$ be a *rearrangement-invariant space* (for the definition and basic properties, see [1]). Rearrangement-invariant spaces are wide generalizations

[1] The author is partially supported by F.C.T. (Portugal) grants POCTI/ 34222/MAT/2000 and PRAXIS/XXI/BPD/22006/99.

of Lebesgue spaces $L^p := L^p(\mathbb{T}, m), 1 \le p \le \infty$, as well as Orlicz and Lorentz spaces. The *domain* of a linear operator T from X to X is denoted by $\mathcal{D}(T)$. The set of all closed (bounded) linear operators from X to X is denoted by $\mathcal{C}(X)$ (respectively, $\mathcal{B}(X)$).

We shall denote by I the identity operator on X and by M_a the operator of multiplication by a measurable (not necessarily bounded) function a on $\mathbb{T}$. The *flip operator* is defined by

$$(U\varphi)(t) := \bar{t}\varphi(\bar{t}) \quad (t \in \mathbb{T}).$$

The *Cauchy singular integral* of a function $\varphi \in L^1$ is given by

$$(S\varphi)(t) := \frac{1}{\pi i} p.v. \int_{\mathbb{T}} \frac{\varphi(\tau)}{\tau - t} d\tau \quad (t \in \mathbb{T}).$$

We will always assume that the rearrangement-invariant space X is reflexive and the *Boyd indices* α_X, β_X of X (see [3]) are nontrivial, i.e., $\alpha_X, \beta_X \in (0, 1)$.

Lemma 1.1. (a) $S, U \in \mathcal{B}(X)$ and $U^2 = I, S^2 = I, SU = -US$;
(b) $M_a \in \mathcal{B}(X)$ if and only if $a \in L^\infty$.

Proof. (a) Since $S, U \in \mathcal{B}(L^p)$ for $p \in (1, \infty)$, by Boyd's interpolation theorem [3], $S, U \in \mathcal{B}(X)$ whenever X has nontrivial Boyd indices. The equality $U^2 = I$ is obvious. The equality $S^2 = I$ is proved in [7, Lemma 6.5]. The equality $SU = -US$ is proved similarly. Part (b) is proved in [10, Theorem 1]. ∎

Put $P_+ := (I + S)/2$ and $P_- := (I - S)/2$. From Lemma 1.1(a) it follows that

$$P_+^2 = P_+, \quad P_-^2 = P_-, \quad U^2 = I, \quad P_+U = UP_-, \quad P_-U = UP_+. \tag{1}$$

For a linear operator T from X to X, we shall use the notation

$$R(T) := P_+ + TP_-.$$

Following [9], we study Fredholmness of singular integral operators with flip of the form $R(A)$, where

$$A := M_b + M_u U,$$

$b \in L^\infty$, but the function u may be unbounded. We refer to [8, Ch. 3–4] for properties of operators in $\mathcal{C}(X)$ and to [2, Ch. 1], [5, Ch. 4] for the definition and properties of (bounded) Fredholm operators on Banach spaces.

The paper is organized as follows. In Section 2, we prove that $R(A)$ is closed whenever $b \in L^\infty$ and u is measurable and a.e. finite. Further we study the invertibility of two auxiliary operators. In Section 3, we define the class $\mathbb{A}$ of, in general, unbounded functions u such that $\log \widetilde{u} \in L^1$ and $u\mathcal{O}_{1/\widetilde{u}}$ belongs to the Douglas algebra $H^\infty + C$. Here $\widetilde{u} := \max\{1, |u|\}$ and $\mathcal{O}_{1/\widetilde{u}}$ is the outer function generated by $1/\widetilde{u}$. Then we prove the relative compactness of the Hankel operator $P_- M_u P_+$ with respect to the operator $M_u P_+$ assuming $u \in \mathbb{A}$. By using this key result, we obtain Fredholm criteria for the operator $R(A)$, where $b \in L^\infty$ and $u \in \mathbb{A}$.

2. Auxiliary results

2.1. *Closedness of* $R(A)$

Lemma 2.1. If a is a measurable a.e. finite function on $\mathbb{T}$, then $M_a \in \mathcal{C}(X)$.

Proof. The idea of the proof is borrowed from [4, Ch. 3, Theorem 9.2]. Clearly,

$$\mathcal{D}(M_a) = \{\varphi \in X : \ a\varphi \in X\}.$$

For each $n \in \mathbb{N}$, let $\mathbb{T}_n := \{t \in \mathbb{T} : |a(t)| \leq n\}$. Then

$$\mathbb{T}_n \subset \mathbb{T}_{n+1}, \quad \bigcup_{n=1}^{\infty} \mathbb{T}_n = \mathbb{T} \setminus \mathbb{T}_\infty, \quad \text{where} \quad \mathbb{T}_\infty := \{t \in \mathbb{T} : a(t) = \infty\}. \tag{2}$$

For every $\varphi \in X$ and $n \in \mathbb{N}$, put $\varphi_n := \chi_{\mathbb{T}_n}\varphi$. Obviously,

$$|\varphi_n| \leq |\varphi|, \quad |a\varphi_n| = |a\chi_{\mathbb{T}_n}\varphi| \leq n|\varphi|, \quad n \in \mathbb{N}.$$

Hence, by the lattice property (see [1, Ch. 1, Definition 1.1]), φ_n and $a\varphi_n$ belong to X whenever $\varphi \in X$ and $n \in \mathbb{N}$. Therefore, $\varphi_n \in \mathcal{D}(M_a)$ for $n \in \mathbb{N}$. Since a is a.e. finite, we have $m(\mathbb{T}_\infty) = 0$. Then $\chi_{\mathbb{T}\setminus\mathbb{T}_n} \to 0$ a.e. on $\mathbb{T}$ as $n \to \infty$. By [1, Ch. 1, Corollary 4.4], the reflexivity of X implies the absolutely continuity of the norm of $\varphi \in X$. Hence, $\|\varphi - \varphi_n\|_X = \|\chi_{\mathbb{T}\setminus\mathbb{T}_n}\varphi\|_X \to 0$ as $n \to \infty$. This means that $\mathcal{D}(M_a)$ is dense in X.

Let us prove that $(M_a)^* = M_{\overline{a}}$. From [1, Ch. 1, Corollaries 4.3–4.4] it follows that $X' = X^*$, where X' is the associate space of X (see [1, Ch. 1, Section 2]). This means that the general form of a functional $G \in X^*$ is given by

$$G(f) = (f, g) := \int_{\mathbb{T}} f\overline{g}\, dm,$$

where $f \in X, g \in X'$ and $\|G\|_{X^*} = \|g\|_{X'}$. Then for all $\varphi \in \mathcal{D}(M_a) \subset X$ and all $g \in \mathcal{D}(M_{\overline{a}}) \subset X'$,

$$(M_a\varphi, g) = \int_{\mathbb{T}} a\varphi \cdot \overline{g}\, dm = \int_{\mathbb{T}} \varphi \cdot \overline{M_{\overline{a}}g}\, dm = (\varphi, M_{\overline{a}}g). \tag{3}$$

Hence, the operators M_a and $M_{\overline{a}}$ are adjoint to each other. Let us show that any other adjoint to M_a is a restriction of $M_{\overline{a}}$. Assume T is an adjoint to M_a, that is,

$$(M_a\varphi, g) = (\varphi, Tg), \quad \varphi \in \mathcal{D}(M_a), \quad g \in \mathcal{D}(T). \tag{4}$$

Above we show that $\chi_{\mathbb{T}_n}\varphi \in \mathcal{D}(M_a)$ for every $\varphi \in X, n \in \mathbb{N}$. Analogously one can prove that $\chi_{\mathbb{T}_n}g \in \mathcal{D}(M_{\overline{a}})$ for every $g \in \mathcal{D}(T) \subset X', n \in \mathbb{N}$. Then from (3) and (4) we get for every $\varphi \in X, g \in \mathcal{D}(T)$, and $n \in \mathbb{N}$,

$$\int_{\mathbb{T}} \varphi\chi_{\mathbb{T}_n} \cdot \overline{Tg - M_{\overline{a}}g}\, dm = \Big(\varphi, \chi_{\mathbb{T}_n}(Tg - M_{\overline{a}}g)\Big) = 0.$$

Therefore, $Tg = M_{\overline{a}}g$ a.e. on $\mathbb{T}_n$ for every $g \in \mathcal{D}(T)$ and every $n \in \mathbb{N}$. Since $m(\mathbb{T}_\infty) = 0$, taking into account (2), we deduce that $Tg = M_{\overline{a}}g$ a.e. on $\mathbb{T}$ for every $g \in \mathcal{D}(T)$. Thus, T is a restriction of $M_{\overline{a}}$, and $M_{\overline{a}}$ is the adjoint to M_a.

By [1, Ch. 1, Corollary 4.4], $(X')^* = X'' = X$. In view of just proved, $M_{\overline{a}}$ is densely defined in X' and $(M_{\overline{a}})^* = M_{\overline{\overline{a}}} = M_a$. Since $M_{\overline{a}}$ is densely defined in X', its adjoint M_a is closed in X (see [8, Ch. 3, Section 5.4]). ■

Lemma 2.2. If $b \in L^\infty$ and u is a measurable a.e. finite function on $\mathbb{T}$, then $R(A), M_u P_+ \in \mathcal{C}(X)$.

Proof. By Lemma 1.1, $M_b, U, P_-, P_+ \in \mathcal{B}(X)$. Lemma 2.1 gives $M_u \in \mathcal{C}(X)$. By using [8, Ch. 3, Section 5.2, Problems 5.6–5.7], we subsequently conclude that the operators M_uU, M_uP_+, $M_b + M_uU$, $(M_b + M_uU)P_-$, $P_+ + (M_b + M_uU)P_-$ are closed. ■

If $T \in \mathcal{C}(X)$, then its domain $\mathcal{D}(T)$ becomes a Banach space with respect to the *graph norm*

$$\|f\|_{\mathcal{D}(T)} := \|f\|_X + \|Tf\|_X$$

(see, e.g., [8, Ch. 4, Section 1.1, Remark 1.4]). Hence, from Lemma 2.2 it follows that $\mathcal{D}(R(A))$ equipped with the graph norm is a Banach space.

Lemma 2.3. If $b \in L^\infty$ and u is a measurable a.e. finite function on $\mathbb{T}$, then

$$\mathcal{D}(R(A)) = \mathcal{D}(M_uUP_-) = \mathcal{D}(P_\pm M_uUP_-) = U\mathcal{D}(M_uP_+). \tag{5}$$

Proof. By Lemma 1.1, the operators $P_\pm$ and $R(M_b)$ are bounded on X. Therefore, using the representation

$$R(A) = R(M_b) + M_uUP_-, \tag{6}$$

we can easily obtain the equalities $\mathcal{D}(R(A)) = \mathcal{D}(M_uUP_-) = \mathcal{D}(P_\pm M_uUP_-)$. The equality $\mathcal{D}(R(A)) = U\mathcal{D}(M_uP_+)$ is proved as in [9, Lemma 2.3]. ■

2.2. *Invertibility of some auxiliary operators*

Lemma 2.4. If $b \in L^\infty$ and u is a measurable a.e. finite function on $\mathbb{T}$, then

$$I - P_+M_uUP_- : \mathcal{D}(R(A)) \to \mathcal{D}(R(A)) \tag{7}$$

is a bounded and continuously invertible operator. Its inverse is $I + P_+M_uUP_-$.

The proof is straightforward.

Consider the usual Hardy spaces $H^p, 1 \le p \le \infty$, over $\mathbb{T}$ (see, e.g., [6]). For a nonnegative function Φ with $\log\Phi \in L^1$ one can define the corresponding *outer function* $\mathcal{O}_\Phi$ by the formula

$$\mathcal{O}_\Phi(z) := \exp\left(\int_{\mathbb{T}} \frac{\tau + z}{\tau - z} \log\Phi(\tau)\, dm(\tau)\right), \quad z \in \mathbb{D} := \{\zeta \in \mathbb{C} : |\zeta| < 1\}.$$

It is analytic in $\mathbb{D}$ and has the crucial property that $|\mathcal{O}_\Phi| = \Phi$ a.e. on $\mathbb{T}$. Moreover, $\mathcal{O}_\Phi \in H^1$ if and only if $\Phi \in L^1$ (see, e.g., [6, p. 62]).

For a measurable function u on $\mathbb{T}$, put $\widetilde{u} := \max\{1, |u|\}$.

Proposition 2.5. (a) If $\log \widetilde{u} \in L^1$, then $\mathcal{O}_{1/\widetilde{u}} \in H^\infty$.

(b) If $\log \widetilde{u} \in L^1$, $F \in H^1$, and $\mathcal{O}_{\widetilde{u}}F \in L^1$, then $\mathcal{O}_{\widetilde{u}}F \in H^1$.

(c) If $\log \widetilde{u} \in L^1$ and $\varphi \in \mathcal{D}(M_u P_+)$, then $\varphi \in \mathcal{D}(M_{\mathcal{O}_{\widetilde{u}}} P_+)$ and

$$P_+ M_{\mathcal{O}_{\widetilde{u}}} P_+ \varphi = M_{\mathcal{O}_{\widetilde{u}}} P_+ \varphi, \quad P_- M_{\mathcal{O}_{\widetilde{u}}} P_+ \varphi = 0. \tag{8}$$

Proof. (a) Obviously, $\log \widetilde{u} \in L^1$ if and only if $\log(1/\widetilde{u}) \in L^1$. Hence, $\mathcal{O}_{1/\widetilde{u}}$ is well defined whenever $\log \widetilde{u} \in L^1$. From the definition of $\widetilde{u}$ it follows that $1/\widetilde{u} \in L^\infty \subset L^1$. Hence, $\mathcal{O}_{1/\widetilde{u}} \in L^\infty \cap H^1 = H^\infty$. Part (a) is proved.

(b) Obviously,

$$\mathcal{O}_{\widetilde{u}} \mathcal{O}_{1/\widetilde{u}} = 1. \tag{9}$$

Therefore, $\mathcal{O}_{\widetilde{u}}F = F/\mathcal{O}_{1/\widetilde{u}}$. The denominator is in H^1, in view of Part (a). By the assumption, the numerator is in H^1 and the fraction is in L^1. Hence, the fraction is in H^1, due to [6, Ch. 5, Exercise 5, p. 75]. Part (b) is proved.

(c) If $\varphi \in \mathcal{D}(M_u P_+)$, then $M_u P_+ \varphi \in X$ and $P_+\varphi \in P_+X \subset H^1$. From properties of outer functions and the definition of $\widetilde{u}$ it follows that

$$|\mathcal{O}_{\widetilde{u}}| = \widetilde{u} \le 1 + |u| \quad \text{a.e. on} \quad \mathbb{T}.$$

Then $|M_{\mathcal{O}_{\widetilde{u}}} P_+ \varphi| \le |P_+\varphi| + |M_u P_+ \varphi|$ a.e. on $\mathbb{T}$. By the lattice property,

$$\|M_{\mathcal{O}_{\widetilde{u}}} P_+ \varphi\|_X \le \|P_+\varphi\|_X + \|M_u P_+ \varphi\|_X. \tag{10}$$

Therefore, $M_{\mathcal{O}_{\widetilde{u}}} P_+ \varphi \in X \subset L^1$. Applying Part (b) to $F := P_+\varphi$, we obtain $M_{\mathcal{O}_{\widetilde{u}}} P_+ \varphi \in X \cap H^1 = P_+X$. From the latter inclusion we get (8). ■

Lemma 2.6. If $b \in L^\infty$ and $\log \widetilde{u} \in L^1$, then the operator

$$R(UM_{\mathcal{O}_{1/\widetilde{u}}}U) : X \to \mathcal{D}(R(A)) \tag{11}$$

is bounded and continuously invertible. Its inverse is

$$R(UM_{\mathcal{O}_{\widetilde{u}}}U) : \mathcal{D}(R(A)) \to X. \tag{12}$$

Proof. The idea of the proof is borrowed from the proof of [9, Lemma 2.6]. If $\log \widetilde{u} \in L^1$, then u is a measurable and a.e. finite function. Hence, by Lemma 2.2, $R(A)$ is closed. Therefore, $\mathcal{D}(R(A))$ is a Banach space with respect to the graph norm.

By Proposition 2.5(a), $\mathcal{O}_{1/\widetilde{u}} \in H^\infty$. Then $M_{\mathcal{O}_{1/\widetilde{u}}}, R(UM_{\mathcal{O}_{1/\widetilde{u}}}U) \in \mathcal{B}(X)$. By using (1), (6), and $\mathcal{O}_{1/\widetilde{u}} \in H^\infty$, we get for $\varphi \in X$,

$$R(A)R(UM_{\mathcal{O}_{1/\widetilde{u}}}U)\varphi = R(M_b)R(UM_{\mathcal{O}_{1/\widetilde{u}}}U)\varphi + M_u M_{\mathcal{O}_{1/\widetilde{u}}} U P_- \varphi. \tag{13}$$

Since $b \in L^\infty$, we have $R(M_b) \in \mathcal{B}(X)$, in view of Lemma 1.1. On the other hand, from properties of outer functions and the definition of $\widetilde{u}$ we get $|u\mathcal{O}_{1/\widetilde{u}}| = |u|/\widetilde{u} \le 1$ a.e. on $\mathbb{T}$. Therefore, by the lattice property (see [1, Ch. 1, Definition 1.1]),

$$\|M_u M_{\mathcal{O}_{1/\widetilde{u}}}\|_{\mathcal{B}(X)} \le 1. \tag{14}$$

Thus, from (13) and (14) we deduce that the operator (11) is bounded and

$$\operatorname{Im} R(UM_{\mathcal{O}_{1/\tilde{u}}}U) \subset \mathcal{D}(R(A)). \tag{15}$$

Let us show that

$$\mathcal{D}(R(A)) \subset \operatorname{Im} R(UM_{\mathcal{O}_{1/\tilde{u}}}U). \tag{16}$$

Assume that $f \in \mathcal{D}(R(A)) \subset X$. By Lemma 2.3, $Uf \in \mathcal{D}(M_u P_+)$. Therefore, by Proposition 2.5(c) and (1), we get

$$P_+UM_{\mathcal{O}_{\tilde{u}}}UP_-f = 0, \quad P_-UM_{\mathcal{O}_{\tilde{u}}}UP_-f = UM_{\mathcal{O}_{\tilde{u}}}UP_-f. \tag{17}$$

Put $\varphi := P_+f + UM_{\mathcal{O}_{\tilde{u}}}UP_-f$. Then from (1), (9), and (17) we obtain

$$R(UM_{\mathcal{O}_{1/\tilde{u}}}U)\varphi = P_+f + UM_{\mathcal{O}_{1/\tilde{u}}}U^2M_{\mathcal{O}_{\tilde{u}}}UP_-f = f.$$

Therefore, $f \in \operatorname{Im} R(UM_{\mathcal{O}_{1/\tilde{u}}}U)$. Hence, (16) holds and

$$R(UM_{\mathcal{O}_{1/\tilde{u}}}U)R(UM_{\mathcal{O}_{\tilde{u}}}U)f = f, \quad f \in \mathcal{D}(R(A)). \tag{18}$$

Analogously to (18), by using Proposition 2.5(a), one can show that

$$R(UM_{\mathcal{O}_{\tilde{u}}}U)R(UM_{\mathcal{O}_{1/\tilde{u}}}U)\varphi = \varphi, \quad \varphi \in X. \tag{19}$$

Equalities (18) and (19) imply that the operator (11) is invertible and its inverse is given by (12). Moreover, we have shown that the operator (11) is bounded (and, therefore, it is closed) and its image coincides with the range space (see (15) and (16)). From above and the closed graph theorem (see, e.g., [8, Ch. 3, Section 5.4, Problem 5.21]) it follows that the operator (12) is bounded. ∎

3. Fredholm theory

3.1. *Relative compactness of Hankel operators*

By analogy with [9] define the class

$$\mathbb{A} := \left\{ u \text{ is measurable}: \quad \log \widetilde{u} \in L^1, \quad u\mathcal{O}_{1/\widetilde{u}} \in H^\infty + C \right\}.$$

Lemma 3.1. If $u \in \mathbb{A}$, then the operator $P_-M_uP_+$ is relatively compact with respect to the operator M_uP_+.

Proof. This lemma is proved by analogy with [9, Lemma 2.1]. Since P_- is bounded on X, we have $\mathcal{D}(M_uP_+) = \mathcal{D}(P_-M_uP_+)$. Let $\{\varphi_n\} \subset \mathcal{D}(M_uP_+)$ be a bounded sequence in X such that the sequence $\{M_uP_+\varphi_n\}$ is also bounded in X. Since P_+ is bounded on X, the sequence $\{P_+\varphi_n\}$ is bounded too. Put $\psi_n := M_{\mathcal{O}_{\tilde{u}}}P_+\varphi_n$. From (10) it follows that

$$\|\psi_n\|_X \le \|P_+\varphi_n\|_X + \|M_uP_+\varphi_n\|_X \le C$$

for some constant $C > 0$ and all $n \in \mathbb{N}$. Thus, $\{\psi_n\}$ is bounded in X. Taking into account (8) and (9), we obtain

$$P_- M_u P_+ \varphi_n = P_- M_{u\mathcal{O}_{1/\tilde{u}}} P_+ M_{\mathcal{O}_{\tilde{u}}} P_+ \varphi_n = P_- M_{u\mathcal{O}_{1/\tilde{u}}} P_+ \psi_n. \tag{20}$$

Since $u \in \mathbb{A}$, we have $u\mathcal{O}_{1/\widetilde{u}} \in H^\infty + C$. By [2, Theorem 2.54], the operator $P_- M_{u\mathcal{O}_{1/\tilde{u}}} P_+$ is compact on L^p for every $p \in (1, \infty)$. Hence, by [11, Corollary 1], it is compact on X. This means that, for the bounded sequence $\{\psi_n\} \subset X$, the corresponding sequence $\{P_- M_{u\mathcal{O}_{1/\tilde{u}}} P_+ \psi_n\}$ contains a convergent subsequence. From (20) we deduce that the sequence $\{P_- M_u P_+ \varphi_n\}$ contains a convergent subsequence. Thus, the operator $P_- M_u P_+$ is relatively compact with respect to the operator $M_u P_+$. ■

3.2. *Fredholm criteria*

Lemma 3.2. If $b \in L^\infty$ and $\log \widetilde{u} \in L^1$, then

(a) the operators

$$R(M_b) : \mathcal{D}(R(A)) \to X, \tag{21}$$

$$R(M_b U M_{\mathcal{O}_{1/\tilde{u}}} U) : X \to X \tag{22}$$

are bounded;

(b) the operator (21) is Fredholm if and only if the operator (22) is Fredholm. If one of these operators is Fredholm, then their indices coincide.

Proof. This statement is proved similarly to [9, Lemma 2.7]. By analogy with (18) one can prove that

$$R(M_b U M_{\mathcal{O}_{1/\tilde{u}}} U) R(U M_{\mathcal{O}_{\tilde{u}}} U)\varphi = R(M_b)\varphi, \quad \varphi \in \mathcal{D}(R(A)). \tag{23}$$

(a) By Proposition 2.5(a), $M_{\mathcal{O}_{1/\tilde{u}}} \in \mathcal{B}(X)$. Due to Lemma 1.1, $M_b \in \mathcal{B}(X)$. Thus, the operator (22) is bounded. On the other hand, by Lemma 2.6, the operator

$$R(U M_{\mathcal{O}_{\tilde{u}}} U) : \mathcal{D}(R(A)) \to X \tag{24}$$

is also bounded. From above and (23) we infer that the operator (21) is bounded too. Part (a) is proved.

(b) By Lemma 2.6, the operator (24) is invertible. Therefore, by [5, Ch. 4, Theorem 6.1], from (23) we conclude that the operators (21) and (22) are Fredholm only simultaneously and their indices coincide. ■

Lemma 3.3. If $b \in L^\infty$ and $u \in \mathbb{A}$, then the operator

$$R(A) : \mathcal{D}(R(A)) \to X \tag{25}$$

is Fredholm if and only if the operator (21) is Fredholm. If one of these operators is Fredholm, then their indices coincide.

Proof. This statement is proved by analogy with [9, Lemmas 2.4 and 2.5]. By Lemma 3.1, the operator $P_-M_uP_+$ is relatively compact with respect to the operator M_uP_+. From this fact and the definition of relative compactness (see [8, Ch. 4, Section 1.3]) one can easily obtain that the operator $P_-M_uUP_- = P_-M_uP_+U$ is relatively compact with respect to the operator $M_uP_+U = M_uUP_-$. Due to Lemma 1.1, $R(M_b)$ is bounded on X. Therefore, it is not difficult to prove that the operator $P_-M_uUP_-$ is relatively compact with respect to the operator $R(A) = R(M_b) + M_uUP_-$. Hence, the operator $P_-M_uUP_-$ is also relatively bounded with respect to the operator $R(A)$ (see [8, Ch. 4, Section 1.3]). Thus, the operator $P_-M_uUP_- : \mathcal{D}(R(A)) \to X$ becomes bounded and compact (see [8, Ch. 4, Section 1.1, Remark 1.4 and Section 1.3, Remark 1.12]). By [5, Ch. 4, Theorem 6.3], the operator (25) is Fredholm if and only if the operator

$$R(A) - P_-M_uUP_- : \mathcal{D}(R(A)) \to X \tag{26}$$

is Fredholm and the indices of (25) and (26) coincide. From (1), (5), and (6) we get for any $\varphi \in \mathcal{D}(R(A))$,

$$(R(A) - P_-M_uUP_-)(I - P_+M_uUP_-)\varphi = R(M_b)\varphi. \tag{27}$$

In view of Lemma 2.4, the operator (7) is bounded and continuously invertible. From above and (27) it follows that the operator (26) (and thus, the operator (25)) is Fredholm if and only if the operator (21) is Fredholm. In that case the indices of the operators (25), (26), and (21) coincide (see [5, Ch. 4, Theorem 6.1]). ■

Now we are in a position to prove the main result of this paper.

Theorem 3.4. If $b \in L^\infty$ and $u \in \mathbb{A}$, then the operator (25) is Fredholm if and only if $u \in L^\infty$ and the operator

$$R(M_b) : X \to X \tag{28}$$

is Fredholm. If one of these operators is Fredholm, then their indices coincide.

Proof. The proof is developed by analogy with [9, Theorem 2.8].

Necessity. By Lemma 3.2(a), the operator (21) is bounded. If the operator (25) is Fredholm, then the operator (21) is Fredholm, due to Lemma 3.3. Therefore, by Lemma 3.2, the operator (22) is bounded and Fredholm. Hence, by [7, Theorem 6.8],

$$\operatorname*{ess\,inf}_{t\in\mathbb{T}} \left| b(t)\mathcal{O}_{1/\widetilde{u}}(1/t) \right| > 0.$$

Then

$$0 < \operatorname*{ess\,inf}_{t\in\mathbb{T}} \left| \mathcal{O}_{1/\widetilde{u}}(1/t) \right| = \operatorname*{ess\,inf}_{t\in\mathbb{T}} \left(1/\widetilde{u}(t) \right) = \left(\operatorname*{ess\,sup}_{t\in\mathbb{T}} \widetilde{u}(t) \right)^{-1}.$$

The latter condition is obviously equivalent to $\widetilde{u} \in L^\infty$. Therefore, $u \in L^\infty$. In this case $R(A)$ is bounded on X. Thus, $\mathcal{D}(R(A)) = X$. So, the operator (28) is Fredholm. The necessity part is proved.

Sufficiency. Assume that $u \in L^\infty$ and the operator (28) is Fredholm. Then the first condition guarantees that $R(A) \in \mathcal{B}(X)$, and so, $\mathcal{D}(R(A)) = X$. This means that the operator (28) coincide with the operator (21). By Lemma 3.3, the operator (25) is Fredholm and its index coincide with the index of the operator (28). ■

Theorem 3.4 generalizes [9, Theorem 2.8]. It is new even for Lebesgue spaces $L^p, 1 < p < \infty$, because the class $\mathbb{A}$ is essentially wider than the class

$$\widetilde{QC}^+ := \Big\{ f \text{ is measurable} : \ \log \widetilde{u} \in L^1, \ u\mathcal{O}_{1/\widetilde{u}} \in (H^\infty + C) \cap (\overline{H^\infty} + C) \Big\},$$

which was considered in [9].

References

1. C. Bennett, R. Sharpley, *Interpolation of Operators*, Pure and Applied Mathematics **129** Academic Press, Inc., Boston, MA, 1988.
2. A. Böttcher, B. Silbermann, *Analysis of Toeplitz Operators*, Springer-Verlag, Berlin, 1990.
3. D. W. Boyd, Indices of function spaces and their relationship to interpolation, *Canad. J. Math.* **21** (1969), 1245–1254.
4. D. E. Edmunds, W. D. Evans, *Spectral Theory and Differential Operators*, Oxford Mathematical Monographs. Oxford Science Publications. The Clarendon Press, Oxford University Press, New York, 1987.
5. I. Gohberg, N. Krupnik, *One-dimensional Linear Singular Integral Equations*, Vol. 1. Operator Theory: Advances and Applications, **53**. Birkhäuser Verlag, Basel, Boston, Berlin, 1992. Russian original: Shtiintsa, Kishinev, 1973.
6. K. Hoffman, *Banach Spaces of Analytic Functions*, Dover Publications, Inc., New York, 1962.
7. A. Karlovich, Singular integral operators with piecewise continuous coefficients in reflexive rearrangement-invariant spaces, *Integral Equations and Operator Theory* **32** (1998) 436–481.
8. T. Kato, *Perturbation theory for linear operators*, Springer-Verlag, Berlin, 1995.
9. V. G. Kravchenko, A. B. Lebre, G. S. Litvinchuk, F. S. Teixeira, Fredholm theory for a class of singular integral operators with Carleman shift and unbounded coefficients, *Math. Nachr.* **172** (1995) 199–210.
10. L. Maligranda, L.-E. Persson, Generalized duality of some Banach function spaces, *Nederl. Akad. Wetensch. Indag. Math.* **51** (1989) 323–338.
11. R. Sharpley, Interpolation theorems for compact operators, *Indiana Univ. Math. J.* **22** (1972/73) 965–984.

Functional Analysis and its Applications
V. Kadets and W. Żelazko

Spectrum as the Support of Functional Calculus

Vladimir V. Kisil [1]

School of Mathematics, University of Leeds, Leeds LS2 9JT, UK

Abstract

We investigate the new definition of analytic functional calculus in the terms of representation theory of $\mathrm{SL}(2,\mathbb{R})$. *We avoid any usage of its algebraic homomorphism property and replace it by the demand to be an intertwining operator. The related notion of spectrum and spectral mapping theorem are given. The construction is illustrated by a simple example of calculus and spectrum of non-normal* $n \times n$ *matrix.*

Key words: Functional calculus, spectrum, intertwining operator, spectral mapping theorem, jet spaces
2000 MSC: 47A60, 46H30

0. Introduction

United in the trinity functional calculus, spectrum, and spectral mapping theorem play the exceptional rôle in functional analysis and could not be substituted by anything else. All traditional definitions of functional calculus are covered by the following rigid template based on *algebra homomorphism* property:

Definition 1. An *functional calculus* for an element $a \in \mathfrak{A}$ is a continuous linear mapping $\Phi : \mathcal{A} \to \mathfrak{A}$ such that

[1] On leave from Odessa University.

(i) Φ is a unital *algebra homomorphism*

$$\Phi(f \cdot g) = \Phi(f) \cdot \Phi(g).$$

(ii) There is an initialisation condition: $\Phi[v_0] = a$ for a fixed function v_0, e.g. $v_0(z) = z$.

Most typical definition of the spectrum is seemingly independent and uses the important notion of resolvent:

Definition 2. A *resolvent* of element $a \in \mathfrak{A}$ is the function $R(\lambda) = (a - \lambda e)^{-1}$, which is the image under Φ of the Cauchy kernel $(z - \lambda)^{-1}$.

A *spectrum* of $a \in \mathfrak{A}$ is the set $\mathbf{sp}a$ of singular points of its resolvent $R(\lambda)$.

Then the following important theorem links spectrum and functional calculus together.

Theorem 3 (Spectral Mapping). *For a function f suitable for the functional calculus:*

$$f(\mathbf{sp}a) = \mathbf{sp}f(a). \tag{1}$$

However the power of the classic spectral theory rapidly decreases if we move beyond the study of one normal operator (e.g. for quasinilpotent ones) and is virtually nil if we consider several non-commuting ones. Sometimes these severe limitations are seen to be irresistible and alternative constructions, i.e. model theory [10], were developed.

Yet the spectral theory can be revived from a fresh start. While three components—functional calculus, spectrum, and spectral mapping theorem—are highly interdependent in various ways we will nevertheless arrange them as follows:

(i) Functional calculus is an *original* notion defined in some independent terms;
(ii) Spectrum (or spectral decomposition) is derived from previously defined functional calculus as its *support* (in some appropriate sense);
(iii) Spectral mapping theorem then should drop out naturally in the form (1) or some its variation.

Thus the entire scheme depends from the notion of the functional calculus and our ability to escape limitations of Definition 1. The first known to the present author definition of functional calculus not linked to algebra homomorphism property was the Weyl functional calculus defined by an integral formula [1]. Then its intertwining property with affine transformations of Euclidean space was proved as a theorem. However it seems to be the only "non-homomorphism" calculus for decades.

The different approach to whole range of calculi was given in [4] and developed in [7] in terms of *intertwining operators* for group representations. It was initially targeted for several non-commuting operators because no non-trivial algebra homomorphism with a commutative algebra of function is possible in this case. However it emerged later that the new definition is a useful replacement for classical one across all range of problems.

In the present note we will support the last claim by consideration of the simple known problem: characterization a $n \times n$ matrix up to similarity. Even that "freshman" question could be only sorted out by the classical spectral theory for a small set of diagonalisable matrices. Our solution in terms of new spectrum will be full and thus unavoidably coincides with one given by the Jordan normal form of matrix. Other more difficult questions are the subject of ongoing research.

1. Another Approach to Analytic Functional Calculus

Anything called "*functional* calculus" uses properties of *functions* to model properties of *operators*. Thus changing our viewpoint on functions we could get another approach to operators. We start from the following observation reflected in the almost any textbook on complex analysis:

Proposition 4. Analytic function theory *in the unit disk* $\mathbb{D}$ *is a manifestation of the mock discrete series representation* ρ_1 *of* $\mathrm{SL}(2,\mathbb{R})$*:*

$$\rho_1(g) : f(z) \mapsto \frac{1}{\alpha - \beta z} f\left(\frac{\bar{\alpha}z - \bar{\beta}}{\alpha - \beta z}\right), \qquad \text{where} \begin{pmatrix} \bar{\alpha} & -\bar{\beta} \\ -\beta & \alpha \end{pmatrix} \in \mathrm{SL}(2,\mathbb{R}). \tag{2}$$

The representation (2) is unitary irreducible when acts on the Hardy space H_2. Consequently we have one more reason to abolish the template Definition 1: H_2 is *not* an algebra. Instead we replace the *homomorphism property* by a *symmetric covariance*:

Definition 5. An *analytic functional calculus* for an element $a \in \mathfrak{A}$ and an $\mathfrak{A}$-module M is a *continuous linear* mapping $\Phi : A(\mathbb{D}) \to A(\mathbb{D}, M)$ such that

(i) Φ is an *intertwining operator*

$$\Phi\rho_1 = \rho_a\Phi$$

between two representations of the $\mathrm{SL}(2,\mathbb{R})$ group ρ_1 (2) and ρ_a defined bellow in (9).

(ii) There is an initialisation condition: $\Phi[v_0] = m$ for $v_0(z) \equiv 1$ and $m \in M$, where M is a left $\mathfrak{A}$-module.

Note that our functional calculus released form the homomorphism condition can take value in any left $\mathfrak{A}$-module M, which however could be $\mathfrak{A}$ itself if suitable. This add much flexibility to our construction.

The earliest functional calculus, which is *not* an algebraic homomorphism, was the Weyl functional calculus and was defined just by an integral formula as an operator valued distribution [1]. In that paper (joint) spectrum was defined as support of the Weyl calculus, i.e. as the set of point where this operator valued distribution does not vanish. We also define the spectrum as a support of functional calculus, but due to our Definition 5 it will means the set of non-vanishing intertwining operators with primary subrepresentations.

Definition 6. A corresponding *spectrum* of $a \in \mathfrak{A}$ is the *support* of the functional calculus Φ, i.e. the collection of intertwining operators of ρ_a with *prime representations* [3, § 8.3].

More variations of functional calculi are obtained from other groups and their representations [4,7].

2. Background in Complex Analysis from SL(2, ℝ) Group

To understand the functional calculus from Definition 5 we need first to realise the function theory from Proposition 4, see [5,6,8,9] for more details.

Elements of $\mathrm{SL}(2,\mathbb{R})$ could be represented by 2×2-matrices with complex entries such that:

$$g = \begin{pmatrix} \alpha & \bar\beta \\ -\beta & \bar\alpha \end{pmatrix}, \qquad g^{-1} = \begin{pmatrix} \bar\alpha & -\bar\beta \\ -\beta & \alpha \end{pmatrix}, \qquad |\alpha|^2 - |\beta|^2 = 1.$$

There are other realisations of $\mathrm{SL}(2,\mathbb{R})$ which may be more suitable under other circumstances, e.g. in the upper half-plane.

We may identify the unit disk $\mathbb{D}$ with the left coset $\mathbb{T}\backslash\mathrm{SL}(2,\mathbb{R})$ for the unit circle $\mathbb{T}$ through the important decomposition $\mathrm{SL}(2,\mathbb{R}) \sim \mathbb{T}\times\mathbb{D}$ with $K = \mathbb{T}$—the only compact subgroup of $\mathrm{SL}(2,\mathbb{R})$:

$$\begin{pmatrix} \alpha & \bar\beta \\ \beta & \bar\alpha \end{pmatrix} = \frac{1}{\sqrt{1-|u|^2}} \begin{pmatrix} e^{i\omega} & 0 \\ 0 & e^{-i\omega} \end{pmatrix} \begin{pmatrix} 1 & u \\ \bar u & 1 \end{pmatrix}, \tag{3}$$

where

$$\omega = \arg\alpha, \qquad u = \bar\beta\alpha^{-1}, \qquad |u| < 1.$$

Each element $g \in \mathrm{SL}(2,\mathbb{R})$ acts by the linear-fractional transformation (the Möbius map) on $\mathbb{D}$ and $\mathbb{T}$ $H_2(\mathbb{T})$ as follows:

$$g^{-1} : z \mapsto \frac{\bar\alpha z - \bar\beta}{\alpha - \beta z}, \qquad \text{where} \quad g^{-1} = \begin{pmatrix} \bar\alpha & -\bar\beta \\ -\beta & \alpha \end{pmatrix}. \tag{4}$$

In the decomposition (3) the first matrix on the right hand side acts by transformation (4) as an orthogonal rotation of $\mathbb{T}$ and $\mathbb{D}$; and the second one—by transitive family of maps of the unit disk onto itself.

The standard linearization procedure [3, § 7.1] leads from Möbius transformations (4) to the unitary representation ρ_1 irreducible on the *Hardy space*:

$$\rho_1(g) : f(z) \mapsto \frac{1}{\alpha - \beta z} f\left(\frac{\bar{\alpha} z - \bar{\beta}}{\alpha - \beta z}\right) \qquad \text{where} \quad g^{-1} = \begin{pmatrix} \bar{\alpha} & -\bar{\beta} \\ -\beta & \alpha \end{pmatrix}. \tag{5}$$

Möbius transformations provide a natural family of intertwining operators for ρ_1 coming from inner automorphisms of $\mathrm{SL}(2, \mathbb{R})$ (will be used later).

We choose [7,8] K-invariant function $v_0(z) \equiv 1$ to be a *vacuum vector*. Thus the associated *coherent states*

$$v(g, z) = \rho_1(g) v_0(z) = (u - z)^{-1}$$

are completely determined by the point on the unit disk $u = \bar{\beta}\alpha^{-1}$. The family of coherent states considered as a function of both u and z is obviously the *Cauchy kernel* [5]. The *wavelet transform* [5,7] $\mathcal{W} : L_2(\mathbb{T}) \rightarrow H_2(\mathbb{D}) : f(z) \mapsto \mathcal{W}f(g) = \langle f, v_g \rangle$ is the *Cauchy integral*:

$$\mathcal{W}f(u) = \frac{1}{2\pi i} \int_{\mathbb{T}} f(z) \frac{1}{u - z}\, dz. \tag{6}$$

Other classical objects of complex analysis (the Cauchy-Riemann equation, the Taylor series, the Bergman space, etc.) can be also obtained [5,8] from representation ρ_1 but are not used and considered here.

3. Representations of $\mathrm{SL}(2, \mathbb{R})$ in Banach Algebras

A simple but important observation is that the Möbius transformations (4) can be easily extended to any Banach algebra. Let $\mathfrak{A}$ be a Banach algebra with the unit e, an element $a \in \mathfrak{A}$ with $\|a\| < 1$ be fixed, then

$$g : a \mapsto g \cdot a = (\bar{\alpha} a - \bar{\beta} e)(\alpha e - \beta a)^{-1}, \qquad g \in \mathrm{SL}(2, \mathbb{R}) \tag{7}$$

is a well defined $\mathrm{SL}(2, \mathbb{R})$ action on a subset $\mathbb{A} = \{g \cdot a \mid g \in \mathrm{SL}(2, \mathbb{R})\} \subset \mathfrak{A}$, i.e. $\mathbb{A}$ is a $\mathrm{SL}(2, \mathbb{R})$-homogeneous space. Let us define the *resolvent* function $R(g, a) : \mathbb{A} \rightarrow \mathfrak{A}$ by the familiar formula $R(g, a) = (\alpha e - \beta a)^{-1}$ then

$$R_1(g_1, \mathsf{a}) R_1(g_2, g_1^{-1}\mathsf{a}) = R_1(g_1 g_2, \mathsf{a}). \tag{8}$$

The last identity is well known in representation theory [3, § 13.2(10)] and is a key ingredient of *induced representations*. Thus we can again linearise (7) (cf. (5)) in the space of continuous functions $C(\mathbb{A}, M)$ with values in a left $\mathfrak{A}$-module M, e.g.$M = \mathfrak{A}$:

$$\rho_a(g_1) : f(g^{-1} \cdot a) \mapsto R(g_1^{-1}g^{-1}, a) f(g_1^{-1}g^{-1} \cdot a) \qquad (9)$$
$$= (\alpha' e - \beta' a)^{-1} f\left(\frac{\bar{\alpha}' \cdot a - \bar{\beta}' e}{\alpha' e - \beta' a}\right).$$

For any $m \in M$ we can again define a K-invariant *vacuum vector* as $v_m(g^{-1} \cdot a) = m \otimes v_0(g^{-1} \cdot a) \in C(\mathbb{A}, M)$. It generates the associated with v_m family of *coherent states* $v_m(u, a) = (ue - a)^{-1}m$, where $u \in \mathbb{D}$.

The *wavelet transform* defined by the same common formula based on coherent states (cf. (6)):

$$\mathcal{W}_m f(g) = \langle f, \rho_a(g) v_m \rangle,$$

is a version of Cauchy integral, which maps $L_2(\mathbb{A})$ to $C(\mathrm{SL}(2, \mathbb{R}), M)$. It is closely related (but not identical!) to the Riesz-Dunford functional calculus: the traditional functional calculus is given by the case:

$$\Phi : f \mapsto \mathcal{W}_m f(0) \qquad \text{for } M = \mathfrak{A} \text{ and } m = e.$$

The both conditions—the intertwining property and initial value—required by Definition 5 easily follows from our construction.

4. Jet Bundles and Prolongations of ρ_1

Spectrum was defined in 6 as the *support* of our functional calculus. To elaborate its meaning we need the notion of a *prolongation* of representations introduced by S. Lie, see [11,12] for a detailed exposition.

Definition 7. [12, Chap. 4] Two holomorphic functions have nth *order contact* in a point if their value and their first n derivatives agree at that point, in other words their Taylor expansions are the same in first $n + 1$ terms.

A point $(z, u^{(n)}) = (z, u, u_1, \ldots, u_n)$ of the *jet space* $\mathbb{J}^n \sim \mathbb{D} \times \mathbb{C}^n$ is the equivalence class of holomorphic functions having nth contact at the point z with the polynomial:

$$p_n(w) = u_n \frac{(w - z)^n}{n!} + \cdots + u_1 \frac{(w - z)}{1!} + u. \qquad (10)$$

For a fixed n each holomorphic function $f : \mathbb{D} \to \mathbb{C}$ has nth *prolongation* (or *n-jet*) $\mathrm{j}_n f : \mathbb{D} \to \mathbb{C}^{n+1}$:

$$\mathrm{j}_n f(z) = (f(z), f'(z), \ldots, f^{(n)}(z)). \qquad (11)$$

The graph $\Gamma_f^{(n)}$ of $\mathrm{j}_n f$ is a submanifold of $\mathbb{J}^n$ which is section of the *jet bundle* over $\mathbb{D}$ with a fibre $\mathbb{C}^{n+1}$. We also introduce a notation J_n for the map $J_n : f \mapsto \Gamma_f^{(n)}$ of a holomorphic f to the graph $\Gamma_f^{(n)}$ of its n-jet $\mathrm{j}_n f(z)$ (11).

One can prolong any map of functions $\psi : f(z) \mapsto [\psi f](z)$ to a map $\psi^{(n)}$ of n-jets by the formula

$$\psi^{(n)}(J_n f) = J_n(\psi f). \tag{12}$$

For example such a prolongation $\rho_1^{(n)}$ of the representation ρ_1 of the group $\mathrm{SL}(2,\mathbb{R})$ in $H_2(\mathbb{D})$ (as any other representation of a Lie group [12]) will be again a representation of $\mathrm{SL}(2,\mathbb{R})$. Equivalently we can say that J_n *intertwines* ρ_1 and $\rho_1^{(n)}$:

$$J_n \rho_1(g) = \rho_1^{(n)}(g) J_n \quad \text{for all } g \in \mathrm{SL}(2,\mathbb{R}).$$

Of course, the representation $\rho_1^{(n)}$ is not irreducible: any jet subspace $\mathbb{J}^k$, $0 \le k \le n$ is $\rho_1^{(n)}$-invariant subspace of $\mathbb{J}^n$. However the representations $\rho_1^{(n)}$ are *primary* [3, § 8.3] in the sense that they are not sums of two subrepresentations.

The following statement explains why jet spaces appeared in our study of functional calculus.

Proposition 8. *Let matrix a be a Jordan block of a length k with the eigenvalue $\lambda = 0$, and m be its root vector of order k, i.e. $a^{k-1}m \neq a^k m = 0$. Then the restriction of ρ_a on the subspace generated by v_m is equivalent to the representation ρ_1^k.*

5. Spectrum and the Jordan Normal Form of a Matrix

Now we are prepared to describe a spectrum of a matrix. Since the functional calculus is an intertwining operator its support is a decomposition into intertwining operators with prime representations (we could not expect generally that these prime subrepresentations are irreducible).

Recall the transitive on $\mathbb{D}$ group of inner automorphisms of $\mathrm{SL}(2,\mathbb{R})$, which can send any $\lambda \in \mathbb{D}$ to 0 and are actually parametrised by such a λ. This group extends Proposition 8 to the complete characterisation of ρ_a for matrices.

Proposition 9. *Representation ρ_a is equivalent to a direct sum of the prolongations $\rho_1^{(k)}$ of ρ_1 in the kth jet space $\mathbb{J}^k$ intertwined with inner automorphisms. Consequently the spectrum of a (defined via the functional calculus $\Phi = \mathcal{W}_m$) labelled exactly by n pairs of numbers (λ_i, k_i), $\lambda_i \in \mathbb{D}$, $k_i \in \mathbb{Z}_+$, $1 \le i \le n$ some of whom could coincide.*

Obviously this spectral theory is a fancy restatement of the *Jordan normal form* of matrices.

Example 10. Let $J_k(\lambda)$ denote the Jordan block of the length k for the eigenvalue λ. On the Fig. 1 there are two pictures of the spectrum for the matrix

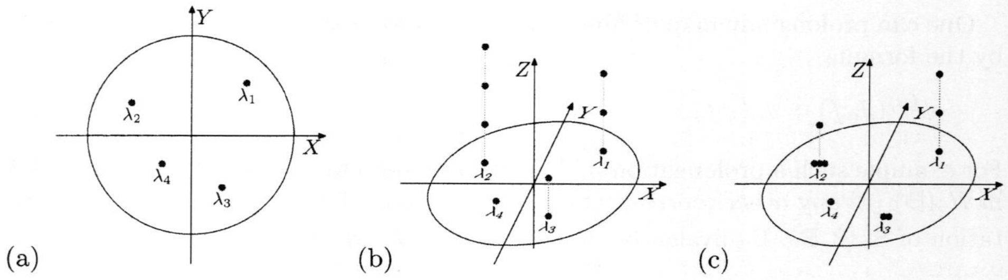

Fig. 1. Classical spectrum (a) of a matrix vs. the new version (b) with its mapping (c).

$$a = J_3(\lambda_1) \oplus J_4(\lambda_2) \oplus J_1(\lambda_3) \oplus J_2(\lambda_4),$$

where

$$\lambda_1 = \frac{3}{4}e^{i\pi/4}, \quad \lambda_2 = \frac{2}{3}e^{i5\pi/6}, \quad \lambda_3 = \frac{2}{5}e^{-i3\pi/4}, \quad \lambda_4 = \frac{3}{5}e^{-i\pi/3}.$$

Part (a) represents the conventional two-dimensional image of the spectrum, i.e. eigenvalues of a, and (b) describes spectrum **sp**a arising from the wavelet construction. The first image did not allow to distinguish a from many other essentially different matrices, e.g. the diagonal matrix

$$\operatorname{diag}(\lambda_1, \lambda_2, \lambda_3, \lambda_4),$$

which even have a different dimensionality. At the same time the Fig. 1(b) completely characterise a up to a similarity. Note that each point of **sp**a on Fig. 1(b) corresponds to a particular root vector, which spans a primary subrepresentation.

6. Spectral Mapping Theorem

As was mentioned in the Introduction a resonable spectrum should be linked to the corresponding functional calculus by an appropriate spectral mapping theorem. The new version of spectrum is based on prolongation of ρ_1 into jet spaces (see Section 4). Naturally a correct version of spectral mapping theorem should also operate in jet spaces.

Let $\phi : \mathbb{D} \to \mathbb{D}$ be a holomorphic map, let us define its action on functions $[\phi_* f](z) = f(\phi(z))$. According to the general formula (12) we can define the prolongation $\phi_*^{(n)}$ onto the jet space $\mathbb{J}^n$. Its associated action $\rho_1^k \phi_*^{(n)} = \phi_*^{(n)} \rho_1^n$ on the pairs (λ, k) is given by the formula:

$$\phi_*^{(n)}(\lambda, k) = \left(\phi(\lambda), \left[\frac{k}{\deg_\lambda \phi}\right]\right), \tag{13}$$

where $\deg_\lambda \phi$ denotes the degree of zero of the function $\phi(z) - \phi(\lambda)$ at the point $z = \lambda$ and $[x]$ denotes the integer part of x.

Theorem 11 (Spectral mapping). *Let ϕ be a holomorphic mapping $\phi : \mathbb{D} \to \mathbb{D}$ and its prolonged action $\phi_*^{(n)}$ defined by* (13)*, then* $\mathbf{sp}\phi(a) = \phi_*^{(n)}\mathbf{sp}a$.

The explicit expression of (13) for $\phi_*^{(n)}$, which involves derivatives of ϕ upto nth order, is known, see for example [2, Thm. 6.2.25], but was not recognised before as form of spectral mapping.

Example 12. Let us continue with Example 10. Let ϕ map all four eigenvalues λ_1, ..., λ_4 of the matrix a into themselves. Then Fig. 1(a) will represent the classical spectrum of $\phi(a)$ as well as a.

However Fig. 1(c) shows mapping of the new spectrum for the case ϕ has *orders of zeros* at these points as follows: the order 1 at λ_1, exactly the order 3 at λ_2, an order at least 2 at λ_3, and finally any order at λ_4.

References

1. Robert F. V. Anderson, The Weyl functional calculus, *J. Functional Analysis* **4** (1969) 240–267.
2. Roger A. Horn and Charles R. Johnson, *Topics in matrix analysis*. Cambridge University Press, Cambridge, 1994, Corrected reprint of the 1991 original.
3. A. A. Kirillov, *Elements of the theory of representations*. Springer-Verlag, Berlin, 1976, Translated from the Russian by Edwin Hewitt, Grundlehren der Mathematischen Wissenschaften, Band 220.
4. Vladimir V. Kisil, Möbius transformations and monogenic functional calculus, *Electron. Res. Announc. Amer. Math. Soc.* **2** (1) (1996) 26–33, (electronic).
5. Vladimir V. Kisil, Analysis in $\mathbb{R}^{1,1}$ or the principal function theory, *Complex Variables Theory Appl.* **40** (2) (1999) 93–118.
6. Vladimir V. Kisil, Two approaches to non-commutative geometry, *Complex Methods for Partial Differential Equations* (H. Begehr, O. Celebi, and W. Tutschke, eds.), Kluwer Academic Publishers, Netherlands, 1999, 219–248.
7. Vladimir V. Kisil, Wavelets in Banach spaces, *Acta Appl. Math.* **59** (1) (1999) 79–109.
8. Vladimir V. Kisil, *Spaces of analytical functions and wavelets—Lecture notes, preprint*, E-print: `arXiv:math.CV/0204018`, 2000–2002, 92 p.
9. Vladimir V. Kisil, *Meeting Descartes and Klein somewhere in a noncommutative space*, Procedsings of ICMP2000, American Mathematical Society, 2002. E-print: `arXiv:math-ph/0112059`, p. 25.
10. N. K. Nikol'skiĭ, *Treatise on the shift operator*. Springer-Verlag, Berlin, 1986, Spectral function theory, With an appendix by S. V. Hruščev [S. V. Khrushchëv] and V. V. Peller, Translated from the Russian by Jaak Peetre.
11. Peter J. Olver, *Applications of Lie groups to differential equations*. second ed., Springer-Verlag, New York, 1993.
12. Peter J. Olver, *Equivalence, invariants, and symmetry*. Cambridge University Press, Cambridge, 1995.

Functional Analysis and its Applications
V. Kadets and W. Żelazko

Generalized Normality in Topological Products and Related Structures[1]

Anatoli P. Kombarov and Andrew N. Yakivchik *

Faculty of Mechanics and Mathematics, Moscow State University, Moscow 119992, Russia

Abstract

We survey recent progress in generalizing the results that provide necessary conditions for (hereditary) normality of topological products, hyperspaces, function spaces and other structures (these results include the well-known theorems by Katětov, Zenor, Noble, Tamano, Keesling–Velichko) to the case where the normality is replaced with a more general property.

Key words: Topological product, Exponential space, Function space, Normal functor, Weak normality, δ-normality, $\mathcal{P}$-normality, Subnormality
2000 MSC: 54B10, 54B20, 54C35, 54D15

Introduction

The property of normality of a topological space, being, unlike weaker separation axioms up to Tychonoff's $T_{3\frac{1}{2}}$, neither productive nor hereditary, is involved in a plenty of remarkable results: a fairly strong and important topological property is often implied by the normality of some topological products, their subspaces, or other structures (hyperspaces etc.); they are recalled in the sequel.

* Corresponding author.
Email addresses: kombarov@mech.math.msu.su (Anatoli P. Kombarov), anyak@mech.math.msu.su (Andrew N. Yakivchik).
[1] The work was supported by Russian Foundation for Basic Research.

The problem that we will discuss here is: *What happens to these results when normality is replaced with a more general property?*

It is worth to note that there are different ways of generalizing the property of normality, particularly because of its *dual* nature: indeed, *equivalently* any two disjoint closed sets are separated by putting them into disjoint open sets, and by mapping them into 0 and 1 by a continuous real-valued function (by virtue of the well-known Urysohn lemma). One can consider the following ways to generalize the normality:

(i) relax the requirement on the separating function (weak normality),
(ii) restrict the class of closed sets to be separated (δ-normality, $\mathcal{P}$-normality),
(iii) allow the separating sets to be more general than the open ones (subnormality).

Below we recollect some recent results in this trend obtained within the last few years. Some open problems are pointed out.

Most of the special topological notions relevant to our subject are recalled in course of the paper and indexed at the end. For the explanation of other terms not defined here, see [6] or [18].

1. Weak normality

Throughout this section, all spaces are assumed to be Tychonoff.

Definition 1 ([1]). Let $\mathcal{P}$ be a class of spaces (resp. Z be a space). A space X is *weakly normal* over the class $\mathcal{P}$ (resp. over Z) if for any two closed sets $A, B \subset X$ such that $A \cap B = \emptyset$ there exists a continuous mapping $f: X \to Y = f(X) \in \mathcal{P}$ (resp. $f: X \to Z$) such that $f(A) \cap f(B) = \emptyset$.

By default, $\mathcal{P}$ is the class of separable metric spaces (resp. $Z = \mathbb{R}^\omega$).

The concept of weak normality was introduced by A.V. Arhangel'skiĭ in connection with the theory of cleavable spaces. A space is *cleavable* (over $\mathcal{P}$, over Z) if *arbitrary* disjoint sets A, B are taken to disjoint sets in $\mathbb{R}^\omega$ (in $Y \in \mathcal{P}$, in Z)]. The (standard) cleavability of a space X has an important application in functional analysis; it is equivalent to the possibility of pointwise approximation of an arbitrary real-valued function on X by a countable family of continuous functions, see [2].

By the Urysohn lemma, every normal space is weakly normal (over $\mathbb{R}$). Also, cleavable spaces are evidently (hereditarily) weakly normal.

Remark 2. Weak normality is preserved *neither* by discrete sums (of more than $\mathfrak{c}$ spaces, see [33]) *nor* by perfect mappings.

1.1. *Tamano's theorem on $X \times \beta X$*

Tamano's theorem of 1960 (from [29]) says: A space X is paracompact iff the product $X \times \beta X$ ($X \times bX$ for some/every compactification bX of X) is normal.

A.N. Yakivchik proved [33] that this result holds true when the normality is relaxed to the weak normality (even over the class of metrizable spaces).

Theorem 3. *For a space X and its (arbitrary) compactification bX, if $X \times bX$ is weakly normal over the class of all metrizable spaces then X is paracompact.*

Recall that the *Hewitt–Nachbin number* $q(X)$ does not exceed τ if the remainder $\beta X \setminus X$ is the union of G_τ-sets in βX; thus, $q(X) \leq \omega$ means the *realcompactness* of X.

Theorem 4 ([33]). *Let X be a pseudocompact space and bX be its compactification. If $X \times bX$ is weakly normal over $\mathbb{R}^\tau$ then $q(X) \leq \tau$.*

The last theorem lets one easily find a space X_τ of weight τ^+ which is not weakly normal over $\mathbb{R}^\tau$; namely, $X_\tau = \tau^+ \times (\tau^+ + 1)$.

Another result resembling Tamano's theorem was found by Y. Yajima [32] who characterized the Lindelöfness of a space X as follows:

Theorem 5. *A space X is Lindelöf iff the subspace $(X \times \beta X) \cup (\beta X \times X)$ of $(\beta X)^2$ is normal.*

Problem 6. Is it possible to characterize Lindelöf property by weak normality?

1.2. *Morita's P-spaces*

K. Morita (1964) proved that, for a space X, the following are equivalent [21]:
(i) $X \times M$ is normal for any metrizable space M;
(ii) X is a normal P-space.
We now recall Morita's definition of P-spaces.

Definition 7 ([21]). A space X is called a *P-space* iff for any family of open sets $G_{s_0 \ldots s_k} \subset G_{s_0 \ldots s_k s_{k+1}} \subset X$, $s_i \in S$, there exist closed sets $F_{s_0 \ldots s_k} \subset G_{s_0 \ldots s_k}$ such that $X = \bigcup_{k \in \omega} G_{s_0 \ldots s_k}$ implies $X = \bigcup_{k \in \omega} F_{s_0 \ldots s_k}$.

It is also appropriate to recollect two covering properties which are of special importance for studying normality and its generalizations.

Definition 8. A space X is *countably paracompact* (resp. *countably metacompact*) if every countable open cover of X has a locally finite (resp. point-finite) open refinement.

An equivalent formulation of the countable paracompactness of a space X is that every countable open cover $\{G_i\}_{i\in\omega}$ of X has a closed refinement $\{F_i\}_{i\in\omega}$ where $F_i \subset G_i$. Hence, every normal P-space is countably paracompact.

Problem 9. Suppose that $X \times M$ is weakly normal for every metrizable space M. Is it true that X is a P-space?

Such a space X may appear to be neither normal nor countably paracompact, as shown by the following

Example 10 ([33]). Let $X = ((\omega_1 + 1) \times (\omega + 1)) \setminus \{(\omega_1, \omega)\}$ be the Tychonoff plank. Then $X \times P$ is weakly normal for every first countable paracompact space P.

1.3. *Products with the unit interval and Dowker spaces*

In 1951, C.H. Dowker [5] proved that, for a space X, the following are equivalent:
(i) X is normal and countably paracompact;
(ii) $X \times [0;1]$ is normal;
(iii) $X \times (\omega + 1)$ is normal.

Recall that a space X which is normal but not countably paracompact (hence $X \times [0;1]$ fails to be normal) is called a *Dowker space*. For a survey of Dowker spaces see, e.g., [26].

Dowker's result cannot be verbally extended to the weak normality. It can easily be observed that for every weakly normal space X and every countable space K the product $X \times K$ is weakly normal. However, there is still a hope to accomplish this without clause (iii), i.e., for the products with a closed interval.

Problem 11. If a space X is (weakly) normal, is then the product $X \times [0;1]$ weakly normal?

A good positive approximation to this problem was found by P. Szeptycki in 1995. The following two results are contained in [28].

Theorem 12. *If X is weakly normal and countably metacompact then $X \times [0;1]$ is weakly normal.*

Theorem 13. *If $Y = \prod_{i\in\omega}(\omega_i + 1)$ with the box product topology (M.E. Rudin's Dowker space, 1971, [25]) then $Y \times [0;1]$ is weakly normal.*

1.4. *Large powers*

N. Noble's theorem of 1971 [22] affirms that the normality is not multiplicative "in large": If X is a space and X^τ is normal for any cardinal τ (for some $\tau \geq \max\{\omega_1, w(X)\}$) then X is compact.
A.P. Kombarov [13] carried over this result to the weak normality.

Theorem 14. *If X^τ is weakly normal for any cardinal τ (for some cardinal $\tau \geq \max\{\omega_1, w(X)\}$) then X is compact.*

The proof is based on the facts that, firstly, the space ω^{ω_1} (A.H. Stone's example) is not weakly normal and, secondly, every countably compact weakly normal space is normal.

1.5. *Hyperspaces*

For a space X, by $\exp(X)$ we mean the hyperspace of non-empty closed subsets of X endowed with the Vietoris topology.
If X is a space such that $\exp(X)$ is normal, then X is compact. This result was proved by J. Keesling (under CH, [12]) and N.V. Velichko (in ZFC, [31]).
It is known [10] that if $\exp(X)$ is normal then X is countably compact. Hence, the next theorem, due to Kombarov [14], is a strengthening of Velichko's theorem:

Theorem 15. *If $\exp(X)$ is weakly normal and X is countably compact then X is compact.*

Actually, in Theorem 15 [Theorem 14] one needs weak normality of $\exp(X)$ over the class of all normal spaces where any countably compact subspace are closed (particular cases are metrizable spaces, normal sequential spaces) [and all points are G_δ].
Nevertheless, the requirement of countable compactness of X in Theorem 15 cannot be omitted: obviously, $\exp(\omega)$ is weakly normal because it admits a one-to-one continuous mapping onto the Cantor set D^ω.

2. Other generalized normality properties

All spaces considered from now on are Hausdorff.

2.1. *δ-normality*

Recall that a (necessarily closed) subset F of a space X is a *regular G_δ-set* if it is the intersection of the closures of countably many open neighborhoods of F in X.

Definition 16 (J. Mack [20]). A space X is *δ-normal* if for any two disjoint closed sets $A, B \subset X$ where A is a regular G_δ-set, there exist disjoint open neighborhoods.

Every countably paracompact space is δ-normal. Moreover, a space X is countably paracompact iff $X \times [0;1]$ is δ-normal [20].

C. Good and I.J. Tree in [8] proved that there exist: 1) a δ-normal Moore space which is not countably paracompact; 2) a δ-normal space which is not countably metacompact.

2.2. *$\mathcal{P}$-normality*

Definition 17 (A.P. Kombarov, 2000, [16]). Let $\mathcal{P}$ be a class of spaces. A space X is *$\mathcal{P}$-normal* iff any two disjoint closed subsets one of which belongs to $\mathcal{P}$ have disjoint open neighborhoods.

The property of $\mathcal{P}$-normality becomes more general as $\mathcal{P}$ is taken narrower. If $\mathcal{P}$ is the class of all countable spaces, this property is the *pseudonormality* introduced by C.W. Proctor (1970, [24]). Compact-normality is equivalent to regularity.

Every Tychonoff δ-normal space is Lindelöf-normal, and hence σ-compact-normal, pseudonormal.

2.3. *Subnormality*

This is a generalized normality property introduced by T.R. Kramer in 1973 [17].

Definition 18. A space is *subnormal* if every two its disjoint closed subsets are contained in disjoint G_δ-sets.

Evidently, subnormality generalizes both normality and perfectness. Similar to the latter and unlike the former, it is inherited by countable unions of closed subsets.

In 1998 Yajima [32] obtained the following generalization of Dowker's theorem:

Theorem 19. *A space X is countably subparacompact (or, equivalently, subnormal and countably metacompact) iff $X \times [0;1]$ is subnormal.*

Recall that a space X is called *countably subparacompact* [17] if every countable open cover of X has a countable closed refinement.

3. Hereditary generalized normality

3.1. *Theorems by Katětov and Zenor and hereditary δ-normality*

First, let us recall these two classical theorems:

If $X \times Y$ is hereditarily normal then either every countable subset in X is closed or Y is perfectly normal (M. Katětov, 1948, [11]).

If $X \times Y$ is hereditarily countably paracompact then either every countable discrete subset in X is closed or Y is perfectly normal (P. Zenor, 1971, [35]).

In 1999, Kombarov [15] proved the following theorem, which is a common generalization of these results.

Theorem 20. *If $X \times Y$ is hereditarily δ-normal then either all countable subsets in X are closed or Y is perfectly normal.*

In the same paper of 1948, M. Katětov [11] proved that if the cube X^3 of a compact space X is hereditarily normal then X is metrizable. For the hereditary normality of the square X^2, this was an open problem very famous and long-standing since, but finally its independence on ZFC has been proven. Counterexamples had been found under MA + ¬CH (P. Nyikos, 1977, [23]) and in CH (G. Gruenhage and P. Nyikos, 1993, [9]); and only recently P. Larson and S. Todorčević [19], using forcing techniques, proved the consistency of the affirmative solution.

Some analogous results for hereditary δ-normality of products were published by Kombarov in 1999–2000.

Theorem 21 ([15]). *If X is a countably compact space and X^3 is hereditarily δ-normal then X is metrizable (and compact).*

Problem 22. Is a compact space X metrizable if X^3 is hereditarily pseudonormal?

Theorem 23 ([16]). (MA + ¬CH) *If X is a countably compact space and X^2 is hereditarily δ-normal then X is compact.*

The latter is deduced from Theorem 20 and the fact that a countably compact perfect regular space is compact under MA + ¬CH.

3.2. *F_σ-δ-normality*

We say that a space X has property F_σ-$\mathcal{P}$ if every F_σ-subset of X has $\mathcal{P}$.

F_σ-δ-normality is a simultaneous generalization of normality and F_σ-countable paracompactness.

Zenor's theorem of 1976 (see [36]) claims: If $X \times Y$ is F_σ-countably paracompact then either all countable discrete subsets of X are closed or Y is normal. It has been generalized as follows:

Theorem 24 (A.P. Kombarov, 1999, [15]). *If $X \times Y$ is F_σ-δ-normal then either all countable subsets of X are closed or Y is normal and countably paracompact.*

This also yields the following generalization of the Keesling–Velichko theorem:

Theorem 25 ([15]). *If $\exp(X)$ is F_σ-δ-normal then the space X is compact.*

Note that δ-normality does not suffice here, because $\exp(\omega_1)$ is δ-normal.

Another application of Theorem 24 concerns spaces $C_p(X)$ of continuous real-valued functions in the topology of pointwise convergence (i.e., inherited from Tychonoff product topology of $\mathbb{R}^X$).

V.V. Tkachuk proved in [30] that if $C_p(X)$ is F_σ-countably paracompact then it is normal.

Theorem 26 ([15]). *If $C_p(X)$ is F_σ-δ-normal then it is normal.*

And, there is another generalization of Noble's theorem:

Theorem 27 ([15]). *If X^τ is F_σ-δ-normal for every cardinal τ (for some $\tau \geq \max\{\omega_1, w(X)\}$) then X is compact.*

3.3. *Katětov's theorems and hereditary subnormality*

For hereditary subnormality of finite products, Yakivchik obtained the following results [34]:

Theorem 28. *Let τ be an uncountable cardinal. If the product $X \times Y$ is hereditarily subnormal then either every subspace of X of cardinality $\leq \tau$ has countable pseudocharacter or every closed subset of Y is a G_τ-set.*

Example 29. For any space P of cardinality ω_1 with at most one non-isolated point (in particular, for the compact space $P = A(\omega_1)$), every finite power P^k is hereditarily subnormal.

Problem 30. Let X be a compact space such that X^2 (X^3, X^k for finite k) is hereditarily subnormal. Is it then true that $w(X) \leq \omega_1$? That $\chi(X) \leq \omega_1$?

It can be observed that a compact space X such that X^2 is hereditarily subnormal cannot have two different uncountable regular cardinals as local characters at points in closed subspaces of X.

Problem 30 has quite recently got almost complete affirmative solution. T. Banakh proved [3] the following result (we formulate its particular case).

Theorem 31. *For a compact space X:*

(i) if X^2 is hereditarily subnormal then the hereditary Lindelöf number $hl(X)$ of X does not exceed ω_1, and hence $\chi(X) \leq \omega_1$;

(ii) if X^3 is hereditarily subnormal then either X is perfectly normal or $w(X) \leq \omega_1$;

(iii) if X^4 is hereditarily subnormal then $w(X) \leq \omega_1$.

It remains unknown whether the hereditary subnormality of X^3 for compact X implies the inequality $w(X) \leq \omega_1$.

Theorem 32. *Suppose that P is a regular space and $P = Q \cup D$ where Q is an F_σ-discrete space (i.e., the union of countably many discrete closed subspaces) and D is discrete expandable in P (i.e., there exists a discrete family $\{V_y\}_{y \in D}$ of open subsets of P where $V_y \ni y$). Then the product $X \times P$ of a space X with P is hereditarily normal in each of the following cases: 1) X is hereditarily Lindelöf; 2) X is metrizable.*

Problem 33. Is Theorem 32 true when X is hereditarily paracompact and perfectly normal?

3.4. *Hyperspaces*

A well-known result of M.M. Choban (1971, see [4]) says: If $\exp(X)$ is hereditarily normal then X is a metrizable compact space.

Some analog of this theorem happens to be true for the case of hereditary weak normality:

Theorem 34 (A.P. Kombarov, 1997, [13]). *If X is a countably compact space and* $\exp(X)$ *is hereditarily weakly normal then X is a perfectly normal hereditarily separable compact space.*

Hence, if X is countably compact and $\exp(\exp(X))$ or $\exp(X \times X)$ is hereditarily weakly normal then X is a metrizable compact space.

Problem 35. Is a compact space X metrizable, if $\exp(X)$ is hereditarily weakly normal?

Theorem 36 (A.P. Kombarov, 2000, [16]). (MA $+ \neg$CH) *If* $\exp(X)$ *is hereditarily Lindelöf-normal then X is a metrizable compact space.*

Problem 37. Does Theorem 36 hold in ZFC?

3.5. *Normal functors*

For the definition of a *normal functor* (of degree n) and a survey of covariant functors, see [27].

M. Katětov's theorem on the cube was extended in 1989 by V.V. Fedorchuk [7] as follows:

Theorem 38. *If X is a compact space and $\mathcal{F}(X)$ is hereditarily normal where $\mathcal{F}$ is a normal functor of degree ≥ 3 then X is metrizable.*

In 2000, the analogous result for the countably paracompact case was proved similarly by T.F. Zhuraev [37]:

Theorem 39. *If X is a compact space and $\mathcal{F}(X)$ is hereditarily countably paracompact where $\mathcal{F}$ is a normal functor of degree ≥ 3 then X is metrizable.*

A simultaneous generalization of the two previous results has been recently obtained by A.P. Kombarov:

Theorem 40. *Let X be a compact space and $\mathcal{F}$ be a normal functor of degree ≥ 3. If $\mathcal{F}(X)$ is hereditarily σ-compact-normal then X is metrizable.*

Problem 41. In the last theorem, is it enough to assume that $\mathcal{F}(X)$ is hereditarily pseudonormal?

If $\exp_3(X)$ is hereditarily σ-compact-normal and X is compact then X is metrizable. It does not suffice to consider countably compact X, because $\exp_3(\omega_1)$ is (even) hereditarily Lindelöf-normal.

3.6. *Hereditarily weakly normal products $X \times Y$*

The results below have visual (rather than logical) relevance to Katětov's theorem.

Theorem 42 (A.N. Yakivchik, 1997, [33]). *Let τ be an infinite cardinal. If $X \times Y$ is hereditarily weakly normal over the class of all spaces with pseudocharacter $\leq \tau$ then either every subset of X with cardinality $\leq \tau$ is closed or every compact subspace of X has tightness $\leq \tau$.*

Theorem 43 ([33]). *If $X \times \tau^+$ is hereditarily weakly normal over the class of all* *spaces with pseudoweight* $\leq \tau$ *then every subset of* X *with cardinality* $\leq \tau$ *is closed.*

References

1. A.V. Arhangel′skiĭ, Divisibility and cleavability of spaces, in: *Recent Developments of General Topology and its Applications. International Conference in memory of Felix Hausdorff*, Math. Res. 67 (Akademie-Verlag, Berlin, 1992) 13–26.
2. A.V. Arhangel'skiĭ, D.B. Shakhmatov, On pointwise approximation of arbitrary functions by countable families of continuous functions, *Trudy Semin. Petrovskogo* **13** (1988) 206–227 (in Russian).
3. T. Banakh, On topological spaces with hereditary subnormal product, preprint.
4. M.M. Choban, Note sur topologie exponentielle, *Fund. Math.* **171** (1971) 27–41.
5. C.H. Dowker, On countably paracompact spaces, *Canad. J. Math.* **3** (1951) 219–224.
6. R. Engelking, *General Topology*. Heldermann-Verlag, Berlin, 1989.
7. V.V. Fedorchuk, On Katětov's theorem on the cube, *Moscow Univ. Math. Bull.* **44** (4) (1989) 102–106.
8. C. Good and I.J. Tree, On δ-normality, *Topology Appl.* **56** (1994) 117–127.
9. G. Gruenhage and P. Nyikos,, Normality in X^2 for compact X, *Trans. Amer. Math. Soc.* **340** (1993) 563–586.
10. V.M. Ivanova, On the theory of spaces of subsets, *Dokl. Akad. Nauk SSSR* **101** (1955) 601–603 (in Russian).
11. M. Katětov, Complete normality of Cartesian products, *Fund. Math.* **35** (1948) 271–274.
12. J. Keesling, On the equivalence of normality and compactness in hyperspaces, *Pacific J. Math.* **33** (1970) 657–667.
13. A.P. Kombarov, Weak normality of subsets of $\exp(X)$, *Topology Appl.* **76** (1997) 157–160.
14. A.P. Kombarov, Weak normality of 2^X and X^τ, *Fund. Prikl. Mat.* **4** (1) (1998) 135–140 (in Russian).
15. A.P. Kombarov, On F_σ-δ-normality and hereditary δ-normality, *Topology Appl.* **91** (1999) 221–226.
16. A.P. Kombarov, On Lindelöf-normal spaces, *Topology Appl.* **107** (2000) 117–122.
17. T.R. Kramer, A note on countably subparacompact spaces, *Pacific J. Math.* **46** (1973) 209–213.
18. K. Kunen and J.E. Vaughan, eds., *Handbook of Set-Theoretic Topology*. North-Holland, Amsterdam, 1984.
19. P. Larson and S. Todorčević, Katětov's problem, *Trans. Amer. Math. Soc.* **354** (2002) 1783–1791 (electronic).
20. J. Mack, Countable paracompactness and weak normality properties, *Trans. Amer. Math. Soc.* **148** (1970) 265–272.
21. K. Morita, Products of normal spaces with metric spaces, *Math. Ann.* **154** (1964) 365–382.

22. N. Noble, Products with closed projections II, *Trans. Amer. Math. Soc.* **160** (1971) 169–183.
23. P. Nyikos, A compact nonmetrisable space P such that P^2 is completely normal, *Topology Proc.* **2** (1977) 359–363.
24. C.W. Proctor, A separable pseudonormal nonmetrizable Moore space, *Bull. Acad. Polon. Sci.* **18** (1970) 179–181.
25. M.E. Rudin, A normal space X for which $X \times I$ is not normal, *Fund. Math.* **73** (1971) 179–186.
26. M.E. Rudin, Dowker spaces, in: *K. Kunen and J.E. Vaughan, eds., Handbook of Set-Theoretic Topology* (North-Holland, Amsterdam, 1984) 761–780.
27. E.V. Shchepin, Functors and uncountable powers of compacta, *Soviet Math. Surveys* **36** (3) (1981) 1–71.
28. P.J. Szeptycki, Weak normality in Dowker spaces, *Topology Proc.* **20** (1995) 289–296.
29. H. Tamano, On paracompactness, *Pacific J. Math.* **10** (1960) 1043–1047.
30. V.V. Tkachuk, Methods in the theory of cardinal invariants and the theory of mappings in application to function spaces, *Siberian Math. J.* **32** (1991) 93–107.
31. N.V. Velichko, On a space of closed subsets, *Siberian Math. J.* **16** (1975) 484–486.
32. Y. Yajima, Analogous results to two classical characterizations of covering properties by products, *Topology Appl.* **84** (1998) 3–7.
33. A.N. Yakivchik, Weakly normal topological spaces and products, *Topology Appl.* **76** (1997) 193–201.
34. A.N. Yakivchik, Subnormality in subspaces of products, *Topology Appl.* **107** (2000) 197–205.
35. P. Zenor, Countable paracompactness in product spaces, *Proc. Amer. Math. Soc.* **30** (1971) 199–201.
36. P. Zenor, Countable paracompactness of F_σ-sets, *Proc. Amer. Math. Soc.* **55** (1976) 201–202.
37. T.F. Zhuraev, Normal functors and the metrizability of compact Hausdorff spaces, *Moscow Univ. Math. Bull.* **55** (4) (2000) 6–9.

Functional Analysis and its Applications
V. Kadets and W. Żelazko

Nonstandard Universe based on Internal Set Theory

Taras Kudryk [a], Wladyslav Lyantse [a] and Vitor Neves [b]

[a]Lviv National University, 1 Universytetska st., 79602 Lviv, Ukraine
[b] Department of Mathematics, University of Aveiro, 3810-193 Aveiro, Portugal

Abstract
This paper is an invitation to look what occurs when tools of Robinson's and Nelson's Nonstandard Analysis act simultaneously. For instance, in this case we have several kinds of infinitesimals in our disposal.

Key words: Standard, internal, external, shadow, infinitesimal, galaxy, hull, Loeb measure
2000 MSC: 03H05, 46S20

1. Preliminaries

We suppose that the reader is familiar with NSA (the Nonstandard Analysis of Robinson, see [9] or [2]) and IST (the Internal Set Theory of Nelson, see [6] or [4], or [5]).

Multiple standardness was studied by E. Gordon and Y. Peraire in a series of interesting papers (see, for instance [3], [7], [8]). Note that none new definition of standardness is introduced in this article. You can see also the following papers [10], [11], or the book [12] discussing similar topics.

To avoid misunderstandings let us fix notation and terminology.

We determine the standard Mathematics as based on the ZFC (Zermelo-Fraenkel with Choice) set theory, but as the background of the whole Mathematics we accept

Email addresses: tkudryk@franko.lviv.ua (Taras Kudryk), wlanc@litech.lviv.ua (Wladyslav Lyantse), vneves@mat.ua.pt (Vitor Neves).

the Internal Set Theory (which includes the ZFC axiomes). Thus "to be standard" means "to be uniquely determined by tools of ZFC (without the predicate st = "to be standard" and related with st axioms of Idealization, Standardization and Transfer) and standard data". This is a recurrent definition with the empty set $\emptyset$ as the beginning. We realize our considerations in some fixed model of set theory. The uniqueness of interpretation allows do not differ a symbol (name of an entity) and the entity itself. We emphasize that the IST recognizes **internal** sets only (i.e. such that are elements of standard sets). Since $x \in \{x\}$ and $\{x\}$ is standard, whenever x is so, each standard item is internal.

Write "$st(x)$" for "x is standard". A predicate of the ZFC-language without occurence of $st(\cdot)$ is called **internal**, whereas one which contains $st(\cdot)$ is called **external**. The subset construction $B = \{x \in A : \ p(x)\}$ in our IST Mathematics is allowed for A and p internal only. Nevertheless one uses the following denotation: ${}^{\mathrm{st}}E := \{x \in E : \ st(x)\}$; ${}^{\mathrm{st}}E$ (the collection of standard elements of the set E) is not a genuine set, but an "external" set. One uses ${}^{\mathrm{st}}E$ to write formulas, for instance, "$x \in {}^{\mathrm{st}}E$", "${}^{\mathrm{st}}E = {}^{\mathrm{st}}F$, which mean respectively "$x \in E \wedge st(x)$", "$(\forall x)(st(x) \Rightarrow x \in E \iff x \in F)$", and so on.

An internal predicate without occurence of nonstandard constants is said to be **standard**. (Obviously, standard predicates belong to the ZFC-language.) For a standard predicate p the Nelson's **transfer** principle holds:

$$\exists\, x\ p(x) \Rightarrow \exists^{\mathrm{st}} x\ p(x), \qquad \forall^{\mathrm{st}} x\ p(x) \Rightarrow \forall\, x\ p(x); \tag{1.1}$$

understand "$\exists\, x$" as "for some internal x" and "$\forall\, x$" as "for any internal x". An essential corollary of (1.1) is the following:

$$st(E) \wedge E \neq \emptyset \Rightarrow {}^{\mathrm{st}}E \neq \emptyset. \tag{1.2}$$

Notice also an essential corollary of the Nelson's idealization principle:

$$st(E) \wedge \operatorname{card} E \geq \operatorname{card} \mathbb{N} \Rightarrow E \setminus {}^{\mathrm{st}}E \neq \emptyset. \tag{1.3}$$

The presence of nontrivial parts ${}^{\mathrm{st}}E$ and $E \setminus {}^{\mathrm{st}}E$ according to (1.2) and (1.3) is an important feature of the Internal Set Theory.

Now a few words about the construction of the Robinson's Nonstandard Universe $\tilde{W}$ and the Transfer Map *. Initial data for these are as follows: S - the set of individuals, Ω - the set of indices, $\mathcal{F}$ - a fixed free ultrafilter on Ω.

We accept the following

1.1 Main assumption. The initial data $S, \Omega, \mathcal{F}$ are *standard* (in the sense of IST). Besides, in order that $\tilde{W}$ and * be nontrivial we assume that $\operatorname{card} \Omega \geq \operatorname{card} S \geq \operatorname{card} \mathbb{N}$.

Recall that the standard universe $\hat{S}$ is the superstructure

$$\hat{S} := \bigcup\nolimits_{n \in \mathbb{N}} S_n, \quad S_0 := S, \quad \forall\, n \in \mathbb{N}\ S_{n+1} := S_n \cup 2^{S_n}.$$

To each "block" S_n of $\hat{S}$ there corresponds a set ${}^*S_n$ defined by the conditions

(i) each function $f \in S_n^{\Omega}$ determines an element $\hat{f} \in {}^*S_n$,

(ii) let $f_1, f_2 \in S_n^{\Omega}$, then $\hat{f}_1 = \hat{f}_2$ iff $f_1 =_{\mathcal{F}} f_2$ (i.e., $\{\omega \in \Omega : f_1(\omega) = f_2(\omega)\} \in \mathcal{F}$),

(iii) let $f \in (2^{S_n})^\Omega$, then $\hat{f}$ is the set of $\hat{g} \in {}^*S_n$ such that $g \in_{\mathcal{F}} f$ (i.e., $\{\omega \in \Omega : g(\omega) \in f(\omega)\} \in \mathcal{F}$),

(iv) for $f \in S_0^\Omega$ $\hat{f} = \{g \in S_0^\Omega : g =_{\mathcal{F}} f\}$ and ${}^*S_0$ is the set of all such $\hat{f}$.

The nonstandard universe $\tilde{W}$ is defined as the bounded ultrapower:

$$\tilde{W} := \hat{S}^\Omega / \mathcal{F} = \bigcup_{n \in \mathbb{N}} {}^*S_n.$$

The transfer map * transforms $\hat{S}$ into $\tilde{W}$ as follows: if $E \in S_n$ and $n \geq 1$, then ${}^*E := \{\hat{f} \in {}^*S_{n-1} : f \in_{\mathcal{F}} E\}$; if $E \in S_0 = S$, then ${}^*E = \hat{f}$ where $\hat{f} \in {}^*S_0$ and $f =_{\mathcal{F}} E$. ${}^*E$ is said to be the **star-version** of $E \in \hat{S}$.

1.2 *Remark.* On account of injectivity of the map * one can identify each individual with its star-version:

$$\forall s \in S_0 \quad {}^*s = s \tag{1.4}$$

But for $n > 0$ the restriction of * to $S_n \setminus S_0$ is not surjective. That is why an $E \in \hat{S} \setminus S$ cannot be identified with its star-version ${}^*E$.

1.3 *Remark.* It may well happen for two individuals that each one is an element of the other. (Recall that all items in Internal Set Theory are sets.) It is necessary to keep in mind that after identification (1.4) for $s, t \in S$ the formula $s \in t$ is meaningless. Indeed, both ${}^*s$ and ${}^*t$ are classes of functions, which excludes the relation ${}^*s \in {}^*t$.

Let α be an assertion of the ZFC-language which states something about $E \in \hat{S}$. Denote by ${}^*\alpha$ the assertion obtained from α by the replacement of E by ${}^*E$. Then the Robinson's **transfer** theorem holds:

$$\alpha \Longleftrightarrow^* \alpha. \tag{1.5}$$

For instance, for $E \in \hat{S} \setminus S$ define ${}^\wedge E := \{{}^*x : x \in E\}$; ${}^\wedge E$ is the pointwise image of E by the map *. Theorem (1.5) implies that $\forall E \in \hat{S} \setminus S\ {}^\wedge E \subseteq {}^*E$, but ${}^*E \setminus {}^\wedge E \neq \emptyset \Longleftrightarrow \operatorname{card} E \geq \operatorname{card} \mathbb{N}$. In the framework of NSA ${}^\wedge E$ is a counterpart of ${}^{\mathrm{st}}E$ of IST.

An essential corollary of assumption 1.1 is the following statement:

1.4 Proposition. *The superstructure $\hat{S}$, the ultrapower $\tilde{W}$, the transfer map $*$ are* **standard** *(in the sense of IST). If $E \in {}^{\mathrm{st}}(\hat{S} \setminus S)$, then ${}^*E$ and ${}^\wedge E$ are standard.*

For instance, ${}^*S_n$ is standard for each $n \in {}^{\mathrm{st}}\mathbb{N}$. Notice that for $n \in \mathbb{N} \setminus {}^{\mathrm{st}}\mathbb{N}$ S_n and ${}^*S_n$ are internal.

1.5 *Warning.* The terminology in Robinson's NSA is somewhat different, namely an $E \in \tilde{W}$ is said to be standard iff it is the star-version of some $F \in \hat{S}$. To avoid misunderstandings, we call an $E \in \tilde{W}$ such that $E = {}^*F$, for some $F \in \hat{S}$, **R-standard** (standard in the sense of Robinson).

2. Numbers

Let us look what numbers are available in the universe $\tilde{W}$ constructed within IST. As usually, by $\mathbb{N}$, $\mathbb{Z}$, $\mathbb{Q}$, $\mathbb{R}$, $\mathbb{C}$ we denote standard sets of natural, integer, rational, real and complex numbers respectively. Their star-versions ${}^*\mathbb{N}, \ldots, {}^*\mathbb{C}$ are called respectively sets of hypernatural, ..., hypercomplex numbers. For convenience assume that $\mathbb{C} \subset S$. This implies

$$^{\mathrm{st}}\mathbb{C} \subset \mathbb{C} = \mathbb{C} \subset {}^*\mathbb{C};$$

and similarly for $\mathbb{N}, \ldots, \mathbb{R}$. Notice that (in our terminology) ${}^*\mathbb{N}, \ldots, {}^*\mathbb{C}$ are standard and ${}^{\mathrm{st}}\mathbb{N}, \ldots, {}^{\mathrm{st}}\mathbb{C}$ are strictly external.

Begin with naturals and hypernaturals. Consider each $n \in \mathbb{N}$ as a (finite) ordinal, i.e. a transitive set, totally ordered by the membership relation $\in$. So for $n \in \mathbb{N}$ we have $n := \{0, \ldots, n-1\}$. Since $\mathbb{N} \subset S$, in view of remark 1.3 this cannot be transferred to ${}^*\mathbb{N}$. But it can be used that the restriction of $\in$ to $\mathbb{N}$ is identical to the ordinary inequality relation $<$. Temporararily denote the last by $\mathcal{I}$. We have $\mathcal{I} \in \hat{S}$, therefore the star-version ${}^*\mathcal{I}$ really exists. Because $\mathcal{I} \subset \mathbb{N} \times \mathbb{N} \subset S \times S$, $\mathcal{I} \subset {}^*\mathcal{I}$ and ${}^*\mathcal{I} \cap (\mathbb{N} \times \mathbb{N}) = \mathcal{I}$. These relation allow to denote $\mathcal{I}$ and ${}^*\mathcal{I}$ by the unique usual symbol $<$. Since $(\mathbb{N}, \leq)$ is a linearly ordered set, by the transfer theorem (1.5) $({}^*\mathbb{N}, \leq)$ is also so. But $({}^*\mathbb{N}, \leq)$ is not totally ordered (though $(\mathbb{N}, \leq)$ is a totally ordered set). The cause is that for $E \in \hat{S} \setminus S$ the star-version of 2^E is not $2^{{}^*E}$, but ${}^*(2^E) = 2^{{}^*E} \cap \tilde{W}$. This involves that some sets $A \subset {}^*\mathbb{N}$ can be simultaneously internal and R-external. For instance, the set ${}^*\mathbb{N} \setminus \mathbb{N}$ is standard, nonempty and does not contain a minimal number. Even for $n \in {}^*\mathbb{N} \setminus \mathbb{N}$ the nonempty internal set $\{k \in {}^*\mathbb{N} \setminus \mathbb{N} : k < n\}$ lacks the minimum.

Call a number $n \in \mathbb{N} \setminus {}^{\mathrm{st}}\mathbb{N}$ **infinite**, and a number $n \in {}^*\mathbb{N} \setminus \mathbb{N}$ **R-infinite** (Robinson's infinite). Of course, if p is R-infinite, then $\forall n \in \mathbb{N}\ n < p$.

The main purpose of natural numbers is to be values of the measure card. Let E be some internal set and $n \in \mathbb{N}$. We write $\operatorname{card} E = n$ iff there exists a bijection $f \in E^n$ (i.e. $\operatorname{dom} f = n = \{0, \ldots, n-1\}$, $\operatorname{im} f = E$). By restricting the map card to $\hat{S} \setminus S$, we get $\operatorname{card} \in \hat{S}$. Thus there exists a star-version ${}^*\mathrm{card} \in \tilde{W}$; ${}^*\mathrm{card}$ as well as card are standard. The domain of ${}^*\mathrm{card}$ is $\tilde{W} \setminus {}^*S$.

2.1 Proposition. *There exists a* **standard** *R-infinite hypernatural number.*

Proof. Since $\mathbb{N}$ and ${}^*\mathbb{N}$ are standard, ${}^*\mathbb{N} \setminus \mathbb{N}$ is so. Because ${}^*\mathbb{N} \setminus \mathbb{N} \neq \emptyset$, by (1.1) ${}^{\mathrm{st}}({}^*\mathbb{N} \setminus \mathbb{N}) \neq \emptyset$. ▷

2.2 *Remark.* A standard $n \in {}^*\mathbb{N} \setminus \mathbb{N}$ can be demonstrated explicitly. For $f, g \in \mathbb{N}^\Omega$ define $f =_{\mathcal{F}} g \equiv \{\omega \in \Omega : f(\omega) = g(\omega)\} \in \mathcal{F}$. Then each $f \in \mathbb{N}^\Omega$ uniquely determines a hypernatural number $\tilde{f}$, namely the equivalence class $\tilde{f} := \{g \in \mathbb{N}^\Omega : g =_{\mathcal{F}} f\}$. If f is **essentially** unbounded, i.e., $(\forall n \in \mathbb{N})\ (\{\omega \in \Omega : f(\omega) > n\} \in \mathcal{F})$, then $\tilde{f} \in {}^*\mathbb{N} \setminus \mathbb{N}$ and $\tilde{f}$ is standard whenever $f \in {}^{\mathrm{st}}(\mathbb{N}^\Omega)$. ▷

2.3 *Remark.* A standard hypernatural number can be even greater than an R-infinite one. Indeed, let $n, p \in {}^{\mathrm{st}}({}^*\mathbb{N} \setminus \mathbb{N})$ and $n < p$. If the set $\{x \in {}^*\mathbb{N} : n < x < p\}$ is infinite, for instance with $p = n^2$, by Nelson's theorem (1.3) it contains nonstandard elements. It is easy to demonstrate such elements explicitly. Standard hypernaturals entirely fill in some galaxies.

It is also interesting to mention the existence of **extra infinite** hypernaturals, i.e. the existence of $p \in {}^*\mathbb{N}$ such that $\forall n \in {}^{st}({}^*\mathbb{N})\ n < p$. This follows from the Idealization principle:

$$\forall\ ^{st\,fin}X\ \exists y\ \forall x \in X\ (x \in {}^*\mathbb{N} \wedge y \in {}^*\mathbb{N} \wedge x < y) \Rightarrow$$
$$\Rightarrow \exists y \forall x \in {}^{st}({}^*\mathbb{N})\ (y \in {}^*\mathbb{N} \wedge x < y).$$

Pass to the reals and hyperreals. The arithmetical operations, the order and their star-versions are denoted identically by $+$, $\cdot$, $<$. For $x, y \in {}^*\mathbb{R}$ define

$$x \approx y \equiv \forall n \in {}^{st}\mathbb{N}\ |x-y| < \frac{1}{n}, \qquad \mathbb{I} := \{x \in \mathbb{R} :\ x \approx 0\}, \tag{2.1}$$
$$x \overset{*}{\sim} y \equiv \forall n \in \mathbb{N}\ |x-y| < \frac{1}{n}, \qquad {}_*\mathbb{I} := \{x \in {}^*\mathbb{R} :\ x \overset{*}{\sim} 0\}, \tag{2.2}$$
$$x \overset{*}{\approx} y \equiv \forall n \in {}^{st}({}^*\mathbb{N}) |x-y| < \frac{1}{n}, \qquad {}_e\mathbb{I} := \{x \in {}^*\mathbb{R} : x \overset{*}{\approx} 0\}.$$

Elements of $\mathbb{I}$ and ${}_*\mathbb{I}$ (${}_*\mathbb{I}$ is not a star-version of $\mathbb{I}$) are called respectively **infinitesimals** and **R-infinitesimals**, and ones of ${}_e\mathbb{I}$ **extra infinitesimals**. ${}_*\mathbb{I}$ is standard in the sense of IST whereas $\mathbb{I}$ and ${}_e\mathbb{I}$ are strictly external. In view of the standard archimedean principle, we have

$$\forall x, y \in \mathbb{R} \quad x \overset{*}{\sim} y \Rightarrow x = y. \tag{2.3}$$

Of course, $\forall x \in {}_*\mathbb{I}\ \forall y \in \mathbb{I} \setminus_* \mathbb{I}\ |x| < |y|$. For $x \in {}^*\mathbb{R}$ define

$$|x| \ll \infty \equiv \exists n \in {}^{st}\mathbb{N}\ |x| < n, \qquad \mathbb{F} := \{x \in \mathbb{R} :\ |x| \ll \infty\},$$
$$|x| <\!* \infty \equiv \exists n \in \mathbb{N}\ |x| < n, \qquad {}_*\mathbb{F} := \{x \in {}^*\mathbb{R} :\ |x| <\!* \infty\},$$
$$|x| < \infty \equiv \exists n \in {}^{st}({}^*\mathbb{N})\ |x| < n, \ {}_e\mathbb{F} := \{x \in {}^*\mathbb{R} :\ |x| < \infty\}.$$

Elements of the strictly external sets $\mathbb{F}$ and ${}_e\mathbb{F}$ respectively are called **finite** and **extra finite**. The set ${}_*\mathbb{F}$ is standard and its elements are called **R-finite**. ${}_*\mathbb{F}$ ($\mathbb{F}$) is a subring of the ordered field ${}^*\mathbb{R}$ ($\mathbb{R}$) and ${}_*\mathbb{I}$ ($\mathbb{I}$) is an ideal of the ordered ring ${}_*\mathbb{F}$ ($\mathbb{F}$). Similarly for ${}_e\mathbb{I}$ and ${}_e\mathbb{F}$. Permanence principles do not reach the sets of extra infinitesimals and extra finite numbers.

It is known that for each $x \in \mathbb{F}$ there exists a unique $y \in {}^{st}\mathbb{R}$ such that $x \approx y$. This y is called the **shadow** of x and denoted by ${}^\circ x$. Also to each $x \in {}_*\mathbb{F}$ there corresponds a unique $y \in \mathbb{R}$ such that $x \overset{*}{\sim} y$. This y is called the **R-shadow** of x and denoted by ${}^\diamond x$. Thus,

$$\forall x \in \mathbb{F} \quad {}^\circ x \in {}^{st}\mathbb{R},\ x \approx {}^\circ x, \quad \forall x \in {}_*\mathbb{F} \quad {}^\diamond x \in \mathbb{R},\ x \overset{*}{\sim} {}^\diamond x. \tag{2.4}$$

2.4 Proposition. *Let $x \in {}^*\mathbb{R}$ and $|x| \ll \infty$. Then there exists a unique $y \in {}^{st}\mathbb{R}$ such that*

$$\forall n \in {}^{st}\mathbb{N} \qquad |x-y| < 1/n. \tag{2.5}$$

Proof. If $x \in {}^*\mathbb{R}$ and $|x| \ll \infty$, then $x \in {}_*\mathbb{F}$ and ${}^\diamond x \in \mathbb{F}$. Therefore we can define $y := {}^\circ({}^\diamond x)$. Evidently this y is a unique standard real which satisfies (2.5). ▷

2.5 *Remark.* If $x \in {}^*\mathbb{R}$ and $|x| \ll \infty$, we write ${}^\circ x$ instead of ${}^\circ({}^\diamond x)$. The reason is as follows: for (2.3), (2.4) we have $\forall x \in \mathbb{R}\ {}^\diamond x = x$, hence for $x \in \mathbb{F}\ {}^\circ({}^\diamond x) = {}^\circ x$. ▷

The maps $x \mapsto {}^\circ x$ and $x \mapsto {}^\diamond x$ defined respectively on $\{x \in {}^*\mathbb{R} : |x| \ll \infty\}$ and ${}_*\mathbb{F}$ have natural extensions. Let $[x]$ denote the integral part of the hyperreal x, i.e. $\forall x \in {}^*\mathbb{R}\ \forall n \in {}^*\mathbb{Z}\ [x] = n \equiv n \leq x < n + 1$.

2.6 Definition. On account of $x - [x] \in {}^*[0,1[$ for $x \in {}^*\mathbb{R}$ put

$$ {}^\circ x := [x] +^\circ (x - [x]), \quad {}^\diamond x := [x] +^\diamond (x - [x]). \tag{2.6}$$

Notice that ${}^*\mathbb{R} = {}^*\mathbb{Z} + {}^*[0,1[$, $\forall x \in {}^*\mathbb{R}\ {}^\circ x \in {}^*\mathbb{Z} + {}^{\mathrm{st}}[0,1[$, ${}^\circ x \approx x$, ${}^\diamond x \in {}^*\mathbb{Z} + [0,1[$, ${}^\diamond x \overset{*}{\sim} x$. The map $x \mapsto {}^\circ x$ is strictly external and $x \mapsto {}^\diamond x$ is standard.

2.7 Proposition. *(External) sets ${}^*\mathbb{R}/\approx$ and ${}^*\mathbb{Z} + {}^{\mathrm{st}}[0,1[$ are ordered additive groups. The extended shadow map is an isomorphism of theirs, i.e. $\forall x, y \in {}^*\mathbb{R}$ $\forall n \in {}^*\mathbb{Z}\ {}^\circ(x+n) = {}^\circ x + n$, ${}^\circ(x+y) = {}^\circ x + {}^\circ y$, $x < y \Rightarrow {}^\circ x \leq {}^\circ y$.*

Proof. Let $x = m + \xi$, where $m \in {}^*\mathbb{Z}$, $\xi \in {}^*[0,1[$. Then by (2.6) ${}^\circ x = m + {}^\circ \xi$. For $n \in {}^*\mathbb{Z}$ we have $[x+n] = m + n$. Therefore ${}^\circ(x+n) = m + n + {}^\circ\xi$, i.e. the first equality holds. Using it it is quite easy to prove the rest. ▷

2.8 *Warning.* In general ${}^\circ(xy) \neq {}^\circ x {}^\circ y$.

2.9 *Remark.* By usual means one can form the set of standard hyperrationals from the set of R-infinite hypernaturals which is even more rich than $\mathbb{Q}$ but that is not the same that ${}^*\mathbb{Q}$.

3. Metric spaces

In the sequel (X, d) will denote a standard metric space such that X is infinite and $\mathbb{C} \cup X \subseteq S$. From this assumption it follows (see (1.4)) that ${}^{\mathrm{st}}X \subset X = \hat{}X \subset {}^*X$. ${}^*X$ and ${}^*d$ as well as X and d are standard. We write d instead of ${}^*d$, because $d \subset {}^*d$. For $x_1, x_2 \in {}^*X$ define (see (2.1), (2.2)) $x_1 \approx x_2 \equiv d(x_1, x_2) \approx 0$, $x_1 \overset{*}{\sim} x_2 \equiv d(x_1, x_2) \overset{*}{\sim} 0$, $x_1 \overset{*}{\approx} x_2 \equiv d(x_1, x_2) \overset{*}{\approx} 0$. Notice that (2.3) implies

$$\forall x_1, x_2 \in X \qquad x_1 \overset{*}{\sim} x_2 \Rightarrow x_1 = x_2. \tag{3.1}$$

3.1 *Remark.* We cannot consider $({}^*X, d)$ as a metric space, because values of d on ${}^*X \times {}^*X$ are hyperreals. Nevertheless ${}^*X$ is a **standard** topological space with the topology ${}^*\tau$ which is the star-version of the topology τ induced by the distance d on X. $({}^*X, d)$ is said to be a **hypermetric space**.

3.2 Definition. For any $x \in {}^*X$ form equivalency classes $h(x) := \{\xi \in X : \xi \approx x\}$, ${}_*h(x) := \{\xi \in {}^*X : \xi \overset{*}{\sim} x\}$, ${}_eh(x) = \{\xi \in {}^*X : \xi \overset{*}{\approx} x\}$. The sets $h(x)$ and ${}_*h(x)$ are called respectively the **halo** and the **R-halo** of x. ${}_eh(x)$ is the **extra halo**.

3.3 *Remark.* Of course, the map ${}^*X \ni x \mapsto {}_*h(x) \in 2^{*X}$ is standard. Therefore, if $x \in {}^{\mathrm{st}}({}^*X)$, then the R-halo ${}_*h(x)$ is standard. But for $x \in {}^*X \setminus X$ it is possible that $h(x) = \emptyset$. Notice that from (3.1) it follows that ${}_*h(x) \cap X = \{x\}$ whenever $x \in X$.

3.4 Definition. Denote by $\hat{X}$ the quotient of ${}^*X$ relative to $\overset{*}{\sim}$: $\hat{X} := {}^*X/\overset{*}{\sim} = \{\hat{x} : x \in {}^*X,\ \hat{x} = {}_*h(x)\}$. For $\hat{x}_1, \hat{x}_2 \in \hat{X}$, $\xi_1 \in \hat{x}_1$, $\xi_2 \in \hat{x}_2$ put $\hat{d}(\hat{x}_1, \hat{x}_2) := {}^\diamond d(\xi_1, \xi_2)$; here ${}^\diamond$ is the R-shadow map (see (2.6)). (The independence of $\hat{d}(\hat{x}_1, \hat{x}_2)$ from $\xi_1 \in \hat{x}_1$, $\xi_2 \in \hat{x}_2$ follows from (3.1).)

3.5 Proposition. *The function $\hat{d}$ is an* **almost distance**, *i.e. it has properties of a genuine distance except that its values are reals modulo* ${}^*\mathbb{N}$.

Proof. is an immediate check-up.

Warning. The quotient $X/\approx$ cannot be formed in the framework of IST although it has a clear naive sense.

3.6 *Remark.* The almost distance $\hat{d}$ defines on $\hat{X}$ a *standard separable* topology $\hat{\tau}$: a set $E \in 2^{\hat{X}}$ is $\hat{\tau}$-**open** iff for each $\hat{x} \in \hat{X}$ there exists a *real* $r > 0$ (at least infinitesimal) such that for $\hat{\xi} \in \hat{X}$ the inequality $\hat{d}(\hat{\xi}, \hat{x}) < r$ implies $\hat{\xi} \in E$. The standard topological space $(\hat{X}, \hat{\tau})$ is called the **ns-hull** of (X, d).

3.7 Definition. Let $f \in {}^*(X^X) = \tilde{W} \cap {}^*X^{{}^*X}$ and $x \in {}^*X$. The function f is said to be **R-continuous** at the point x iff

$$\forall \xi \in {}^*X \quad \xi \overset{*}{\sim} x \Rightarrow f(\xi) \overset{*}{\sim} f(x). \tag{3.2}$$

(Notice that it is well known that an R-standard function f is continuous at a R-standard point $x \in {}^*X$ iff it is R-continuous at this point.) Let $f \in {}^*(X^X)$ be R-continuous at each point $x \in {}^*X$. For $x \in {}^*X$ and $\xi \in \hat{x} := {}_*h(x)$ put

$$\hat{f}(\hat{x}) := \widehat{f(\xi)}. \tag{3.3}$$

It is clear that (3.3) determines $\hat{f}$ as a map $\hat{X}$ into $\hat{X}$. This map is called the **ns-hull** of f.

3.8 Proposition. *The ns-hull $\hat{f}$ of an R-continuous function $f \in {}^*(X^X)$ is continuous in respect to the topology $\hat{\tau}$.*

Proof. Let ε be a positive real. For $x \in {}^*X$ consider the set $E_{x,\varepsilon}$ of positive reals δ such that $\hat{d}(\hat{x}, \hat{\xi}) < \delta$ implies $\hat{d}(\hat{f}(\hat{\xi}), \hat{f}(\hat{x})) < \varepsilon$, i.e. from ${}^\diamond d(\xi, x) < \delta$ it follows that ${}^\diamond d(f(\xi), f(x)) < \varepsilon$. For (3.2) the set $E_{x,\varepsilon}$ contains each positive R-infinitesimal δ. By R-overspill it contains some positive real δ. ▷

Consider another variant of continuity. Let (X, d), (Y, d) be standard metric spaces such that $\mathbb{C} \cup X \cup Y \subseteq S$. For a function $f \in Y^X$ we have $f \subseteq {}^*f$, that is why we write f instead of ${}^*f$.

3.9 Proposition. *Suppose that* $f \in {}^{\mathrm{st}}(Y^X)$ *and*

$$\forall x, \xi \in {}^*X \qquad \xi \approx x \Rightarrow f(\xi) \overset{*}{\sim} f(x). \tag{3.4}$$

Then f is constant on each connected component of X.

Proof. For a fixed $x_0 \in {}^{\mathrm{st}}X$ denote $E := \{\xi \in X : f(\xi) = f(x_0)\}$. Then E is standard and closed. If $x, \xi \in X$ and $\xi \approx x$ then by (3.1) and (3.4) $f(\xi) = f(x)$. Therefore for each $x \in {}^{\mathrm{st}}E$ $h(x) \subset E$, hence E is open. ▷

3.10 Proposition. *Let* $f \in \tilde{W} \cap {}^*Y^{{}^*X}$, $x \in {}^*X$. *Suppose that* $\forall x \in {}^*X$ $\xi \overset{*}{\sim} x \Rightarrow f(\xi) \approx f(x)$. *Then f is continuous at the point x in the usual sense, i.e. for each positive real* ε *there exists a positive real* δ *such that for any* $\xi \in {}^*X$ *the inequality* $d(\xi, x) < \delta$ *implies* $d(f(\xi), f(x)) < \varepsilon$.

Proof. For a real $\varepsilon > 0$ put $E := \{r \in {}^*\mathbb{R} : r > 0 \text{ and } d(\xi, x) < r \Rightarrow d(f(\xi), f(x)) < \varepsilon\}$. By the principle of internal definition $E \in \tilde{W}$. Since $\{r \in {}_*\mathbb{I} : r > 0\} \subset E$, by the R-overspill E contains a ball of the space ${}^*X$ with center x and radius $\delta \mathrel{*{>}} 0$. ▷

4. Galaxies

Define some more equivalencies on ${}^*X$. For $x_1, x_2 \in {}^*X$ put $x_1 \overset{F}{\sim} x_2 \equiv d(x_1, x_2) \ll \infty$, $x_1 \overset{*F}{\sim} x_2 \equiv d(x_1, x_2) \mathrel{<*} \infty$, $x_1 \overset{eF}{\sim} x_2 \equiv d(x_1, x_2) < \infty$ and for each fixed $x \in {}^*X$ let $G(x) := \{\xi \in X : \xi \overset{F}{\sim} x\}$, ${}_*G(x) := \{\xi \in {}^*X : \xi \overset{*F}{\sim} x\}$, ${}_eG(x) := \{\xi \in {}^*X : \xi \overset{eF}{\sim} x\}$.

4.1 Definition. $G(x)$ and ${}_*G(x)$ are called respectively the **galaxy** and the **R-galaxy** of the point x. ${}_eG(x)$ is the **extra galaxy**.

4.2 *Remark.* The map $x \mapsto {}_*G(x)$ is standard. In particular, if $x \in {}^{\mathrm{st}}({}^*X)$, then ${}_*G(x)$ is a standard set. For $x \in X$ $G(x) \subset {}_*G(x)$ and ${}_*G(x) \cap X = X$. If $x \in {}^*X \backslash X$, then it is possible that $G(x) = \emptyset$. ▷

4.3 Definition. Let $\mathcal{G}$ be some R-galaxy. Define $\hat{\mathcal{G}}$ as the image of $\mathcal{G}$ under the quotient map ${}^*X \ni x \mapsto {}_*h(x) := \hat{x}$. It is easy to check that the restriction of the almost distance $\hat{d}$ to $\hat{\mathcal{G}}$ is a usuall distance: values of $\hat{d}_{|\hat{\mathcal{G}}}$ are reals. The metric space $(\hat{\mathcal{G}}, \hat{d})$ is called the **ns-hull** of the galaxy $\mathcal{G}$.

4.4 *Remark.* If $\mathcal{G} := {}_*G(x)$ for some standard $x \in {}^*X$, then the metric space $(\hat{\mathcal{G}}, \hat{d})$ is **standard**. This space is **complete** (by the saturation principle of NSA). It is easy to check that $\hat{\mathcal{G}}$ is an open and closed subspace of the topological space $(\hat{X}, \hat{\tau})$, i.e. $\hat{\mathcal{G}}$ is a connected component of $\hat{X}$.

4.5 Proposition. *Suppose that for any two points* x_1, x_2 *of a standard metric space* (X, d) *there exists an isometric map* $f \in X^X$, $\operatorname{im} f = X$, *such that* $f(x_1) =$

x_2. Then for any two R-galaxies $\mathcal{G}_1$, $\mathcal{G}_2$ their ns-hulls $(\hat{\mathcal{G}}_1, \hat{d})$, $(\hat{\mathcal{G}}_2, \hat{d})$ are isometric metric spaces.

Proof. Let $\mathcal{G}_1 = {}_*G(x_1)$, $\mathcal{G}_2 = {}_*G(x_2)$ and let f be an isometric map X on X such that $f(x_1) = x_2$. Since $f \subset {}^*f$, we write f instead of ${}^*f$. For the NSA-transfer principle (1.5) f is an isometry ${}^*X$ on ${}^*X$. Therefore f is R-continuous, hence has the ns-hull $\hat{f}$. For arbitrary $\xi_1, \xi_2 \in {}^*X$ we have $\hat{d}(\hat{f}(\hat{\xi}_1), \hat{f}(\hat{\xi}_2)) = {}^\diamond d(f(\xi_1), f(\xi_2)) = {}^\diamond d(\xi_1, \xi_2)$, whence

$$\forall \xi_1, \xi_2 \in {}^*X \quad \hat{d}(\hat{f}(\hat{\xi}_1), \hat{f}(\hat{\xi}_2)) = \hat{d}(\hat{\xi}_1, \hat{\xi}_2). \tag{4.1}$$

In particular, the restriction of $\hat{f}$ to an arbitrary ns-hull $\mathcal{G}$ is an isometric map $\mathcal{G}$ on $\mathcal{G}$. Let $\xi \in \mathcal{G}_1$, then by (4.1) $\hat{d}(\hat{x}_2, \hat{f}(\hat{\xi})) = \hat{d}(\hat{f}(\hat{x}_1), \hat{f}(\hat{\xi})) = \hat{d}(\hat{x}_1, \hat{\xi})$, i.e. $\hat{d}(\hat{x}_2, \hat{f}(\xi)) \in \mathbb{R}$. We see that $\forall \hat{\xi} \in \hat{\mathcal{G}}_1 \ \hat{f}(\hat{\xi}) \in \hat{\mathcal{G}}_2$. Evidently $\hat{f}^{-1}$ exists and is an isometric map $\hat{\mathcal{G}}_2$ on $\hat{\mathcal{G}}_1$. ▷

4.6 *Corollary.* For a standard normed space X and points $x_1, x_2 \in X$ put $\forall \xi \in X$ $f(\xi) = \xi + x_2 - x_1$. The shift f is an isometry X onto X such that $f(x_1) = x_2$. Therefore the ns-hulls $\hat{\mathcal{G}}_1$, $\hat{\mathcal{G}}_2$ of arbitrary galaxies $\mathcal{G}_1$, $\mathcal{G}_2$ of the space X are isometric. Thus it is sufficient to investigate the **principal** galaxy $\mathcal{G} = {}_*G(x)$, $x \in X$, only. ▷

5. A standard R-infinite-dimensional Hilbert space

Let m be a fixed standard R-infinite number: $m \in {}^{\mathrm{st}}({}^*\mathbb{N} \setminus \mathbb{N})$. Denote by $\mathbb{C}^m$ the linear space of complex valued functions x defined on $m = \{0, 1, \ldots, m-1\}$. Arithmetical operations in $\mathbb{C}^m$ are defined pointwise.

5.1 Definition. A function $x \in \mathbb{C}^m$ is said to be **integrable** iff for any real $\varepsilon > 0$ there exists a finite set $\Delta \subset m$ (obviously, "ε is real" means "$\varepsilon \in \mathbb{R}$", and "$\Delta$ is finite" means "card$\Delta \in \mathbb{N}$") such that for arbitrary finite $E_1, E_2 \subset m$ we have

$$\Delta \subset E_1 \cap E_2 \Rightarrow \left| \sum\nolimits_{t \in E_1} |x(t)| - \sum\nolimits_{t \in E_2} |x(t)| \right| < \varepsilon. \tag{5.1}$$

The subspace of integrable functions $x \in \mathbb{C}^m$ is denoted by $\ell_1(m)$ or shortly by ℓ_1.

5.2 *Remark.* For $x \in \mathbb{C}^m$ put $\operatorname{supp} x := \{t \in m : \ x(t) \neq 0\}$. The support $\operatorname{supp} x$ of $x \in \ell_1$ is **enumerable**: $\forall x \in \ell_1$ $\operatorname{card}\operatorname{supp} x \leq \operatorname{card}\mathbb{N}$. Indeed, if $\operatorname{card}\operatorname{supp} x > \operatorname{card}\mathbb{N}$, then there exists a real $\varepsilon > 0$ such that $\operatorname{card}\{t \in m : \ |x(t)| > \varepsilon\} > \operatorname{card}\mathbb{N}$. Thus x does not satisfy condition (5.1). ▷

5.3 Definition. For $x \in \ell_1$ by $\int x$ denote the sum of the absolutely convergent series $\sum_{t \in \operatorname{supp} x} x(t)$.

5.4 *Remark.* If $x \in \ell_1$, then for any real $\varepsilon > 0$ there is a finite set $\Delta \subset m$ such that for any finite set $E \subset m$ $\Delta \subset E \Rightarrow \left| \int x - \sum_{t \in E} x(t) \right| < \varepsilon$. Notice that the integral $\int x$ is a linear positive functional on ℓ_1, bounded relative to the norm $\|x\|_1 := \int |x| = \sum_{t \in \operatorname{supp} x} |x(t)|$.

5.5 Definition. By $\ell_2(m)$ or shortly by ℓ_2 we denote the space of square integrable functions $x \in \mathbb{C}^m$ with the inner product

$$(x|y) := \int x\overline{y} = \sum\nolimits_{t<m} x(t)\overline{y(t)}. \tag{5.2}$$

5.6 *Remark.* ℓ_2 is a standard nonseparable (complete) Hilbert space. The Kronecker deltas $(\delta_s)_{s \in m}$ form an orthonormal basis of ℓ_2: $\forall x \in \ell_2 \ x = \sum_{s \in m} x(s)\delta_s$.

There is a natural standard decomposition of ℓ_2 as an orthogonal sum of its separable subspaces. At first let us decompose m.

5.7 Definition. The restriction of the equivalence $\overset{*F}{\sim}$ to $m := \{0, 1, \dots, m-1\}$ determines the quotient set $\mu := m/\overset{*F}{\sim}$. For each $m' \in \mu$ put $\ell_2(m') = \{x \in \ell_2 : \operatorname{supp} x \subset m'\}$.

5.8 Proposition. *For each $m' \in \mu$ $\ell_2(m')$ is a closed separable (i.e. $\ell_2(m')$ has orthonormal bases of cardinality* $\operatorname{card} \mathbb{N}$, *for instance $(\delta_s)_{s \in m'}$) subspace of ℓ_2 and $\ell_2 = \ell_2(m) = \bigoplus_{m' \in \mu} \ell_2(m')$.*

Proof. For $\nu \in m$ let m_ν denote the equivalence class $m_\nu := \{t \in m : t \overset{*F}{\sim} \nu\}$. Observe that $m_0 = \mathbb{N}$, $m_{m-1} = (m-1) - \mathbb{N}$ and for $\nu \in m \backslash (m_0 \cup m_{m-1})$ $m_\nu = \nu + \mathbb{Z}$. This proves our assertion. ▷

6. Shift on $\ell_2(m)$

Denote by $\mathfrak{Z}$ the shift operator defined on $\ell_2(\mathbb{Z})$ by $\forall x \in \ell_2(\mathbb{Z})\ \forall t \in \mathbb{Z}\ \mathfrak{Z}x(t) = x(t+1)$. Notice that $\mathfrak{Z}$ is unitary. For $x \in \ell_2(\mathbb{Z})$ and $0 \le \varphi < 2\pi$ put $\hat{x}(\varphi) = \sum_{t \in \mathbb{Z}} x(t)e^{-it\varphi}$. Then $\forall t \in \mathbb{Z}\ x(t) = \frac{1}{2\pi}\int_0^{2\pi} \hat{x}(\varphi)e^{it\varphi}\,d\varphi$ and $\widehat{\mathfrak{Z}x}(\varphi) = e^{i\varphi}\hat{x}(\varphi)$. Thus the spectrum of $\mathfrak{Z}$ is the circle $\{e^{i\varphi} : 0 \le \varphi < 2\pi\}$.

Now let $\mathfrak{Y}$ be the shift operator of the space $\ell_2(m)$ (notation as in section 5) with the following **periodicity condition**: $\forall x \in \ell_2(m)$

$$\forall t \in m \qquad \mathfrak{Y}x(t) = \begin{cases} x(t+1) & \text{for } t < m-1, \\ x(0) & \text{for } t = m-1. \end{cases} \tag{6.1}$$

6.1 Proposition. *The operator $\mathfrak{Y}$ is unitarily equivalent to a (nonenumerable) orthogonal sum of copies $\mathfrak{Z}_{m'}$, $m' \in \mu = m/\overset{*F}{\sim}$, of the operator $\mathfrak{Z}$: $\mathfrak{Y} \sim \bigoplus_{m' \in m} \mathfrak{Z}_{m'}$. In particular, the spectrum of $\mathfrak{Y}$ is purely continuous and fills in the circle $\{e^{i\varphi} : 0 \le \varphi < 2\pi\}$ with the multiplicity* $\operatorname{card} \mu$.

Proof. Let $m' \in \mu \setminus (m_0 \cup m_{m-1})$. Then the subspace $\ell_2(m')$ of $\ell_2(m)$ is invariant under $\mathfrak{Y}$. Denote $\mathfrak{Y}' = \mathfrak{Y}_{|\ell_2(m')}$ and for $\nu \in m'$, $t \in \mathbb{Z}$, $x \in \ell_2(\mathbb{Z})$ put $\mathfrak{Z}^\nu x(t) = x(t+\nu)$. Then $\mathfrak{Z}^\nu$ is a unitary map $\ell_2(\mathbb{Z}) \to \ell_2(m')$ which transforms $\mathfrak{Z}$ in $\mathfrak{Y}'$: $\mathfrak{Z}^\nu \mathfrak{Z} = \mathfrak{Y}'\mathfrak{Z}^\nu$.

Now let $m'' = m_0 \cup m_{m-1}$ and let $\mathfrak{Y}'' = \mathfrak{Y}_{\ell_2(m'')}$. The map $t \mapsto m-1+t$ transforms $\mathbb{Z}_- := \{t \in \mathbb{Z} : t < 0\}$ onto m_{m-1}. Observe that $m_0 = \mathbb{N}$. Therefore the operator $\mathfrak{Z}_0 := \mathbb{I}_{\ell_2(m_0)} + \mathfrak{Z}^{m-1}$ is a unitary map $\ell_2(\mathbb{Z}) \to \ell_2(m'')$ which transforms $\mathfrak{Z}$ in $\mathfrak{Y}''$: $\mathfrak{Z}_0\mathfrak{Z} = \mathfrak{Y}''\mathfrak{Z}_0$. Because $\mathfrak{Y} = (\oplus\mathfrak{Y}') \oplus \mathfrak{Y}''$, our assertion is proven. ▷

6.2 *Remark.* Another point of view on the shift $\mathfrak{Y}$ defined by (6.1) is natural, too. Consider $\ell_2(m)$ as a Hilbert space (not over $\mathbb{C}$ but) over the field $^*\mathbb{C}$ of hypercomplex numbers. Now the sum in $(x|y) = \sum_{t\in m} x(t)\overline{y(t)}$ (see (5.2)) is simply the star-version of the sum of a finite set of summands. Under this treatment $\mathfrak{Y}$ is also standard, unitary and its spectrum consists of eigenvalues $\rho_0, \ldots, \rho_{m-1}$ which are roots of the equation $\rho^m = 1$. The normed eigenfunction x which corresponds to the eigenvalue ρ is the following: $\forall t \in m\ x(t) = \frac{1}{\sqrt{m}}\rho^t$.

7. Shift and Loeb measure

Consider another approach to the same subject. Let as before $m \in {}^{\mathrm{st}}({}^*\mathbb{N} \setminus \mathbb{N})$. Denote by ℓ the normed counting measure defined on the R-finite algebra $\mathfrak{A}_0 = \tilde{W} \cap 2^m$ by $\forall A \in \mathfrak{A}_0\ \ell A = \frac{1}{m}({}^*\mathrm{card})A$; notice that $\ell A \in {}^*\mathbb{Q}$ and $\ell m = 1$. To the R-finite measure space $(m, \mathfrak{A}_0, \ell)$ there corresponds the standard Loeb measure space $(m, \mathfrak{A}, \mathcal{L})$.

The starting condition is $\mathfrak{A}_0 \subset \mathfrak{A}$; $\forall A \in \mathfrak{A}_0\ \mathcal{L}A = {}^\diamond\ell A$. The σ-algebra $\mathfrak{A}$ and the σ-additive measure $\mathcal{L}$ are obtained by the usual standard procedure of extension. The relation between $\mathcal{L}-$ and $\ell-$ integrals is as follows. A function $\xi \in \tilde{W} \cap ({}^*\mathbb{C})^m$ is said to be S-integrable iff $\int_m |\xi|\, d\ell := \frac{1}{m}\sum_{t\in m} |\xi(t)| < *\infty$ and for $n \in {}^*\mathbb{N} \setminus \mathbb{N}$ $\frac{1}{m}\sum_{|\xi(t)|>n} |\xi(t)| \overset{*}{\sim} 0$. A function $\xi \in \tilde{W} \cap ({}^*\mathbb{C})^m$ is said to be a lifting for a function $x \in ({}^*\mathbb{C})^m$ iff $x(t) = {}^\diamond\xi(t)$ $\mathcal{L}$-a.e. It is proven that a function $x \in ({}^*\mathbb{C})^m$ is ($\mathcal{L}$-measurable and) $\mathcal{L}$-integrable iff it has an S-integrable lifting. If it is so, then $\int_m x\, d\mathcal{L} = {}^\diamond\frac{1}{m}\sum_{t\in m} |\xi(t)|$.

After this introduction denote by H the Hilbert space $L_2(m, \mathfrak{A}, \mathcal{L})$ with the scalar product $\forall x, y \in H\ (x|y) = \int_m x(t)\overline{y(t)}\, \mathcal{L}(dt)$. If ξ and η are square S-integrable liftings for x and y, then $(x|y) = {}^\diamond(\xi|\eta)$, where $(\xi|\eta) = \frac{1}{m}\sum_{t\in m} \xi(t)\overline{\eta(t)}$. Now define the shift $\mathfrak{Y}$ on H by $\forall x \in H\ \forall t \in (m-1)\ \mathfrak{Y}x(t) = x(t+1)$; this is meaningless for $t = m-1$, but never mind: the Loeb measure of an one-point set is: $\mathcal{L}\{t\} = {}^\diamond\frac{1}{m} = 0$. Moreover, because $\mathcal{L}$ is σ-additive, we have $\mathcal{L}\mathbb{N} = \mathcal{L}(m - \mathbb{N}) = \mathcal{L}(t + \mathbb{Z}) = 0$. Besides, define on the linear space $H_0 = \tilde{W} \cap ({}^*\mathbb{C})^m$ the shift $\mathfrak{Y}_0$ by $\forall t \in (m-1)\ \mathfrak{Y}_0\xi(t) = \xi(t+1)$ and $\mathfrak{Y}_0\xi(m-1) = \xi(0)$. Let $\xi \in H_0$ be a lifting for $x \in H$. Then $\mathfrak{Y}_0\xi$ is a lifting for $\mathfrak{Y}x$, i.e. $\mathfrak{Y}x(t) = {}^\diamond\mathfrak{Y}_0\xi(t)$ $\mathcal{L}$-a.e. We see that

$\forall x \in H\ \mathfrak{Y}x \in H$. Because $\mathfrak{Y}_0$ is a unitary map of H_0 relative to scalar product (6.1), $\mathfrak{Y}$ is a unitary map $H \to H$. Denote

$$_*\sigma := \{\rho \in {}^*\mathbb{C} :\ \rho^m = 1\}, \quad \sigma := \{r \in \mathbb{C} :\ |r| = 1\}.$$

7.1 Proposition. *The circle σ is the spectrum of the shift $\mathfrak{Y}$ in $H = L_2(m, \mathfrak{A}, \mathcal{L})$. For $\rho \in {}_*\sigma$ denote $\forall t \in m\ x_\rho(t) := {}^\diamond(\rho^t)$. Then x_ρ is an eigenfunction of $\mathfrak{Y}$ corresponding to the eigenvalue $r = {}^\diamond\rho$. The family $(x_\rho)_{\rho\in{}_*\sigma}$ is complete in H.*

Proof. The set $_*\sigma$ is the spectrum of the shift $\mathfrak{Y}_0$ in $\tilde{W} \cap ({}^*\mathbb{C}^m)$ and for $\rho \in {}_*\sigma$ the function ξ_ρ: $\forall t \in m\ \xi_\rho(t) = \rho^m$ is an eigenfunction of $\mathfrak{Y}_0$ corresponding to the eigenvalue ρ. Therefore $\mathfrak{Y}x_\rho(t) = {}^\diamond\mathfrak{Y}_0\xi_\rho(t) = {}^\diamond(\rho\xi_\rho(t)) = rx_\rho(t)$, where $r = {}^\diamond\rho$. Suppose that $h \in H$ and $\forall r \in \sigma\ (h|x_\rho) = 0$. Because the set m, the map ${}^*\mathbb{C} \ni z \mapsto {}^\diamond z \in \mathbb{C}$, and the set $_*\sigma$ are standard, the family $(x_\rho)_{\rho\in{}_*\sigma}$ is standard, too. Therefore it is sufficient to prove that $h = 0$ assuming $h \in {}^{\text{st}}H$. But $h \in {}^{\text{st}}H$ has a lifting $g \in {}^{\text{st}}H_0$. We have $(g|\xi_\rho) \overset{*}{\sim} (h|x_\rho) = 0$. Hence $\forall \rho \in {}^{\text{st}}{}_*\sigma\ (g|\xi_\rho) = 0$. By transfer, $\forall x \in {}_*\sigma\ (g|\xi_\rho) = 0$, whence $g = 0$ and $h = 0$. ▷

References

1. S. Albeverio, J. Fenstad, R. Høegh-Krohn, T. Lindstrøm, *Nonstandard methods in stochastic analysis and mathematical physics*. Orlando, Academic Press, 1986.
2. M. Davis, *Applied Nonstandard Analysis*. Wiley, New York, 1977.
3. E. Gordon, Relatively standard elements in E.Nelson's Internal Set Theory, *Sib. Math. J.* **30** (1) (1989) 89–95 (Russian).
4. R. Lutz, M. Goze, Nonstandard Analysis: a practical guide with applications, *Lect. Notes in Math.* **881** (1981).
5. W. Lyantse, T. Kudryk, *Introduction to Nonstandard Analysis*. Mathematical Studies, Monograph Series, VNTL Publishers, L'viv, **3**, 1997.
6. E. Nelson, A new approach to nonstandard analysis, *Bull. Amer. Math. Soc.* **83** (6) (1977) 1165–1198.
7. Y. Peraire, Un nouvelle théorie des infinitésimaux, *C.R.Acad.Sc.Paris (Serie A)* **301** (5) (1985) 157-159.
8. Y. Peraire, Théorie relative des ensembles internes, *Osaka J. Math.* **29** (1992) 267-297.
9. A. Robinson, *Non-Standard Analysis*. North-Holland, Amsterdam, 1966.
10. V. Kanovei, M. Reeken, Internal approach to external sets and universes, parts 1-3, *Studia Logica*, **55** (2) (1995) 229-257; **55** (3) (1995) 347-376; **56** (3) (1996) 293-322.
11. V. Kanovei, M. Reeken, Mathematics in a nonstandard world.II, *Math. Japon.* **45** (3) (1997) 555-571.
12. E. Gordon, A. Kusraev and S. Kutateladze, *Infinitesimal Analysis*. Kluwer, 2002.

V. Kadets and W. Żelazko

The Beurling Theorem for Entire Functions of Finite Order

Konstantin G. Malyutin

Deparment of the mathematics, Ukrainian Academy of Banking, Sumy, Ukraine

Nazim Sadyk

Department of mathematics, Istanbul state university, Turkey

Abstract

We extend the result of Beurling on the closure in H^p of the linear manifold $F(z) \cdot \{polynomials\ of\ z\}$ to the classes of entire functions of finite order.

Key words: Beurling theorem, entire functions of finite order
2000 MSC: 30D20, 30D55

The following theorem (for $p = 2$ proven by Beurling, the general case was considered by T. Srinivasan and J.K. Wang [4]) is well known.

Beurling theorem. *Let $F = I_F Q_F \in H^p$, $p > 0$, where I_F is the inner function of F. Then the closure in H^p of the linear manifold $F(z) \cdot \{polynomials\ of\ z\}$ is $I_F \cdot H^p$.*

In this paper we generalize the Beurling theorem to the classes of entire functions of finite order.

Denote by E the set of all entire functions on the complex plane $\mathbb{C}$. For real constants $\rho > 0$, $\sigma > 0$, we define the Banach space

$$E_\rho^\sigma := \{f : f \in E, ||f||_{\rho,\sigma} = \sup_{z \in \mathbb{C}} |f(z)| \exp(-\sigma|z|^\rho) < \infty\}$$

and set $E_\rho^\infty = \bigcup_{\sigma>0} E_\rho^\sigma$. The set E_ρ^∞ is a linear locally convex space with the topology of inductive limit. The sequence of entire functions $\{f_n\}$ converges in the space E_ρ^∞

Email address: kgm@academy.sumy.ua (Konstantin G. Malyutin).

as $n \to \infty$ to a function f if and only if $\{f_n\}$ converges to a function f uniformly on every compact subset of $\mathbb{C}$ and there exists $K > 0$ such that $|f_n(z)| \le \exp(K|z|^\rho)$ for all $z \in \mathbb{C}$. The space E_ρ^∞ is a topological algebra with respect to the product of the functions.

Linear continuous functional l on the space E_ρ^∞ is such that

$$l(f) = \sum_{k=0}^{\infty} c_k b_k, \quad f(z) = \sum_{k=0}^{\infty} b_k z^k, \quad f \in E_\rho^\infty,$$

where $\lim\limits_{k\to\infty} k^{-1/\rho}|c_k|^{1/k} = 0$ [2].

Lemma. *Let $f = \in E_\rho^\infty$. Then there exists a sequence of polynomials $P_n(z)$ such that $P_n(z)$ converge in E_ρ^∞ as $n \to \infty$ to f.*

Proof. Let $f(z) = \sum\limits_{k=0}^{\infty} b_k z^k$. Then we set

$$P_n(z) = \sum_{k=0}^{n} b_k z^k .$$

Let l be a linear continuous functional on E_ρ^∞. Then $l(f - P_n) = \sum\limits_{k=n+1}^{\infty} c_k b_k$ converge as $n \to \infty$ to zero.

Let f be any entire function, let $\{a_n\}$ be the set of all zeros of the function f, and let $I_f(z)$ be canonical product of the set $\{a_n\}$ (if the set $\{a_n\}$ is empty then $I_f(z) \equiv 1$) [3]. We shall say that $I_f(z)$ is the interior factor for f. The function $Q_f(z) = f(z)/I_f(z)$ is called the exterior factor of f.

Theorem. *Let $f = I_f Q_f \in E_\rho^\infty$. Then the closure in E_ρ^∞ of the linear manifold $f(z) \cdot \{polynomials\ of\ z\}$ is $I_f \cdot E_\rho^\infty$.*

Proof. a) $I_f \cdot E_\rho^\infty$ includes this closure. Let $g \in E_\rho^\infty$, and let $\{P_n(z)\}$ be the sequence of polynomials such that fP_n converge in E_ρ^∞ as $n \to \infty$ to g. Let a_k be a zero of the function $f(z)$. Since fP_n converges to g uniformly on every compact set, the $f(a_k)P_n(a_k)$ converges as $n \to \infty$ to $g(a_k)$. Since we have $f(a_k)P_n(a_k) = 0$ for all $k = 1, 2, \ldots$, $g(a_k) = 0$, $k = 1, 2, \ldots$, follows then. We just prove $G := g/I_f \in E_\rho^\infty$.

The following argument we shall call "standard argument". Let $C(\xi, \rho)$ be a disc of radius ρ about ξ. Let $\{C(\xi_n, \rho_n)\}$ be a sequence of discs. The number

$$L = \limsup_{r\to\infty} \frac{1}{r} \sum_{|\xi_n| \le r} \rho_n \tag{1}$$

is called the upper density of the set $\bigcup\limits_{n=1}^{\infty} C(\xi_n, \rho_n)$ [3].

If $f \in E_\rho^\infty$ then for all $\varepsilon > 0$ there exists $K_\varepsilon > 0$ such that

$$\ln|f(re^{i\theta})| \geq -K_\varepsilon r^\rho ,$$

for all $re^{i\theta} \notin C_\varepsilon$, where C_ε is a set of upper density ε and K_ε is the constant dependent on ε [1].

Thus, there exists a set C_ε of upper density ε such that

$$\ln|G(re^{i\theta})| \leq K_\varepsilon r^\rho \tag{2}$$

for all $re^{i\theta} \notin C_\varepsilon$.

Let $z = re^{i\theta} \in C_\varepsilon$ and let ρ_n be the radius of the disc centered at ξ_n, denoted $C(\xi_n, \rho_n)$, such that $z \in C(\xi_n, \rho_n)$. It follows from (1) that $\rho_n \leq \frac{1+\varepsilon}{1-\varepsilon} r$. By the maximum modulus principle the relation (2) is true (probably with other constant) for all $z \in \mathbb{C}$.

b) The closure of $f(z) \cdot \{\text{polynomials of } z\}$ includes $I_f \cdot E_\rho^\infty$. It is equivalent to prove that the linear manifold $Q_f(z) \cdot \{\text{polynomials of } z\}$ is dense set in E_ρ^∞. Really, if $g \in E_\rho^\infty$ and $fP_n \to g$ in E_ρ^∞ then $g = I_f G$ as in a). We prove that G belongs to the closure in E_ρ^∞ of the manifold $Q_f(z) \cdot \{\text{polynomials of } z\}$. There exists $K > 0$ such that $|f(z)P_n(z)| \leq \exp[K_1 r^\rho]$ for all $z = re^{i\theta}$. By "standard argument", we find $K_1 > 0$ such that

$$|Q_f(z)P_n(z)| \leq \exp[K_1 r^\rho] .$$

By Stieltyes-Vitali theorem, the sequence $\{Q_f P_n\}$ converges uniformly on every compact set, it is clear to the function G and $G \in E_\rho^\infty$.

Let g be any function of E_ρ^∞. Then $g/G_f \in E_\rho^\infty$. Let $\{P_n(z)\}$ be the sequence of polynomials converges in E_ρ^∞ to $g(z)$. Then $Q_f P_n$ convergent to $Q_f g$.

Acknowledgements. The authors would like to thank the Referee for pointing out mistakes in earlier versions of the paper.

References

1. A.F. Grishin, Continuity and asymptotic continuity of subharmonic functions, *Mat. Fiz., Anal., Geom.* **1** (1994) 193-215 (Russian).
2. A.F. Leontyev, *Generalization of series of exponents.* Nauka, Moscow, 1981 (Russian).
3. B.Ya. Levin, *Distribution of zeros of entire functions.* GITTL, Moscow, 1956 (Russian). Revised English transl.: Amer. Math. Soc., Providence, Rh. I., 1980.
4. T. Srinivasan, J.K. Wang, On closed ideals of analytic functions, *Proc. Amer. Math. Soc.* **16** (1965) 49-52.

V. Kadets and W. Żelazko

Dvoretzky's Theorem by Gaussian Method

Ivan Matsak [a] **and Anatolij Plichko** [b]

[a]Kyiv State University of Technology and Design,
2 Str. Nemirovich-Danchenko, Kyiv 04011 Ukraine
[b]Institute of Mathematics, Cracow University of Technology, Cracow, Poland

Abstract
A complete proof of the Dvoretzky theorem, accessible to graduate students, is given.

Key words: Dvoretzky theorem, Gaussian random variables, Gaussian measures
2000 MSC: 46B20, 46B09, 28C20, 46G12

In this note we give a complete proof of the well known Dvoretzky theorem on the almost spherical (or rather ellipsoidal) sections of convex bodies. Our proof follows Pisier [18],[19]. It is accessible to graduate students. In the references we list papers containing other proofs of Dvoretzky's theorem.

1. Gaussian random variables

Let ξ be a real valued random variable (r.v.) on a probability space $(\Omega, \mathcal{B}, \mathbf{P})$. Put $\mathbf{E}\xi := \int_\Omega \xi(\omega) d\mathbf{P}$. If $\xi \geq 0$, then $\mathbf{E}\xi = \int_0^\infty \mathbf{P}\{\xi \geq t\}dt = \int_0^\infty (1 - \mathbf{P}\{\xi < t\})dt$ [6], 15.6.

Definition 1. An r.v. g with a distribution function

$$\mathbf{P}\{g < t\} = \frac{1}{\sqrt{2\pi}} \int_{-\infty}^{t} \exp(-\frac{u^2}{2})du \ , \ \ t \in \mathbb{R} \ ,$$

is called *a standard Gaussian r.v.* Its *characteristic function* is (see f.e. [3], 17.2)

$$\phi(t) := \mathbf{E}\exp(itg) = \exp(-\frac{t^2}{2}) \ .$$

Email addresses: `infor1@valtek.kiev.ua` (Ivan Matsak), `aplichko@kspu.kr.ua` (Anatolij Plichko).

Proposition 1. *Let $(g_j)_1^n$ be independent standard Gaussian r.v. and $\sum_1^n a_j^2 = 1$. Then $\sum_1^n a_j g_j$ is a standard Gaussian r.v. as well.*

Proof. Let us find the characteristic function of r.v. $\sum_1^n a_j g_j$. As is known ([3], 15.12), the characteristic function of a sum of independent r.v. equals to the product of the characteristic functions of summands. So

$$\mathbf{E}\exp(it\sum_1^n a_j g_j) = \prod_{j=1}^n \mathbf{E}\exp(ita_j g_j) =$$

$$\prod_1^n \exp(-\frac{a_j^2 t^2}{2}) = \exp(-\frac{t^2}{2}\sum_1^n a_j^2) = \exp(-\frac{t^2}{2})$$

i.e we obtain the characteristic function of standard Gaussian r.v. However, the distribution of an r.v. is uniquely determined by its characteristic function [3], 10.3. Hence, $\sum_1^n a_j g_j$ is a standard Gaussian r.v. □

Proposition 2. *There exists a constant $c > 0$ such that for any independent standard Gaussian r.v. $(g_j)_1^n$*

$$\mathbf{E}\max_{j\le n}|g_j| \ge c(\ln n)^{1/2}.$$

Proof. By definition,

$$\mathbf{E}\max_{j\le n}|g_j| = \int_0^\infty [1 - \mathbf{P}\{\max_{j\le n}|g_j| < t\}]dt =$$

$$\int_0^\infty [1 - \mathbf{P}\{|g_1| < t, \ldots, |g_n| < t\}]dt =$$

(since g_j are independent and identically distributed) =

$$\int_0^\infty [1 - (\mathbf{P}\{|g| < t\})^n]dt =$$

$$\int_0^\infty [1 - (1 - \mathbf{P}\{|g| > t\})^n]dt \ge (\text{for any } a > 0) \ge$$

$$\int_0^a [1 - (1 - \mathbf{P}\{|g| > t\})^n]dt \ge a[1 - (1 - \mathbf{P}\{|g| > a\})^n].$$

Now

$$\mathbf{P}\{|g| > a\} = \sqrt{\frac{2}{\pi}}\int_a^\infty \exp(-\frac{t^2}{2})dt \ge$$

$$\sqrt{\frac{2}{\pi}}\int_a^{a+1} \exp(-\frac{t^2}{2})dt \ge \sqrt{\frac{2}{\pi}}\exp(-\frac{(a+1)^2}{2}) .$$

Take $a = (\ln n)^{1/2}$. Then for sufficiently large n

$$\mathbf{P}\{|g| > a\} > \sqrt{\frac{2}{\pi}}\exp(-\frac{((\ln n)^{1/2}+1)^2}{\sqrt{2}\cdot\sqrt{2}}) \ge \sqrt{\frac{2}{\pi}}\exp(-\frac{\ln n}{\sqrt{2}}) \ge \frac{1}{n} .$$

Hence, for some $c > 0$

$$\mathbf{E} \max_{j \le n} |g_j| > (\ln n)^{1/2}[1 - (1 - \frac{1}{n})^n] \ge (\ln n)^{1/2} \cdot (1 - \frac{1}{e}) \ge c \cdot (\ln n)^{1/2}$$

for sufficiently large n . □

2. Standard Gaussian random vectors and Gaussian measures

A random vector $\overrightarrow{g} = (g_j)_1^n$ with independent standard Gaussian components g_j we call *a standard Gaussian random vector.*

Theorem 1. *Let $\overrightarrow{g}$ be a standard Gaussian random vector and let U be an orthogonal matrix in $\mathbb{R}^n$. Then $U\overrightarrow{g}$ is a standard Gaussian random vector as well.*

Proof. Let $\phi(\overrightarrow{t}) := \mathbf{E}\exp(i\langle \overrightarrow{t}, \overrightarrow{g}\rangle) = \exp(-\frac{1}{2}\sum_{j=1}^n t_j^2)$ be the *characteristic function* of $\overrightarrow{g}$. Let us find the characteristic function $\phi_U(\overrightarrow{t})$ of $U\overrightarrow{g}$

$$\phi_U(\overrightarrow{t}) = \mathbf{E}\exp(i\langle \overrightarrow{t}, U\overrightarrow{g}\rangle) = \mathbf{E}\exp(i\langle U^*\overrightarrow{t}, \overrightarrow{g}\rangle) =$$
$$= \phi(U^*\overrightarrow{t}) = \exp(-\frac{1}{2}\langle U^*\overrightarrow{t}, U^*\overrightarrow{t}\rangle) = \exp(-\frac{1}{2}\langle UU^*\overrightarrow{t}, \overrightarrow{t}\rangle) .$$

However, U is an orthogonal matrix, so $UU^* = I$. Hence,

$$\phi_U(\overrightarrow{t}) = \phi(\overrightarrow{t}) = \exp(-\frac{1}{2}\sum_{j=1}^n t_j^2) .$$

But the distribution of a random vector is uniquely determined by its characteristic function ([3], 10.6). Therefore, $U(\overrightarrow{g})$ is a standard Gaussian random vector. □

Definition 2. A measure $\gamma(A) := \mathbf{P}\{g \in A\}$ ($A \subset \mathbb{R}$ is Borel) is called a *Gaussian measure* (generated by a standard Gaussian r.v. g).

Of course, different standard Gaussian r.v. generate the same Gaussian measure γ.

A Gaussian random vector $\overrightarrow{g}$ generates a *Gaussian measure* $\gamma_n(A) = \mathbf{P}\{\overrightarrow{g} \in A\}$ in $\mathbb{R}^n$. The fundamental property of Gaussian measure is its invariance under rotations. More exactly, Theorem 1 implies

Corollary. *Let a Gaussian random vector $\overrightarrow{g}$ generate a Gaussian measure γ_n and let U be an orthogonal matrix in $\mathbb{R}^n$. Then $U\overrightarrow{g}$ generates the same measure γ_n.*

Lemma 1. $\int_{\mathbb{R}} \exp(at)d\gamma(t) = \exp\frac{a^2}{2}$.

Proof.

$$\int_{\mathbb{R}} \exp(at) d\gamma(t) = \int_{\mathbb{R}} \exp(at) \frac{1}{\sqrt{2\pi}} \exp(-\frac{t^2}{2}) dt =$$

$$\frac{1}{\sqrt{2\pi}} \int_{\mathbb{R}} \exp(\frac{a^2 - (t-a)^2}{2}) dt = \exp(\frac{a^2}{2}) \frac{1}{\sqrt{2\pi}} \int_{\mathbb{R}} \exp(-\frac{t^2}{2}) dt = \exp\frac{a^2}{2} . \square$$

Proposition 3. *Let* $\overrightarrow{f} = (f_1, \dots, f_n) \in \mathbb{R}^n$. *Then*

$$\int_{\mathbb{R}^n} \exp\langle \overrightarrow{f}, \overrightarrow{t} \rangle d\gamma_n(\overrightarrow{t}) = \exp(\frac{1}{2} \| \overrightarrow{f} \|_2^2) .$$

Proof. Let $\overrightarrow{t} = (t_1, \dots, t_n)$. Then

$$\int_{\mathbb{R}^n} \exp\langle \overrightarrow{f}, \overrightarrow{t} \rangle d\gamma_n(\overrightarrow{t}) = \int_{\mathbb{R}^n} \exp(\sum_1^n f_i t_i) d\gamma_n(\overrightarrow{t}) =$$

$$\text{(by the Fubini theorem)} == \prod_1^n \int_{\mathbb{R}} \exp(f_i t_i) d\gamma(t_i) =$$

$$\text{(by Lemma 1)} = \prod_1^n \exp(\frac{1}{2} f_i^2) = \exp(\frac{1}{2} \| \overrightarrow{f} \|_2^2) . \square$$

3. Some mathematical analysis

Definition 3. A vector function $x(\theta) = (x_1(\theta), \dots, x_n(\theta))$ from $\mathbb{R}$ to $\mathbb{R}^n$ is said to be *differentiable in a point* θ if there are usual derivatives $x_1'(\theta), \dots, x_n'(\theta)$. A vector function $x'(\theta) = (x_1'(\theta), \dots, x_n'(\theta))$ is called a *derivative* of $x(\theta)$. A mapping $F(x)$ from $\mathbb{R}^n$ to $\mathbb{R}$ is called *differentiable in a point* x if there exist a linear functional (denote it by $F'(x)$ and its using to $y \in \mathbb{R}^n$ by $\langle F'(x), y \rangle$) such that

$$F(y) - F(x) = \langle F'(x), y - x \rangle + o(\|y - x\|)$$

if $\|y - x\| \to 0$.

Proposition 4. *Let a vector function* $x(\theta)$ *be differentiable in a point* θ *and let a map* $F : \mathbb{R}^n \to \mathbb{R}$ *be differentiable in* $x(\theta)$. *Then the usual function* $F(x(\theta))$ *is differentiable in* θ, *moreover*

$$[F(x(\theta))]' = \langle F'(x(\theta)), x'(\theta) \rangle .$$

Proof. Indeed,

$$F(x(\vartheta)) - F(x(\theta)) = \langle F'(x(\theta)), x(\vartheta) - x(\theta) \rangle + o(\|x(\vartheta) - x(\theta)\|)$$

Now, divide by $\vartheta - \theta$ and pass to limit. $\square$

Proposition 5. *Let $\Phi(t)$ be a convex measurable scalar function and let $f(x)$ be an integrable function on a measurable space (A, Σ, μ). Then*

$$\Phi[\int_A f(x)d\mu] \leq \frac{1}{\mu(A)} \int_A \Phi[\mu(A) \cdot f(x)]d\mu .$$

Proof. Let, first, f be a simple function taking values t_i on sets A_i. Then

$$\Phi[\int_A f(x)d\mu] = \Phi[\sum \frac{\mu(A_i)}{\mu(A)}\mu(A)t_i]$$

$$\leq \sum \frac{\mu(A_i)}{\mu(A)}\Phi[\mu(A)t_i] = \frac{1}{\mu(A)} \int_A \Phi[\mu(A) \cdot f(x)]d\mu .$$

To finish the proof one can use a passage to limit. □

Recall also the *Markov (or Chebyshev) inequality* $\int_A f(x)d\mu \geq t \cdot \mu\{x : f(x) > t\}$ which (in the case of strictly increasing function *Phi*) we can write in the form

$$\mu\{x : f(x) > t\} = \mu\{x : \Phi(f(x)) > \Phi(t)\} \leq \frac{\int_A \Phi(f(x))d\mu}{\Phi(t)} . \quad (1)$$

4. Sobolev inequality

Theorem 2. ("Sobolev inequality"). *Let $F : \mathbb{R}^n \to \mathbb{R}$ be a differentiable map and let Φ be a convex measurable function in $\mathbb{R}$. Then*

$$\int_{\mathbb{R}^n} \Phi[F(x) - \int_{\mathbb{R}^n} F(y)d\gamma_n(y)]d\gamma_n(x) \leq \int_{\mathbb{R}^n}\int_{\mathbb{R}^n} \Phi[\frac{\pi}{2}\langle F'(x), y\rangle]d\gamma_n(y)d\gamma_n(x) .$$

Proof. For fixed elements $x, y \in \mathbb{R}^n$ and $\theta \in [0, \frac{\pi}{2}]$ put $x(\theta) = x \cdot \sin\theta + y \cdot \cos\theta$. Then, by Proposition 4,

$$F(x) - F(y) = F(x(\frac{\pi}{2})) - F(x(0)) = \int_0^{\frac{\pi}{2}} [F(x(\theta))]'d\theta = \int_0^{\frac{\pi}{2}} \langle F'(x(\theta)), x'(\theta)\rangle d\theta .$$

This equality and Proposition 5 give

$$\Phi[F(x) - F(y)] = \Phi[\int_0^{\frac{\pi}{2}} \langle F'(x(\theta)), x'(\theta)\rangle d\theta] \leq \frac{2}{\pi}\int_0^{\frac{\pi}{2}} \Phi[\frac{\pi}{2}\langle F'(x(\theta)), x'(\theta)\rangle]d\theta .$$

Integrating the last inequality over x and y we obtain

$$\int_{\mathbb{R}^n}\int_{\mathbb{R}^n} \Phi[F(x) - F(y)]d\gamma_n(y)d\gamma_n(x) \leq \quad (2)$$

$$\int_{\mathbb{R}^n}\int_{\mathbb{R}^n} \frac{2}{\pi}\int_0^{\frac{\pi}{2}} \Phi[\frac{\pi}{2}\langle F'(x(\theta)), x'(\theta)\rangle]d\theta d\gamma_n(y)d\gamma_n(x) .$$

Now observe that $x'(\theta) = x\cos\theta - y\sin\theta$, that for every fixed θ the map $(x,y) \to (x(\theta), x'(\theta))$ is a rotation in $\mathbb{R}^n \times \mathbb{R}^n$ an that (*this is the crucial point*) $\gamma_n \times \gamma_n$ is invariant with respect to this map. Thus the right-hand side of (2) is equal to

$$\int_{\mathbb{R}^n}\int_{\mathbb{R}^n} \Phi[\frac{\pi}{2}\langle F'(x), y\rangle] d\gamma_n(y) d\gamma_n(x) \ . \qquad (3)$$

By Proposition 5,

$$\int_{\mathbb{R}^n} \Phi[F(x) - \int_{\mathbb{R}^n} F(y) d\gamma_n(y)] d\gamma_n(x) \le \int_{\mathbb{R}^n}\int_{\mathbb{R}^n} \Phi[F(x) - F(y)] d\gamma_n(y) d\gamma_n(x) \ . \quad (4)$$

From (4),(2) and (3) we get the theorem. □

5. Gaussian random elements in normed spaces. Second weak moment

Definition 4. We say that X is a *standard Gaussian random element (r.e.) in a normed space* E if there are linearly independent elements $(e_i)_1^n$ in E and independent scalar standard Gaussian r.v. $(g_i)_1^n$ such that $X = \sum_1^n g_i e_i$.

Definition 4 is connected with the definition of standard Gaussian random vector (Section 2) in the following way. Consider a map $T : \mathbb{R}^n \to E$:

$$T(a_1, \dots, a_n) = \sum_1^n a_i e_i \ . \qquad (5)$$

This map generates a standard random vector $T^{-1}X(\omega) = (g_1(\omega), \dots, g_n(\omega))$ in $\mathbb{R}^n$.

Proposition 6. *Let $(X_k)_1^m$ be independent identically distributed standard Gaussian r.e. in a normed space E and let $\sum_1^m a_k^2 = 1$. Then $\sum_1^m a_k X_k$ is a standard Gaussian r.e. with the same distribution as X_k.*

Proof. Let $X_k = \sum_{i=1}^n g_{ki} e_i$. Then

$$\sum_1^m a_k X_k = \sum_{k=1}^m a_k \sum_{i=1}^n g_{ki} e_i = \sum_{i=1}^n (\sum_{k=1}^m a_k g_{ki}) e_i \ .$$

Since $(g_{ki})_{k,i}$ all are independent and (by Proposition 1) for every i , $\sum_{k=1}^m a_k g_{ki}$ is a standard Gaussian r.v., hence $\sum_{k=1}^m a_k X_k$ is a standard Gaussian r.e. with the same distribution as X_k. □

Definition 5. *The second weak moment* of r.e. X in a normed space E is, by definition,

$$\sigma = \sigma(X) := \sup_{\|f\|=1, f\in E^*} (\mathbf{E} f^2(X))^{1/2} \ .$$

Proposition 7. *Let $X = \sum_1^n g_i e_i$ be a standard Gaussian r.e. in a normed space E and let T be defined by* (5). *Then $\sigma = \|T\|$.*

Proof. Obviously, without restriction of generality, one can suppose $\text{lin}(e_i)_1^n = E$. Let (f_i) be the biorthogonal to (e_i) functionals. Then for $f \in E^*$ we have $f = \sum_1^n c_i f_i$ and

$$\sigma = \sup_{\|f\|=1} (\mathbf{E}\langle \sum_1^n g_i e_i, \sum_1^n c_i f_i \rangle^2)^{1/2} = \sup_{\|f\|=1} (\mathbf{E}(\sum_1^n g_i c_i)^2)^{1/2} =$$

$$\sup_{\|f\|=1} (\sum_1^n c_i^2 \mathbf{E} g_i^2)^{1/2} = \sup_{\|f\|=1} (\sum_1^n c_i^2)^{1/2} = \sup_{\|f\|=1} \|T^* f\| = \|T^*\| = \|T\| \ . \square$$

6. Tail behavior (the concentration of measure phenomenon)

Theorem 3. (Maurey-Pisier; [19], Theorem 4.7). *Let $X = \sum_1^n g_i e_i$ be a standard Gaussian r.e. in a normed space E. Then for any $t > 0$*

$$\mathbf{P}\{|\|X\| - \mathbf{E}\|X\|| > t\} \le 2\exp(-\frac{2t^2}{\pi^2\sigma^2}) \ . \tag{6}$$

Proof. Let T be the operator of Section 5. Then (6) is equivalent to

$$\gamma_n\{x \in \mathbb{R}^n : |\|Tx\| - \int_{\mathbb{R}^n} \|Tx\| d\gamma_n| > t\} \le 2\exp(-\frac{2t^2}{\pi^2\sigma^2}) \ . \tag{7}$$

Put $F(x) = \|Tx\|$ and suppose temporary that $\|\cdot\|$ is differentiable. Since for any $x, y \in \mathbb{R}^n$

$$|F(x) - F(y)| \le \|T(x-y)\| \le \|T\|\|x-y\|_2 \ ,$$

$\|F'(x)\|_2 \le \sigma$ for every $x \in \mathbb{R}^n$ (recall, $\sigma = \|T\|$!). Thus, by Proposition 3, for any $\lambda \in \mathbb{R}$

$$\int_{\mathbb{R}^n} \exp[\frac{\lambda\pi}{2}\langle F'(x), y\rangle] d\gamma_n(y) = \exp[\frac{1}{2}(\frac{\lambda\pi}{2})^2 \|F'(x)\|_2^2] \le \exp[\frac{1}{2}(\frac{\lambda\pi}{2})^2\sigma^2] \ . \tag{8}$$

Taking in Theorem 2 $\Phi(t) = \exp(\lambda t)$, we get

$$\begin{aligned} &\int_{\mathbb{R}^n} \exp[\lambda\{F(x) - \int_{\mathbb{R}^n} F(y) d\gamma_n(y)\}] d\gamma_n(x) \le \\ &\quad \int_{\mathbb{R}^n}\int_{\mathbb{R}^n} \exp[\lambda\frac{\pi}{2}\langle F'(x), y\rangle] d\gamma_n(y) d\gamma_n(x) \le \\ &\quad (\text{by } (8)) \le \int_{\mathbb{R}^n} \exp[\frac{1}{2}(\frac{\lambda\pi\sigma}{2})^2] d\gamma_n(x) = \\ &\qquad \exp[\frac{1}{2}(\frac{\lambda\pi\sigma}{2})^2] \ . \end{aligned} \tag{9}$$

Using the Markov inequality (1), taking there $\Phi(t) = \exp(\lambda t)$, $f(x) = F(x) - \int_{\mathbb{R}^n} F(y) d\gamma_n(y)$ and $\mu = \gamma_n(x)$, we get from (9)

$$\gamma_n\{x : F(x) - \int F(y) d\gamma_n(y) > t\} \le \exp[\frac{1}{2}(\frac{\lambda\pi\sigma}{2})^2 - \lambda t] \ .$$

Putting $\lambda = \frac{4t}{(\pi\sigma)^2}$ we obtain

$$\gamma_n\{x : F(x) - \int_{\mathbb{R}^n} F(y) d\gamma_n(y) > t\} \le \exp[-\frac{2t^2}{(\pi\sigma)^2}] \,.$$

Clearly, the same inequality holds for $-F$, so we finally obtain (7), and hence (6), for differentiable norms.

To finish the proof, observe that in the finite dimensional space $\mathrm{lin}(e_i)_1^n$ one can approximate any norm by differentiable one as exact as need. □

Remark. The estimation (6) is an important result in theory of Banach valued Gaussian r.e. It demonstrates that the distribution of $\|X\|$ is concentrated near $\mathbf{E}\|X\|$ as dense as the distribution of Gaussian r.v. $\sigma \cdot g$ near 0. In a somewhat different form:

$$\mathbf{P}\{|\|X\| - \mathbf{m}\|X\|| > t\} \le 2(1 - \Phi(\frac{t}{\sigma})) \,, \tag{10}$$

(where $\mathbf{m}\|X\|$ is the median of $\|X\|$ and $\Phi(t)$ is the distribution function of g), the concentration phenomenon is presented in books Ledoux-Talagrand ([14], 3.1) and Lifshits ([15], Chapt.12).

In fact, the concentration phenomenon in the form (10) was contained as far as in Borell [2] and Sudakov-Tsirelson [21].

In connection with Theorem 3, note also papers Fernique [7], Landau-Shepp [13] and Skorokhod [20], where for the first time were established exponential integrability of the norm for Gaussian r.e., and Yurinskii [24], where for the first time was proposed the martingale approach to estimations of probabilities of large deviations for the norm of sums of independent r.e. in Banach spaces.

7. Lévy's inequality

Recall that an r.e. X in a normed space E is called *symmetric* if $\mathbf{P}\{X \in A\} = \mathbf{P}\{X \in -A\}$ for any Borel set $A \subset E$.

Theorem 4. (Lévy's inequality; [11], II.3). *Let $(X_i)_1^n$ be symmetric independent r.e. in a normed space E and $S = \sum_1^n X_i$. Then for every $t > 0$*

$$2\mathbf{P}\{\|S\| > t\} \ge \mathbf{P}\{\max_{i \le n} \|X_i\| > t\} \,.$$

Proof. Put $\tau = \tau(\omega) = \min\{i : \|X_i\| > t\}$. Then

$$\mathbf{P}\{\|S\| > t\} \ge \sum_{i=1}^{n} \mathbf{P}\{\|S\| > t \,,\, \tau = i\} \,. \tag{11}$$

Now since $(X_1, \dots, X_n)$ and $(-X_1, \dots, -X_{i-1}, X_i, -X_{i+1}, \dots, -X_n)$ are identically distributed and τ depends on $\{\|X_i\|\}_1^n$ only, we have

$$\mathbf{P}\{\|S\| > t \,,\, \tau = i\} \;=\; \mathbf{P}\{\|X_i - R_i\| > t \,,\, \tau = i\} \,,$$

where $R_i = S - X_i$. This equality and (11) imply

$$2\mathbf{P}\{\|S\| > t\} \geq \sum_{i=1}^{n} \mathbf{P}\{\|S\| > t \ , \ \tau = i\} + \mathbf{P}\{\|X_i - R_i\| > t \ , \ \tau = i\} \ . \tag{12}$$

By triangle inequality

$$2\|X_i\| \leq \|X_i + R_i\| + \|X_i - R_i\| = \|S\| + \|X_i - R_i\| \ .$$

Hence

$$\mathbf{P}\{\|S\| > t \ , \ \tau = i\} + \mathbf{P}\{\|X_i - R_i\| > t \ , \ \tau = i\} \geq$$
$$\mathbf{P}\{\|S\| > t \bigvee \|X_i - R_i\| > t \ , \ \tau = i\} \geq \mathbf{P}\{\|S\| + \|X_i - R_i\| > 2t \ , \ \tau = i\} \geq$$
$$\mathbf{P}\{2\|X_i\| > 2t \ , \ \tau = i\} = \mathbf{P}\{\|X_i\| > t \ , \ \tau = i\} = \mathbf{P}\{\tau = i\} \ .$$

This inequality and (12) imply

$$2\mathbf{P}\{\|S\| > t\} \geq \sum_{i=1}^{n} \mathbf{P}\{\tau = i\} = \mathbf{P}\{\max_{i \leq n} \|X_i\| > t\} \ . \ \square$$

Corollary *There exists $c > 0$ such that for any normed space E and any standard Gaussian r.e. $X = \sum_1^n g_i e_i$ in E*

$$\mathbf{E}\|X\| \geq c \cdot \min_{i \leq n} \|e_i\| \cdot (\ln n)^{1/2} \ .$$

Proof. Evidently, standard Gaussian r.e. are symmetric. Hence one can use Lévy's inequality taking $S = X$ and $X_i = g_i e_i$. We have for any t

$$2\mathbf{P}\{\|X\| > t\} \geq \mathbf{P}\{\max_{i \leq n} \|g_i e_i\| > t\} \ .$$

Integrating over t we receive

$$\mathbf{E}\|X\| \geq \frac{1}{2}\mathbf{E} \max_{i \leq n} \|g_i e_i\| \geq \frac{1}{2} \min_{i \leq n} \|e_i\| \cdot \mathbf{E} \max_{i \leq n} |g_i| \geq$$
$$(\text{Proposition 2}) \geq c \cdot \min_{i \leq n} \|e_i\| \cdot (\ln n)^{1/2} \ . \ \square$$

8. Dvoretzky-Rogers theorem

Theorem 5. (Dvoretzky-Rogers, [5]). *Let $\| \cdot \|$ be a norm on $\mathbb{R}^N$ and let D be the ellipsoid of maximal volume inscribed in the unit $\|\cdot\|$-ball B. Then there exists a basis $(e_i)_1^N$, orthonormal with respect to D, such that $1 \geq \|e_i\| \geq 2^{-(N-1)/(N-i)}$, $i = 1, \ldots, N-1$.*

Proof. We choose the basis (e_i) inductively in the following way.

Let e_1 be a vector in D with maximal norm (clearly $\|e_1\| = 1$).

Given $e_1, \ldots, e_i$, choose e_{i+1} in $D \cap (e_1, \ldots, e_i)^{\perp}$ with maximal possible norm ($\perp$ is the orthogonal complement with respect to D).

Then for any $x \in \mathrm{lin}(e_i, \dots, e_N) \cap D$

$$\|x\| \le \|e_i\| \,. \tag{13}$$

Now consider the ellipsoid

$$U = \{\sum_{j=1}^{N} a_j e_j : \frac{\sum_1^{i-1} a_j^2}{a^2} + \frac{\sum_i^N a_j^2}{b^2} \le 1\} \,.$$

Of course, this ellipsoid depends on i and scalars a and b which we shall choose late on.

If $\sum_{j=1}^N a_j e_j \in U$ then $\sum_1^{i-1} a_j e_j \in aD$ and thus $\| \sum_1^{i-1} a_j e_j \| \le a$. In the same way, $\| \sum_{j=i}^N a_j e_j \| \le b$ and thus, by (13), $\| \sum_i^N a_j e_j \| \le b\|e_i\|$.

Choosing $a = 1/2$, $b = 1/(2\|e_i\|)$, we get that $U \subset B$. On the other hand

$$\mathrm{vol}U = \frac{1}{2^{i-1}} \cdot \frac{1}{(2\|e_i\|)^{N-i}} \mathrm{vol}D$$

so necessarily

$$\frac{1}{2^{i-1}} \cdot \frac{1}{(2\|e_i\|)^{N-i}} \le 1$$

or $\|e_i\| \ge 2^{-(N-1)/(N-i)}$. □

Corollary 1. *Let E be an N-dimensional normed space and $\overline{N} = [\frac{N}{2}]$. Then there exist $(e_i)_1^{\overline{N}} \subset E$ such that $\|e_i\| \ge \frac{1}{2}$ and for any $(a_i)_1^{\overline{N}} \subset \mathbb{R}$*

$$\| \sum_1^{\overline{N}} a_i e_i \| \le (\sum_1^{\overline{N}} a_i^2)^{1/2} \,. \tag{14}$$

Corollary 2. *For every N-dimensional normed space E there exists a subspace $\overline{E} \subset E$ of dimension $\overline{N} = [\frac{N}{2}]$ and $\overline{E}$-valued standard Gaussian r.e. X with $\mathbf{E}\|X\| = 1$ and $\sigma^2(X) \le c/\ln N$, where $c > 0$ is an absolute constant.*

Proof. Let $\overline{X} = \sum_1^{\overline{N}} g_i e_i$ where (g_i) are independent Gaussian r.v. and (e_i) are from Corollary 1. Then, by (14),

$$\sigma(\overline{X}) := \sup_{\|f\|=1} (\mathbf{E} f^2(\overline{X}))^{1/2} = \sup_{\|f\|=1} (\sum_{i=1}^{\overline{N}} f^2(e_i))^{1/2} = \sup_{\|a\|_2=1} \| \sum_{i=1}^{\overline{N}} a_i e_i \|^{1/2} \le 1 \,.$$

On the other hand, by Corollary of Theorem 4

$$\mathbf{E}\|\overline{X}\| \ge c_1 (\ln \overline{N})^{1/2} \ge c_2 (\ln N)^{1/2} \,.$$

Now put $X = \overline{X}/\mathbf{E}\|\overline{X}\|$. □

9. Two geometric lemmas

Lemma 2. *Let $\|\cdot\|$ be any norm on $\mathbb{R}^n$ with unit ball B and unit sphere S. Let $\delta > 0$. There is a δ-net $A \subset S$ with cardinality*

$$\operatorname{card} A \le (1 + \frac{2}{\delta})^n \ . \tag{15}$$

Proof. Let A be a maximal subset of S such that $\|a-b\| \ge \delta$ for all $a, b \in A$, $a \ne b$. Clearly, by maximality, A is a δ-net of S. To majorize $\operatorname{card} A$ we note that balls $a + \frac{\delta}{2}B$, $a \in A$ are disjoint and included into $(1+\frac{\delta}{2})B$. Therefore

$$\sum_{a\in A} \operatorname{vol}(a + \frac{\delta}{2}B) \le \operatorname{vol}((1+\frac{\delta}{2})B) = (1+\frac{\delta}{2})^n \operatorname{vol} B \ ;$$

hence

$$\operatorname{card} A \cdot (\frac{\delta}{2})^n \operatorname{vol} B \le (1+\frac{\delta}{2})^n \operatorname{vol} B \ .$$

This inequality implies (15). □

Lemma 3. *For each $\varepsilon > 0$ there is a $\delta = \delta(\varepsilon)$, $0 < \delta < 1$, with the following property. Let $\|\cdot\|$ be an arbitrary norm on $\mathbb{R}^n$ with the unit sphere S. Let A be a δ-net in S and let $x_1, \dots, x_n$ be elements of a normed space E. If for every $a = (a_1, \dots, a_n) \in A$*

$$1 - \delta \le \|\sum_1^n a_k x_k\| \le 1 + \delta \ ,$$

then for every $a \in S$

$$(1+\varepsilon)^{-1} \le \|\sum_1^n a_k x_k\| \le 1 + \varepsilon; .$$

Proof. There is a^0 in A such that $\|a - a^0\| \le \delta$ hence $a = a^0 + \lambda_1 a'$ with $|\lambda_1| \le \delta$ and $a' \in S$. Continuing this process we obtain $a = a^0 + \lambda_1 a^1 + \lambda_2 a^2 + \cdots$ with $a^j \in A$ and $|\lambda_j| \le \delta^j$. It follows that

$$\|\sum_1^n a_k x_k\| \le \sum_{j\ge 0} \delta^j \|\sum_{k=1}^n a_k^j x_k\| \le \frac{1+\delta}{1-\delta} \ .$$

Similarly

$$\|\sum_1^n a_k x_k\| \ge \|\sum_1^n a_k^0 x_k\| - \frac{\delta(1+\delta)}{1-\delta} \ge 1 - \delta - \frac{\delta(1+\delta)}{1-\delta} = \frac{1-3\delta}{1-\delta} \ .$$

Hence, if $\delta > 0$ is chosen small enough so that

$$\frac{1-3\delta}{1-\delta} \ge \frac{1}{1+\varepsilon} \quad \text{and} \quad \frac{1+\delta}{1-\delta} \le 1 + \varepsilon \ ,$$

we obtain the announced result. Note that one can find a suitable δ depending only on ε (and *independent* on n). □

10. Dvoretzky's theorem

Theorem 6. (Dvoretzky). *For each $\varepsilon > 0$, there is a number $\eta = \eta(\varepsilon) > 0$ with the following property. Every normed space E of dimension N contains a subspace of dimension $n = [\eta \ln N]$ which is $(1+\varepsilon)^2$–isomorphic to l_2^n.*

Let us present first the idea of proof. We take independent copies $X_1, \dots, X_n$ of r.e. X from Corollary 2 of Dvoretzky-Rogers theorem, $n \approx \ln N$, which is determined on a probability space $(\Omega, \mathcal{B}, \mathbf{P})$. Let $\Omega(n, \varepsilon)$ be the set of all ω in Ω such that for every $a = (a_1, \dots, a_n)$ in the unit sphere S of Euclidean space $\mathbb{R}^n$

$$(1+\varepsilon)^{-1} \le \|\sum_1^n a_k X_i(\omega)\| \le 1 + \varepsilon .$$

We will show that $\mathbf{P}\{\Omega(n,\varepsilon)\} > 0$ provided that n is not too large and precisely provided $n \le \eta/\sigma^2$. This clearly yields Theorem 6.

Now the **Proof.** Let $\overline{E}$ be the subspace and X be an r.e. from Corollary 2 of Theorem 5. Let $X_1, \dots, X_n$ be independent copies of X (n we choose late on). Then for any $a = (a_1, \dots, a_n)$ in S the r.e. $\sum_1^n a_k X_k$ has the same distribution as X (Proposition 6).

Thus we have $\mathbf{E}\|\sum_1^n a_k X_k\| = 1$. Therefore, by Theorem 3, for any $\delta > 0$

$$\mathbf{P}\{|\|\sum_1^n a_k X_k\| - 1| > \delta\} \le 2\exp(-\frac{2\delta^2}{\pi^2\sigma^2}) < 2\exp(-\frac{\delta^2}{\sigma^2}) .$$

Let A be as in Lemmas 2 and 3 with $\|a\| = \|a\|_2$. Then preceding inequality implies

$$\mathbf{P}\{\exists a \in A : |\|\sum_1^n a_k X_k\| - 1| > \delta\} \le 2(\mathrm{card} A)\exp(-\frac{\delta^2}{\sigma^2}) \le$$

$$\text{(by Lemma 2)} \le 2(1+\frac{2}{\delta})^n \exp(-\frac{\delta^2}{\sigma^2}) \le \tag{16}$$

$$2\exp(\frac{2n}{\delta})\exp(-\frac{\delta^2}{\sigma^2}) = 2\exp(\frac{2n}{\delta} - \frac{\delta^2}{\sigma^2}) .$$

Now let $\delta = \delta(\varepsilon)$ be the function of ε, given by Lemma 3. Let we choose n so that

$$\frac{2n}{\delta} \le \frac{\delta^2}{2\sigma^2} . \tag{17}$$

Then the probability (16) is not greater than $2\exp(-\frac{\delta^2}{2\sigma^2})$. Clearly, we can always assume that σ is small enough (say $\sigma < \delta/2$), otherwise there is nothing to prove. Hence we can assume that the right side of (16) < 1. We than obtain that with *positive* probability, for every $a \in A$

$$|\|\sum a_k X_k(\omega)\| - 1| < \delta$$

i.e.

$$1 - \delta \le \|\sum_1^n a_k X_k(\omega)\| \le 1 + \delta \,.$$

By Lemma 3 (recall, we choose $\delta = \delta(\varepsilon)!$), we conclude that with positive probability for all $a \in S$

$$(1+\varepsilon)^{-1} \le \|\sum_1^n a_k X_k(\omega)\| \le 1 + \varepsilon \,.$$

Therefore, there exists $\omega_0 \in \Omega$ such that for $x_k = X_k(\omega_0)$ and for every $a \in S$

$$(1+\varepsilon)^{-1} \le \|\sum_1^n a_k x_k\| \le 1 + \varepsilon \,.$$

By homogenity of the norm it means that $\mathrm{lin}(x_k)_1^n$ is $(1+\varepsilon)^2$-isomorphic to l_2^n.

To satisfy (17) put

$$n = [\frac{\delta^3}{4\sigma^2}] = \eta \ln N$$

(recall, $\sigma^2 \le c/\ln N$, by Corollary 2 of Theorem 5). □

Acknowledgement. The first version of this note was written when the second author visited Politecnico di Milano. He express his thanks to P.Terenzi for hospitality and valuable remarks. The authors also express they thanks to M.Ostrovskii for valuable references.

References

1. Y. Benyamini and J. Lindenstrauss, *Geometric Nonlinear Functional Analysis.* V.I, Colloq. Publ. Amer. Math. Soc., Rhode Island, 2000.
2. C. Borell, The Brunn-Minkowski inequality in Gauss space , *Invent. Math.* **30** (1975) 207-216.
3. H. Cramér, *Mathematical Methods of Statistics.* Princeton Univ. Press, N.J., 1946.
4. A. Dvoretzky, *Some results on convex bodies and Banach spaces.* Proc. Int. Symp. on Linear Spaces, Jerusalem, 1961, 123-160.
5. A. Dvoretzky and C.A. Rogers, Absolute and unconditional convergence in normed linear spaces, *Proc. Nat. Acad. Sci. USA* **36** (1950) 192-197.
6. W. Feller, *An Introduction to Probability Theory and its Applications.* V.II, John Wiley and Sons, New York, 1966.
7. K. Fernique, Intégrabilité des vecteurs gaussiens, *C.R. Acad. Sci. Paris, ser. A* **270** (1970) 1698-1699.
8. T. Figiel, *A short proof of Dvoretzky's theorem on almost spherical sections,* Compositio Math., 33 (1976), 297-301.
9. T. Figiel, J. Lindenstrauss and V.D. Milman, The dimension of almost spherical sections of convex bodies, *Acta Math.* **139** (1977) 53-94.

10. Y. Gordon, On Dvoretzky's theorem and extensions of Slepian's lemma, *Israel Semin. Geom. Aspects Funct. Anal., Technion preprint series* No. MT-636 (1983) (4) II.1.1-II.1.25.
11. J.P. Kahane, *Some Random Series of Functions*. D.C. Heath and Co, Lexington, 1968.
12. J.L. Krivine, Sous-espaces de dimension finie des espaces de Banach reticules, *Ann. of Math.* **104** (1) (1976) 1-29.
13. H.J. Landau and L.A. Shepp, On the supremum of a Gaussian process, *Sankhya* **32** (1970) 369-378.
14. M. Ledoux and M. Talagrand, *Probability in Banach Spaces. Isoperimetry and Processes*. Springer, Berlin, 1991.
15. M.A. Lifshits, *Gaussian Random Functions*. TViMS, Kyiv, 1995 (Russian).
16. V.D. Milman, New proof of the theorem of Dvoretzky on sections of convex bodies, *Funkts. Anal. Prilozh.* **5** (5) (1971) 28-37 (Russian).
17. V.D. Milman and G. Schechtman, Asymptotic theory of finite dimensional normed spaces. *Lect. Notes in Math.* **1200** (1986).
18. G. Pisier, *Probabilistic methods in the geometry of Banach spaces*, Lecture Notes in Math., **1206** (1986) 167-241.
19. G. Pisier, *The Wolume of Convex Bodies and Banach Space Geonetry*. Cambridge Univ. Press, Cambridge e.a., 1989.
20. A.V. Skorokhod, A note on Gaussian measures in a Banach space, *Teor. Veroyatn. Prim.* **15** (1970) 519-520 (Russian).
21. V.N. Sudakov and B.S. Tsirel'son, Extremal properties of half-spaces for spherically invariant measures, *Zap. Nauch. Sem. L.O.M.I.* **41** (1974) 14-24 (Russian).
22. A. Szankowski, On Dvoretzky's theorem on almost spherical sections of convex bodies, *Israel J. Math.* **17** (1974) 325-338.
23. L. Tzafriri, On Banach spaces with unconditional bases, *Israel J. Math.* **17** (1974) 84-93.
24. V.V. Yurinskii, Exponential bounds for large deviations, *Teor. Veroyatn. Prim.* **19** (1974) 152-153 (Russian).

Functional Analysis and its Applications
V. Kadets and W. Żelazko

On the Inverse Problem of Scattering Theory for a Differential Operator of the Second Order

Khanlar R. Mamedov, Hamza Menken

Mersin University-Turkey

Abstract

Direct and inverse problem of scattering theory (IPST) are investigated for the differential equation

$$-y'' + q(x)y = \lambda^2 y$$

on half line containing a spectral parameter in the boundary condition

$$y'(0) + (\alpha_0 + i\alpha_1\lambda + \alpha_2\lambda^2)y(0) = 0.$$

Key words: Sturm-Liouville operator, inverse problem of scattering theory, scattering function, scattering data
2000 MSC: 34B24, 34A55

1. Introduction

Consider the Sturm-Liouville equation on the half line

$$-y'' + q(x)y = \lambda^2 y \quad (0 < x < \infty) \tag{1}$$

with the spectral parameter dependent boundary condition

$$y'(0) + (\alpha_0 + i\alpha_1\lambda + \alpha_2\lambda^2)y(0) = 0. \tag{2}$$

Email addresses: `hanlar@aport.ru;hanlar@mersin.edu.tr` (Khanlar R. Mamedov,), `hmenken@mersin.edu.tr` (Hamza Menken).

Here $q(x)$ is real-valued function satisfying the condition

$$\int_0^\infty (1+x)\,|q(x)|\,dx < \infty, \tag{3}$$

and $\alpha_0, \alpha_1, \alpha_2$ are real numbers with $\alpha_1 \geq 0$, $\alpha_2 \geq 0$.

In this paper, we study the inverse problems of scattering theory (IPST) for the boundary problem of (1), (2). Before we discuss our problem we shall recall some works which has the spectral parameter dependent boundary condition. The boundary problems with spectral parameter dependent boundary condition for second order (or higher order) of differential equations are interesting with their applications, and we refer to [4], [13], [14]. The many spectral properties of such this boundary problems were investigated with different methods by the many authors. About this subject it was given more informations and references in [18], and also in [17]-[5].

The other interesting problem with spectral parameter dependent boundary condition for equation (1) is the inverse problems of spectral analysis, so that it is the construction of the differential operator associated the equation (1) with respect to some spectral characteristics (to two spectres, spectral function, scattering data, etc.). In finite interval and half line, this problem with spectral parameter dependent boundary condition were investigated with different methods in [1]-[6]. In [16], on the half line the inverse problem was investigated with respect to spectral function for the boundary problem

$$-y'' + q(x)y = \lambda y,$$

$$y(0) = \theta(\lambda)y'(0)$$

where $\theta(\lambda)$ is an arbitrary analytic function whose imaginary part is non-negative and $q(x)$ is a real-valued function. Similar problem was investigated in [6] in case

$$\theta(\lambda) = a + b\lambda + \sum_{k=1}^{\infty} \frac{1+\lambda_k\lambda}{\lambda_k - \lambda}\rho_k.$$

where the numbers a, b, λ_k, ρ_k $(k = 1, 2, ...)$ are real coefficients and $\rho_k > 0$. Also, in the similar cases, on the finite interval too, it is obtained results with respect to two spectres and spectral function, and now the problem on half line that we study is about the inverse problem of scattering theory. In [11],[12] it was considered the boundary problem without any spectral parameter in the boundary condition

$$-y'' + q(x)y = \lambda^2 y, \quad (0 \leq x < \infty) \tag{4}$$

$$y(0) = 0 \tag{5}$$

where $q(x)$ is a real valued function satisfying the condition (3), and the inverse problem of scattering theory of the boundary problem (4), (5) was solved. That the boundary value problem (4), (5) has bounded solutions $u(\lambda, x)$ for $-\infty < \lambda < \infty$ and $\lambda = i\lambda_k$ $(k = \overline{1,n})$, moreover, as $x \to \infty$

$$u(\lambda, x) = e^{-i\lambda_k x} - S(\lambda)e^{i\lambda x} + o(1), \; (-\infty < \lambda < \infty)$$

and

$$u(i\lambda_k, x) = m_k e^{-i\lambda_k x}(1 + o(1)), \ (k = \overline{1, n}),$$

respectively. Thus, the collection $\{S(\lambda)\ (-\infty < \lambda < \infty); \lambda_k;\ m_k\ (k = \overline{1,n})\ \}$ provides a complete description of the behavior at infinity of all radial wave functions. The collection of quantity $\{S(\lambda)\ (-\infty < \lambda < \infty); \lambda_k;\ m_k\ (k = \overline{1,n})\ \}$ that specify the behavior all radial wave functions $u(\lambda, x)$ at infinity, will be referred to as scattering data of the boundary value problem (4), (5). In terms of the scattering data, the potential $q(x)$ is recover uniquely (see [17], p. 217). In [8], the inverse problem of scattering theory of the boundary problem

$$-y'' + q(x)y = \lambda^2 y, \quad (0 \le x < \infty)$$

$$y'(0) = hy(0)$$

was solved where h is any real number. Similar problems for the boundary problem (1), (2) they are considered in [9] and in this paper. The some properties of scattering data were investigated in [10].

According to [12], for any λ from closed upper-half plane, the equation (1) has a solution $e(\lambda, x)$ described by

$$e(\lambda, x) = e^{i\lambda x} + \int_x^\infty K(x,t)e^{i\lambda t}dt \tag{6}$$

and for the kernel function $K(x,t)$ the inequality

$$|K(x,t)| \le \frac{1}{2}\sigma(\frac{x+t}{2})\exp\left\{\sigma_1(x) - \sigma_1(\frac{x+t}{2})\right\}$$

holds where

$$\sigma(x) \equiv \int_x^\infty |q(t)|\,dt, \quad \sigma_1(x) \equiv \int_x^\infty \sigma(t)dt.$$

Moreover,

$$K(x,x) = \frac{1}{2}\int_x^\infty q(t)dt.$$

The solution $e(\lambda, x)$ is an analytic function of λ in the upper half plane $Im\ \lambda \ge 0$ and is continuous on the real line. The following estimates hold through the half plane $Im\ \lambda \ge 0$,

$$\left|e(\lambda, x) - e^{i\lambda x}\right| \le \left\{\sigma_1(x) - \sigma_1(x + \frac{1}{|\lambda|})\right\}\exp\left\{-Im\ \lambda x + \sigma_1(x)\right\} \tag{7}$$

and

$$\left|e'(\lambda, x) - i\lambda e^{i\lambda x}\right| \le \sigma(x)\exp\left\{-Im\ \lambda x + \sigma_1(x)\right\} \tag{8}$$

for real $\lambda \ne 0$, the functions $e(\lambda, x)$ and $e(-\lambda, x)$ form a fundamental system of solutions of the equation (1) and their Wronskian is equal to $2i\lambda$:

$$W\left\{e(\lambda, x), e(-\lambda, x)\right\} = e'(\lambda, x)e(-\lambda, x) - e(\lambda, x)e'(-\lambda, x) = 2i\lambda. \tag{9}$$

2. Main Results

Let $\omega(x,\lambda)$ be a solution of the equation (1) satisfying the initial-value conditions

$$\omega(\lambda,0)=1,\quad \omega'(\lambda,0)=-(\alpha_0+i\alpha_1\lambda+\alpha_2\lambda^2).$$

Theorem 1. *The identity*

$$\frac{2i\lambda\omega(\lambda,x)}{e'(\lambda,0)+(\alpha_0+i\alpha_1\lambda+\alpha_2\lambda^2)e(0,\lambda)}=\overline{e(x,\lambda)}-S(\lambda)e(\lambda,x) \tag{10}$$

holds for all $\lambda\neq 0$, *where*

$$S(\lambda)=\frac{\overline{e'(\lambda,0)}+(\alpha_0+i\alpha_1\lambda+\alpha_2\lambda^2)\overline{e(\lambda,0)}}{e'(\lambda,0)+(\alpha_0+i\alpha_1\lambda+\alpha_2\lambda^2)e(\lambda,0)}$$

and

$$\overline{S(\lambda)}=S(-\lambda).$$

Proof. Since two function $e(-\lambda,x)$ and $e(\lambda,x)$ form a fundamental system of solutions to equation (1) for all $\lambda\neq 0$, we can write

$$\omega(x,\lambda)=c_1(\lambda)e(\lambda,x)+c_2(\lambda)\overline{e(\lambda,x)}$$

where

$$c_1(\lambda)=\frac{\overline{e'(\lambda,0)}+(\alpha_0+i\alpha_1\lambda+\alpha_2\lambda^2)\overline{e(\lambda,0)}}{-2i\lambda},$$

and

$$c_2(\lambda)=\frac{e'(\lambda,0)+(\alpha_0+i\alpha_1\lambda+\alpha_2\lambda^2)e(\lambda,0)}{2i\lambda}.$$

Then,

$$\omega(x,\lambda)=\frac{\overline{e'(\lambda,0)}+(\alpha_0+i\alpha_1\lambda+\alpha_2\lambda^2)\overline{e(\lambda,0)}}{-2i\lambda}e(\lambda,x)+$$
$$+\frac{e'(\lambda,0)+(\alpha_0+i\alpha_1\lambda+\alpha_2\lambda^2)e(\lambda,0)}{2i\lambda}\overline{e(\lambda,x)} \tag{11}$$

Let $E(\lambda,0)=e'(\lambda,0)+(\alpha_0+i\alpha_1\lambda+\alpha_2\lambda^2)e(\lambda,0)$. Since $q(x)$ is real, it follows that $e(-\lambda,0)=\overline{e(\lambda,0)}$, and hence that $E(\lambda,0)\neq 0$ for all real $\lambda\neq 0$. To prove this we assume that there is a non-zero $\lambda_0\in(-\infty,\infty)$ such that

$$E(\lambda_0,0)=e'(\lambda_0,0)+(\alpha_0+i\alpha_1\lambda+\alpha_2\lambda^2)e(\lambda_0,0)=0$$

or

$$e'(\lambda_0,0)=-(\alpha_0+i\alpha_1\lambda+\alpha_2\lambda^2)e(\lambda_0,0).$$

From the formula (9) we get

$$e'(\lambda_0,0)\overline{e(\lambda_0,0)}-e(\lambda_0,0)\overline{e'(\lambda_0,0)}=2i\lambda_0$$

or

$$-2i\alpha_1\lambda_0\left|e(\lambda_0,0)\right|^2=2i\lambda_0.$$

Since $\alpha_1 \geq 0$ we have a construction, hence we have that $E(\lambda, 0) \neq 0$ for all real $\lambda \neq 0$. According to the formula (11) we obtain

$$\frac{2i\lambda\omega(\lambda, x)}{e'(\lambda, 0) + (\alpha_0 + i\alpha_1\lambda + \alpha_2\lambda^2)e(\lambda, 0)} = \overline{e(\lambda, x)} - S(\lambda)e(\lambda, x)$$

where

$$S(\lambda) = \frac{\overline{e'(\lambda, 0)} + (\alpha_0 + i\alpha_1\lambda + \alpha_2\lambda^2)\overline{e(\lambda, 0)}}{e'(\lambda, 0) + (\alpha_0 + i\alpha_1\lambda + \alpha_2\lambda^2)e(\lambda, 0)}$$

and

$$\overline{E(\lambda, 0)} = e'(-\lambda, 0) + (\alpha_0 - i\alpha_1\lambda + \alpha_2\lambda^2)e(-\lambda, 0) = E(-\lambda, 0),$$

$$S(\lambda) = \overline{S(-\lambda)}.$$

This proves the theorem. ■

Theorem 2. *The function $E(\lambda, 0)$ may have only a finite number of zeros on the half plane* Im $\lambda > 0$, *they are all simple and lie on the imaginary.*

Proof. From the proof of Theorem 1, since $E(\lambda, 0) \neq 0$ for all real $\lambda \neq 0$, the point $\lambda = 0$ is the possible real zero of the function $E(\lambda, 0)$. Since the function $E(\lambda, 0)$ is analytic in the upper half plane $Im\ \lambda > 0$ we have that the zeros of $E(\lambda, 0)$ are at most countable. Now to show that the set of the zeros of $E(\lambda, 0)$ is bounded we assume, by way of contradiction, that this set is bounded, so that there exist the numbers λ_k such that these numbers satisfy the relation $E(\lambda_k, 0) = 0$ for $Im\ \lambda_k > 0$ and $|\lambda_k| \to \infty$ or $e'(\lambda_k, 0) = -(\alpha_0 + i\alpha_1\lambda + \alpha_2\lambda^2)e(\lambda_k, 0)$. From the relation (8) and last statement we have that $\lim_{k\to\infty} e(\lambda_k, 0) = 0$ as $|\lambda_k| \to \infty$. On the other hand, according to (7) it follows that $\lim_{k\to\infty} e(\lambda_k, 0) = 0$. This contradiction shows that the set $\{\lambda_k\}$ is bounded. Hence, the zeros of the function $E(\lambda, 0)$ is bounded and is form at most countable set having 0 the only possible limit point.

In [9], it was shown that the zeros of the function $E(\lambda, 0)$ are simple and are purely imaginary.

Now we shall prove that the function $E(\lambda, 0)$ has finitely many zeros in the half plane $Im\ \lambda > 0$. This is obvious if $E(0, 0) \neq 0$, because under this assumption, the set of zeros cannot have limit points. To verify that the number of zeros of $E(\lambda, 0)$ is finite in the general case too, we show that the distance between neighboring zeros is bounded away from zero.

We let δ denote the infimum of the distances between two neighboring zeros of $E(\lambda, 0)$, and show next that $\delta > 0$. Otherwise, we could exhibit a sequence of zeros, $\left\{i\widetilde{\lambda}_k\right\}$ and $\{i\lambda_k\}$ of the zeros the function $E(\lambda, 0)$, such that $\lim_{k\to\infty} (\widetilde{\lambda}_k - \lambda_k) = 0$, $\widetilde{\lambda}_k > \lambda_k \geq 0$, and $\max_k \widetilde{\lambda}_k < M$. Then it follows from estimate (9) that, for A large enough, the inequality $e(i\lambda, x) > \frac{1}{2}e^{-\lambda x}$ holds uniformly with respect to $x \in [A, \infty)$ and $\lambda \in [0, \infty)$ whence

$$\int_A^\infty e(i\widetilde{\lambda}_k, x)\overline{e(i\lambda_k, x)}dx > \frac{e^{-2AM}}{8M}. \tag{12}$$

On the other hand, according to the formula (11) in [9] we have

$$0 = (\lambda_k - \widetilde{\lambda}_k)\int_0^\infty e(i\widetilde{\lambda}_k, x)\overline{e(i\lambda_k, x)}dx + \alpha_2((\lambda_k - \widetilde{\lambda}_k)\cdot$$

$$\cdot e(i\widetilde{\lambda}_k, 0)\overline{e(i\lambda_k, 0)} + \alpha_1 e(i\widetilde{\lambda}_k, 0)\overline{e(i\lambda_k, 0)}. \tag{13}$$

Letting $k \to \infty$, we get

$$\alpha_1 \left|e(i\lambda_k, 0)\right|^2 = 0.$$

This relation gives a contradiction if $\alpha_1 > 0$. If $\alpha_1 = 0$, from the formula (13) we write

$$0 = \int_0^\infty e(i\widetilde{\lambda}_k, x)\overline{e(i\lambda_k, x)}dx + e(i\widetilde{\lambda}_k, 0)e(i\lambda_k, 0) =$$

$$= \int_0^A e(i\widetilde{\lambda}_k, x)\left[\overline{e(i\lambda_k, x) - e(i\widetilde{\lambda}_k, x)}\right] dx + \int_0^A e(i\widetilde{\lambda}_k, x)\overline{e(i\widetilde{\lambda}_k, x)}dx +$$

$$+ \int_A^\infty e(i\widetilde{\lambda}_k, x)\overline{e(i\lambda_k, x)}dx + e(i\widetilde{\lambda}_k, 0)e(i\lambda_k, 0).$$

Taking limit as $k \to \infty$, we get

$$0 \geq \varliminf_{k\to\infty} \int_A^\infty e(i\widetilde{\lambda}_k, x)\overline{e(i\lambda_k, x)}dx.$$

But this contradict with the inequality (12). Thus, the assumption is not true, so that $\delta > 0$. Thus the function $E(\lambda, 0)$ has finitely many zeros. This completes the proof. ■

Theorem 3. *The function $1 - S(\lambda)$ is the Fourier transform of a function $F_S(x)$ of the form*

$$F_S(x) = F_S^{(1)}(x) + F_S^{(2)}(x),$$

where $F_S^{(1)}(x) \in L_1(-\infty, \infty)$, *whereas* $F_S^{(2)}(x) \in L_2(-\infty, \infty)$ *and* $\sup\limits_{-\infty<x<\infty} \left|F_S^{(2)}(x)\right| < \infty$.

Proof. From the formula (6) it follows that

$$e(\lambda, 0) = 1 + \int_0^\infty K(0, t)e^{i\lambda t}dt,$$

$$e'(\lambda,0) = i\lambda - K(0,0) + \int_0^\infty K_x(0,t)e^{i\lambda t}dt.$$

We shall use the following notations for shortly:

$$q_0 = K(0,0); \quad \varphi_0(\lambda) = \alpha_0 + i\alpha_1\lambda + \alpha_2\lambda^2 - q_0;$$

$$K_1(t) = K_x(0,t) + \alpha_0 K(0,t); \quad K_2(t) = \alpha_1 K(0,t) + \alpha_2 K_t(0,t);$$

$$K_3(t) = \alpha_1 K(0,t) - \alpha_2 K_t(0,t)$$

and

$$\widetilde{K}_j(-\lambda) = \int_0^\infty K_j(t)e^{i\lambda t}dt, \quad j = 1,2,3.$$

Then we have

$$1 - S(\lambda) = \frac{2i\lambda(1+\alpha_2 q_0) + \widetilde{K}_1(-\lambda) + i\lambda\widetilde{K}_2(-\lambda) - \widetilde{K}_1(\lambda) - i\lambda\widetilde{K}_3(\lambda)}{\varphi_0(\lambda) + i\lambda(1+\alpha_2 q_0) + \widetilde{K}_1(-\lambda) + i\lambda\widetilde{K}_2(-\lambda)}. \tag{14}$$

Every one of the functions

$$\widetilde{f}_1(\lambda) = \frac{i\lambda(1+\alpha_2 q_0)}{\varphi_0(\lambda)}, \quad \widetilde{f}_2(\lambda) = \frac{\widetilde{K}_1(-\lambda) + i\lambda\widetilde{K}_2(-\lambda)}{\varphi_0(\lambda)},$$

$$\widetilde{f}_3(\lambda) = \frac{\widetilde{K}_1(\lambda) + i\lambda\widetilde{K}_3(\lambda)}{\varphi_0(\lambda)}.$$

is the Fourier transformation of a summable function. Hence we have

$$1 - S(\lambda) = \frac{\widetilde{f}(\lambda)}{1+\widetilde{K}(-\lambda)}$$

where

$$\widetilde{f}(\lambda) = 2\widetilde{f}_1(\lambda) + \widetilde{f}_2(\lambda) - \widetilde{f}_3(\lambda),$$
$$\widetilde{K}(-\lambda) = \widetilde{f}_1(\lambda) + \widetilde{f}_2(\lambda).$$

We can rewrite the formula (14) as form

$$1 - S(\lambda) = \widetilde{f}(\lambda)\left[\left\{1 + (1-\widetilde{h}(\lambda N^{-1})\widetilde{K}(-\lambda)\right\}^{-1} - 1\right] +$$

$$+\widetilde{f}(\lambda) - \widetilde{f}(\lambda)\left\{\frac{1}{1+\left\{1-\widetilde{h}(\lambda N^{-1})\widetilde{K}(-\lambda)\right\}} - \frac{1}{1+\widetilde{K}(-\lambda)}\right\} \tag{15}$$

where

$$\widetilde{h}(\lambda) = \begin{cases} 1, & if\ |\lambda| \le 1 \\ 2-|\lambda|, & if\ 1 \le \lambda \le 2 \\ 0, & if\ |\lambda| > 2 \end{cases}$$

is the Fourier transform of the function $h(x) \in L_1(-\infty,\infty)$. Also, $h(\lambda N^{-1})$ is the Fourier transform of the function $h_N(x) = Nh(xN)$, and

$$\lim_{N\to\infty} \|f(x) - h_N * f(x)\|_{L_1} = 0 \tag{16}$$

for all $f(x) \in L_1(-\infty,\infty)$, where $h_N * f(x)$ is the convolution of functions $h_N(x)$ and $f(x)$ from $L_1(-\infty,\infty)$. Note that the convolution $h_N * f(x)$ of functions $h_N(x)$ and $f(x)$ from $L_1(-\infty,\infty)$ is defined as $h_N * f(x) = \int\limits_{-\infty}^{\infty} h_N(x-t)f(t)dt$. In general, recall that the Fourier transform of the convolution

$$f * g(x) = \int\limits_{-\infty}^{\infty} f(x-t)g(t)dt$$

of two functions from $L_1(-\infty,\infty)$ equals the product $\widetilde{f}(\lambda)\widetilde{g}(\lambda)$ of their Fourier transforms, and the norm of the convolution does not exceed the product of norms:

$$\|f * g\|_{L_1} \le \|f\|_{L_1} \|g\|_{L_1}.$$

Consequently, if $\|f\|_{L_1} < 1$, then the series

$$-f(x) + f * f(x) - f * f * f(x) + ...$$

converges in the metric of $L_1(-\infty,\infty)$, its sum belongs to this space and its Fourier transform is equal to

$$-\widetilde{f}(\lambda) + \left\{\widetilde{f}(\lambda)\right\}^2 - \left\{\widetilde{f}(\lambda)\right\}^3 + ... = \left\{1 + \widetilde{f}(\lambda)\right\}^{-1} - 1.$$

We conclude, from (16) and the previous argument that for N large enough, the function $\left[1 + \left\{1 - \widetilde{h}(\lambda N^{-1})\right\}\widetilde{f}(-\lambda)\right]^{-1} - 1$ is the Fourier transform of function from $L_1(-\infty,\infty)$. It follows that the sum of the first two terms in the right-hand side of (15) is also the Fourier transform of a summable function $F_S^{(1)}(x)$. Finally, since $\widetilde{h}(\lambda N^{-1}) = 0$ for $|\lambda| > 2N$, the third term in the same formula vanishes for $|\lambda| > 2N$ and is bounded. As such, it is the Fourier transform of a bounded function $F_S^{(2)}(x) \in L_2(-\infty,\infty)$, and the theorem is proved. ∎

Using the identity (10), Theorem 2 and Theorem 3, as in [9], we obtain Gelfand-Levitan-Marcenko's basic equation which has important role in the solution of IPST

$$F(x+y) + K(x,y) + \int_x^{\infty} K(x,t)F(t+y)dt = 0, \quad (x < y < \infty)$$

where

$$F(x) = \sum_{k=1}^{n} m_k^2 e^{-\lambda_k x} + \frac{1}{2\pi}\int_{-\infty}^{\infty} (1 - S(\lambda))e^{i\lambda x}dx$$

and "the norming constants" are

$$m_k^{-2} = \int_0^{\infty} |e(x,i\lambda_k)|^2 dx + \frac{\alpha_1 + 2\alpha_2\lambda_k}{2\lambda_k} |e(0,i\lambda_k)|^2.$$

The collection $\{S(\lambda)\ (-\infty < \lambda < \infty); \lambda_k;\ m_k\ (k = \overline{1, n})\ \}$ which is defined as above is defined as the scattering data of the boundary problem (1), (2). Theorems on uniquely construction of the boundary problem (1), (2) with respect to the scattering data are given in [9].

References

1. P. A. Binding, P. J. Browne, B. A. Watson, Inverse spectral problems for Sturm-Liouville equations with eigenparameter dependent boundary conditions, *J. London Math. Soc.* **2** (62) (2000) 161-182.
2. P. A. Binding, P. J. Browne, B. A. Watson, Sturm-Liouville problems with boundary conditions rationally dependent on the eigenparameter I, *Proc. Edinburg Math. Soc.* **45** (2002) 631-645.
3. P. A. Binding, P. J. Browne, B. A. Watson, Sturm-Liouville problems with boundary conditions rationally dependent on the eigenparameter II, *Journal of Comp. and Appl. Math.* **148** (2002) 147-168.
4. C. T Fulton, Singular eigenvalue problems with eigenvalue-parameter contained in the boundary conditions, *Proc. Soc. Edinburgh Sec. A* **87** (1980) 1-34.
5. N. Y. Kapustin, E. I. Moiseev, A remark on the convergence problem for spectral expansions corresponding to classical problem with a spectral parameter in the boundary condition, *Diff. Equations* **37** (12) (2001) 1677-1683.
6. V. I. Kopılov, Levitan-Gasymov inverse spectral theorem for Sturm-Liouville problem with a spectral parameter in the boundary condition, *Functional Analysis, Linear Spaces*. Ulyanovsk (1983) 73-82.
7. N. B. Kerimov, Kh. R. Mamedov, On a boundary value problem with a spectral parameter in the boundary conditions, *Sib. Math. J.* **40** (2) (1999) 281-290; translation from *Sib. Mat. Zh.* **40** (2) (1999) 325-335.
8. B. M. Levitan, The inverse scattering problem of quantum theory, *Math. Notes* **17** (4) (1975) 611-624.
9. Kh. R. Mamedov, Uniqueness of solution of the inverse problem of scattering theory with a spectral parameter in the boundary condition, *Math. Notes* **74** (1) (2003).
10. Kh. R. Mamedov, H. Menken, Levinson's type formula for boundary problem with a spectral parameter in the boundary condition, *Ukr. Math. J.* (to appear).
11. V. A. Marchenko, Reconstruction of the potential energy from the phases of the scattered waves, *Dokl. Akad. Nauk SSSR* **104** (5) (1955) 695-698.
12. V. A Marchenko, *Strum-Liouville Operators and Applications*. Birkhauser Verlag (1986).
13. A. G. Megrabov, Inverse problems of scattering for planar waves with free boundary or dependent non-homogeneous surface, *Proc. Geom. Math. Problems* **4**, Novosibirsk (1979).
14. S. V. Meleshko, Yu. V. Pokornyj, On a vibrational boundary value problem, *Diff. Uravn.* **23** (8) (1987) 1466-1467.

15. V. N. Pivovarchik, Direct and inverse three-point Sturm-Liouville problem with parameter-dependent boundary conditions, *Asymptotic Anal.* **26** (2001) 219-238.
16. E. A Pocheykina-Fedotova, On the inverse problem of boundary problem for second order differential equation on the half line, *Izv. Vuzov* **17** (1972) 75-84.
17. E. M. Russakovskii, The matrix Sturm-Liouville problem with spectral parameter in the boundary conditions, Algebraic and operator aspects, *Trans. Moscow. Math. Soc.* **57** (1996) 159-184.
18. A. A. Shkalikov, Boundary problems for ordinary differential equations with parameter in the boundary conditions, *J. Soviet. Math.* **33** (1986) 1311-1342.

V. Kadets and W. Żelazko

On the Automorphisms and Multipliers Related to Generalized Integration Operators

Swietłana Minczewa-Kamińska [a]

[a]Institute of Mathematics, University of Rzeszów, Rejtana 16 A, 35-310 Rzeszów, Poland

Abstract

We give necessary and sufficient conditions for multipliers of a convolution of the generalized integration operator to be topological automorphisms in the space of continuous functions. Moreover, an equivalent form of the conditions is found in terms of cyclic elements of the considered integration operator.

Key words: Generalized integration operator, convolution of generalized integration operator, commutant, multiplier, cyclic element.
2000 MSC: 45D05, 45E10, 45P05, 46E15

1. Introduction

Due to the Titchmarsh theorem, the convolutional algebra of continuous functions on the positive half-line has no divisor of zero which makes it possible to construct the field of the convolution quotients, called Mikusiński operators, the objects interesting both for their mathematical elegance and numerous applications (see [8]).

Though the Titchmarsh theorem is not true on the whole real line, the space of continuous functions (on an arbitrary interval Δ containing 0, bounded or not) remains an annihilators-free convolutional algebra with respect to the Duhamel convolution and there is an interesting interplay between the Duhamel convolution, the multipliers (in the sense of Larsen [6]) and continuous automorphisms of this algebra commuting with the Volterra integration operator (see [2] and [9]).

Together with a general notion of the Dimovski convolution introduced in [1], embracing the Duhamel convolution as a particular case, it appeared a problem of mutual relations between this notion, the generalized integration operator and the multipliers of this convolution in various classes of functions. Several papers (see e.g. [2], [9], [3]) were devoted to a study of many aspects of this problem.

In this paper, we are interested in giving a convolutional representation of multipliers of the Dimovski convolution in the space $C(\Delta)$ of continuous functions on Δ. The main result (Theorem 12) is a characterization of the topological automorphisms on $C(\Delta)$ among multipliers of the Dimovski convolution. We also give (see Theorem 14) necessary and sufficient conditions for continuous functions to be cyclic elements of the generalized integration operator in $C(\Delta)$.

We begin with some general definitions. Let X be a linear space over the field $\mathbb{C}$ of complex numbers and let $L\colon\ X \to X$ be a linear operator. A bilinear, commutative and associative operation $\circledast\colon\ X \times X \to X$ is said to be *a convolution of the linear operator* L if

$$L(x \circledast y) = (Lx) \circledast y, \qquad x, y \in X$$

and the pair $(X, \circledast)$ is called *a convolutional algebra.* The convolutional algebra $(X, \circledast)$ is *annihilators-free* if $x \circledast y = 0$ for all $x \in X$ implies $y = 0$. An element d of $(X, \circledast)$ is *a divisor of zero* of the convolution $\circledast$ if $d \neq 0$ and $d \circledast a = 0$ for a certain $a \in X, a \neq 0$. A linear operator $L\colon\ X \to X$ is said to be *a convolutional operator* if there exists an $r \in X$ such that

$$Lx = r \circledast x, \qquad x \in X, \tag{1}$$

where $\circledast$ is a convolution of L.

Definition 1. A linear operator $M:\ X \to X$ is said to be a multiplier of the convolutional algebra $(X, \circledast)$ if $M(x \circledast y) = (Mx) \circledast y$ for all $x, y \in X$.

Due to [7], the multipliers of annihilators-free convolutional algebra $(X, \circledast)$ are linear operators on X and form a commutative ring.

By [6], if a linear topological space X has the closed graph theorem property, the convolutional algebra $(X, \circledast)$ is annihilators-free and the convolution $\circledast$ is separately continuous, then every multiplier of $(X, \circledast)$ is a continuous operator.

Definition 2. An operator $A\colon\ X \to X$, where X is a linear topological space, is called a topological automorphism of X if it is a linear operator and a homeomorphism on X.

Let X_D be a subspace of a linear space X and let $D:\ X_D \to X$ be a linear operator. A linear operator $R:\ X \to X_D$ is said to be a *right inverse* of D, if $DRx = x$ for each $x \in X$. According to Dimovski [1], if a convolutional operator L given by (1) is a right inverse operator of D, then each multiplier M of the convolutional algebra $(X, \circledast)$ has the form

$$Mx = D(m \circledast x), \qquad \text{where } m := Mr. \tag{2}$$

In this paper we will assume that $X = C(\Delta)$, where Δ is an arbitrary (finite, infinite, closed, open, or half-open) interval containing zero, i.e. X is a Banach space in case Δ is a compact interval and is a Fréchet space in case Δ is a noncompact interval.

2. Dimovski convolutions in $C(\Delta)$

We will study the right inverse operator of the differentiation $D = d/dt$ with a given boundary value condition, i.e. the operator $L = L_\Phi$ which assigns to each $f \in C(\Delta)$ the solution $y = Lf$ of the boundary value problem

$$y' = f, \qquad \Phi(y) = 0,$$

where Φ is a nonzero linear continuous functional on $C(\Delta)$. By [1, p. 69], a linear continuous operator $R: \ C(\Delta) \to C(\Delta)$ is a right inverse of the differentiation $D = d/dt$ iff it admits a representation of the form

$$Rf(t) = \int_0^t f(\tau)\,d\tau + \Upsilon(f)$$

with a continuous linear functional Υ on $C(\Delta)$. Consequently, the operator L_Φ is represented in the form:

$$L_\Phi f(t) = \int_0^t f(\tau)\,d\tau - \Phi\left\{\int_0^t f(\tau)\,d\tau\right\}, \tag{3}$$

under the condition that the restriction of Φ to $C^1(\Delta)$ is $I - L_\Phi D$ (necessary for Φ to be so-called *defining projector* of L_Φ - see [1, p. 40-41]). The linear continuous operator L_Φ given by (3) will be called the *generalized integration operator*.

By the *Dimovski convolution*, we mean the operation $\overset{\Phi}{*}$: $C(\Delta) \times C(\Delta) \to C(\Delta)$ defined for a given nonzero linear continuous functional Φ on $C(\Delta)$ and $f, g \in C(\Delta)$ by the formula:

$$(f \overset{\Phi}{*} g)(t) := \Phi_x\left\{I_x^t(f, g)\right\}, \tag{4}$$

where the symbol Φ_x means that Φ acts on x only and

$$I_x^t(f, g) := \int_x^t f(x + t - \tau) g(\tau)\,d\tau. \tag{5}$$

In [1, p. 62], Dimovski introduced the above convolution $\overset{\Phi}{*}$ and proved that $\overset{\Phi}{*}$ is a bilinear, commutative, associative and separately continuous operation on $C(\Delta)$

and the algebra $(C(\Delta), \overset{\Phi}{*})$ is annihilators-free. One can prove that $\overset{\Phi}{*}$ is a convolution of the generalized integration operator L_Φ given by (3).

Example 3. If $\Phi(f) = f(0)$, the generalized integration operator L_Φ given by (3) coincides with the Volterra integration operator l given for $f \in C(\Delta)$ by

$$lf(t) := \int_0^t f(\tau)\, d\tau$$

and the Dimovski convolution $\overset{\Phi}{*}$ in $C(\Delta)$ coincides with the Duhamel convolution $*$ defined for $f, g \in C(\Delta)$ by

$$(f * g)(t) := \int_0^t f(t-\tau)g(\tau)\, d\tau. \tag{6}$$

All multipliers of the Duhamel convolution in $C(\Delta)$ and all automorphisms among them are described in [2].

Example 4. In case $\Phi = E$, where $E(f) := \frac{1}{2}(f(0) + f(1))$ and $[0,1] \subset \Delta$, the generalized integration operator L_Φ, called in this case the Euler integration operator, is of the form

$$L_E f(t) = \int_0^t f(\tau)\, d\tau - \frac{1}{2}\int_0^1 f(\tau)\, d\tau$$

and the convolution $\overset{E}{*}$ in $C(\Delta)$ is given by

$$(f \overset{E}{*} g)(t) = \frac{1}{2}\left[\int_0^t f(t-\tau)g(\tau)\, d\tau - \int_t^1 f(1+t-\tau)g(\tau)\, d\tau\right].$$

Example 5. In case $\Phi = B$, where $B(f) := \int_0^1 f(\tau)\, d\tau$ and $[0,1] \subset \Delta$, the generalized integration operator L_Φ, called in this case the Bernoulli integration operator, is given by

$$L_B f(t) = \int_0^t f(\tau)\, d\tau - \int_0^1 (1-\tau) f(\tau)\, d\tau$$

and the convolution $\overset{B}{*}$ in $C(\Delta)$ is of the form

$$(f \overset{B}{*} g)(t) = \int_0^1 \left\{ \int_x^t f(x+t-\tau)g(\tau)\,d\tau \right\} dx.$$

3. Multipliers of Dimovski convolution

Let us normalize $\Phi \in (C(\Delta))'$, $\Phi \neq 0$, by imposing on Φ the condition

$$\Phi\{1\} = 1. \tag{7}$$

Then operator L_Φ defined in (3) is a convolutional operator of the form

$$L_\Phi f = \{1\} \overset{\Phi}{*} f, \qquad f \in C(\Delta),$$

where $\{1\}$ denotes the constant function on Δ. By the Riesz-Markov theorem (see [5], Theorem 4.10.1), Φ has a compact support, i.e. there are a compact interval $[\alpha, \beta] \subset \Delta$ and a function ω of bounded variation on $[\alpha, \beta]$ such that

$$\Phi(f) = \int_\alpha^\beta f(x)\,d\omega(x).$$

The case where ω is a jump-function is described in [3]. We will assume further on that $\omega \in AC[\alpha, \beta]$ and $\nu := \omega' \in BV[\alpha, \beta]$. In this case (see [9], p. 50) there exists a $\Psi \in (C(\Delta))'$, $\Psi \neq 0$, such that

$$\Phi(f) = \Psi_x \left\{ \int_0^x f(\tau)\,d\tau \right\} = \Psi(lf), \qquad f \in C(\Delta). \tag{8}$$

Then the convolution $\overset{\Phi}{*}$ given by (4) maps $C(\Delta)\times C(\Delta)$ in $C^1(\Delta)$ and the derivative of the convolution $(f \overset{\Phi}{*} g)$ defined in (4) is of the form

$$\begin{aligned}(f \overset{\Phi}{*} g)'(t) = {} & f(t)\Phi(g) + g(t)\Phi(f) \\ & -\Psi\{1\}(f * g)(t) + \Psi_x\left\{I_x^t(f,g)\right\},\end{aligned} \tag{9}$$

where $*$ is the Duhamel convolution defined by (6) and I_x^t is given by (5) (see [1, p. 67]).

Lemma 6. *([1, p. 75]) The operation* $\tilde{*} : C(\Delta) \times C(\Delta) \to C(\Delta)$ *given by*

$$(f \tilde{*} g)(t) := f(t)\Phi(g) + g(t)\Phi(f) - \Psi\{1\}(f * g)(t) + \Psi_x\left\{I_x^t(f,g)\right\} \tag{10}$$

is a convolution of the generalized integration operator of the form

$$L^\Psi f = lf - \Psi(l^2 f) \tag{11}$$

with $\Psi \in (C(\Delta))'$. The operator L^{Ψ} is a convolutional operator of the form $L^{\Psi} f = r \tilde{*} f$ with $r := l\{1\} - \Psi(l^2\{1\})$ and $(C(\Delta), \tilde{*})$ is a convolutional algebra with a unit, i.e. $\{1\} \tilde{*} f = f$ for $f \in C(\Delta)$.

Now we will characterize all multipliers of the convolution $\overset{\Phi}{*}$ defined by (4) with Φ of the form (8).

Theorem 7. *A linear continuous operator $M : C(\Delta) \to C(\Delta)$ is a multiplier of the Dimovski convolution $\overset{\Phi}{*}$ given by (4) with Φ of the form (8) iff*

$$\begin{aligned} Mf(t) &= f(t)\Phi(m) + m(t)\Phi(f) \\ &\quad -\Psi\{1\}(m * f)(t) + \Psi_x\left\{I_x^t(m, f)\right\}, \end{aligned} \tag{12}$$

where $m := M\{1\} \in C(\Delta)$.

Proof. It is clear that every operator M of the form (12) satisfies the identity $M(f \overset{\Phi}{*} g) = (Mf) \overset{\Phi}{*} g$ for $f, g \in C(\Delta)$. On the other hand, if M is a multiplier of the convolution $\overset{\Phi}{*}$ of the form (4), then equations (2) and (9) imply (12). □

Corollary 8. *A linear continuous operator $M : C(\Delta) \to C(\Delta)$ is a multiplier of the convolution $\overset{\Phi}{*}$ with Φ given by (8) iff $Mf = m \tilde{*} f$ where $m := M\{1\} \in C(\Delta)$ and $\tilde{*}$ is defined by (10).*

Lemma 9. *If the multiplier M of the convolution $\overset{\Phi}{*}$, given by (12) with $\Psi \in (C(\Delta))'$, satisfies the condition $\Psi(lm) = 0$, then*

$$\Psi(lMf) = 0, \qquad f \in C(\Delta). \tag{13}$$

Proof. It follows from the assumptions that

$$\Psi(lMf) = -\Psi_t \Psi_x \left\{ \int_0^t m(\tau)\, d\tau \int_0^x f(\sigma)\, d\sigma \right\} + \Psi_t \Psi_x \{h(t,x)\},$$

where

$$h(t,x) = \int_0^t m(\tau) \left[\int_{t-\tau}^{t+x-\tau} f(\sigma)\, d\sigma \right] d\tau - \int_0^x m(\tau) \left[\int_{x-\tau}^{t+x-\tau} f(\sigma)\, d\sigma \right] d\tau.$$

Since $h(t,x) = -h(x,t)$, the assertion follows from the Fubini theorem. □

Define the operator T, for $f \in C(\Delta)$, by

$$Tf(t) := m(t)\Phi(f) - \Psi\{1\}(m * f)(t) + \Psi_x\left\{I_x^t(m,f)\right\}. \tag{14}$$

Lemma 10. *Let Δ be a compact interval. Then T, given by (14) with $m \in C(\Delta)$, is a compact operator in the space $C(\Delta)$.*

Proof. The assertion follows due to the Arzela-Ascoli theorem, applied to each of the addends in the representation (14) of the operator T (see [9]). $\square$

From Lemma 10 one can easily conclude the following assertion:

Corollary 11. *Every operator T of the form (14) with $m \in C(\Delta)$ is compact in $C(\Delta)$ also in case Δ is a noncompact interval.*

Theorem 12. *A multiplier M of the convolutional algebra $(C(\Delta), \overset{\Phi}{*})$ with Φ given by (8) is a topological automorphism on $C(\Delta)$ iff the following conditions hold:*

(*a*) *$M\{1\}$ is not a divisor of zero of the convolution $\overset{\Phi}{*}$;*
(*b*) *$\Psi(lm) \neq 0$, where $m := M\{1\}$.*

Proof. Due to Lemma 2 in [2], it suffices to show that conditions (a) and (b) are necessary and sufficient for M of the form (12) to be invertibile.

Assume that M is a one-to-one linear mapping and condition (a) is not fulfilled, i.e. $M\{1\} \overset{\Phi}{*} f = 0$ for some non-zero function $f \in C(\Delta)$. But then the operator M defined in (12) is not injective, contrary to the assumption. If (b) does not hold, then (13) follows by Lemma 9, which contradicts (7) and so the surjectivity of the operator M. Assume (a) and (b). By (a), (9) and (7), it follows that M is an injection. To prove that M is a surjection consider the equation

$$m \tilde{*} f = g. \tag{15}$$

Since M is an injection and T given by (14) is compact, equation (15) has a unique solution f for a given $g \in C(\Delta)$, according to (b) and the Fredholm alternative. $\square$

4. Cyclic elements of L in $C(\Delta)$

In this part, we suppose that $\Phi \in (C(\Delta))'$ is of the form (8) with $\Psi \in (C(\Delta))'$ and denote, for simplicity, the generalized integration operator L_Φ by L. We will use the convolution $\overset{\Phi}{*}$ to characterize explicitly the cyclic elements (in $C(\Delta)$) of the operator L.

We start from a general definition:

Definition 13. For an arbitrary linear continuous operator S on $C(\Delta)$, a function $k \in C(\Delta)$ is said to be a *cyclic element* of the operator S if the span of $\{S^n k\}_{n=1}^{\infty}$ is dense in $C(\Delta)$.

Theorem 14. *A function $k \in C(\Delta)$ is a cyclic element of the operator L in the space $C(\Delta)$ iff k is not a divisor of zero of the convolution $\overset{\Phi}{*}$ and $\Psi(lk) \neq 0$.*

Proof. First assume that k is a cyclic element of L and a divisor of zero of the convolution $\overset{\Phi}{*}$. Then there exists a $g \in C(\Delta)$, $g \neq 0$ such that $k \overset{\Phi}{*} g = 0$. Hence $L(k \overset{\Phi}{*} g) = 0$ and so $(L^n k) \overset{\Phi}{*} g = 0$ for all $n \in \mathbb{N}_0 := \mathbb{N} \bigcup \{0\}$. Consequently, $f \overset{\Phi}{*} g = 0$ for every $f \in C(\Delta)$, but this contradicts the identity $Lg = \{1\} \overset{\Phi}{*} g$.

Let now k be a cyclic element of L and $\Psi(lk) = 0$. Let A_n be the Appell polynomials, i.e. $A_n(t) := L^n\{1\}$ for $t \in (-\infty, \infty)$ and $n \in \mathbb{N}_0$. Since $Lk = \{1\} \overset{\Phi}{*} k$, we have $L^{n+1}k = A_n(t) \overset{\Phi}{*} k$ for $n \in \mathbb{N}_0$ or, due to formula (9),

$$\begin{aligned} L^n k(t) = {} & A_n(t)\Phi(k) + k(t)\Phi(A_n) \\ & -\Psi\{1\}(A_n * k)(t) + \Psi_x\left\{I_x^t(A_n, k)\right\} \end{aligned} \tag{16}$$

for $t \in (-\infty, \infty)$. This is an equation analogous to (12). By Lemma 9, the condition $\Psi(lk) = 0$ implies $\Phi(L^n k) = 0$ for $n = \mathbb{N}_0$. Therefore, $\{1\}$ cannot be approximated by means of linear combinations of $L^n k$, in view of (7). This proves the necessity of the condition $\Psi(lk) \neq 0$.

Let now $k \in C(\Delta)$ satisfy the condition $\Psi(lk) \neq 0$ and be not a divisor of zero of the convolution $\overset{\Phi}{*}$. We will show that k is a cyclic element of L in $C(\Delta)$.

Fix $f \in C(\Delta)$. We look for a function $g \in C(\Delta)$ such that

$$f(t) = \Phi(k)g(t) + k(t)\Phi(g) - \Psi\{1\}(k * g)(t) + \Psi_x\left\{I_x^t(k, g)\right\}. \tag{17}$$

The operator T, given by

$$Tg(t) := k(t)\Phi(g) - \Psi\{1\}(k * g)(t) + \Psi_x\left\{I_x^t(k, g)\right\},$$

is of the form (14). By Lemma 10 and Corrolary 11, the operator T is compact. Since $\Psi(lk) \neq 0$, equation (17) is a Fredholm integral equation of the second kind. But k is not a divisor of zero of the convolution $\overset{\Phi}{*}$, so the function $g \equiv 0$ is the unique solution of the homogeneous equation $k \tilde{*} g = 0$, which guarantees the existence of a unique solution g of (17).

Let $\{g_n\}_{n=0}^{\infty}$ be a sequence of polynomials converging almost uniformly to g in $C(\Delta)$, which can be represented in the form $g_n = \sum_{j=0}^{n} \alpha_{jn} A_j$ for $n \in \mathbb{N}_0$, where A_j are the Appell polynomials. Let $f_n := \sum_{j=0}^{n} \alpha_{jn} L^j k$. By (16), we can write

$$f_n(t) = \Phi(k)g_n(t) + k(t)\Phi[g_n(t)] - \Psi\{1\}(k * g_n)(t) + \Psi_x\left\{I_x^t(k, g_n)\right\}.$$

Hence the sequence $\{f_n\}_{n=0}^{\infty}$ converges almost uniformly to the function f in $C(\Delta)$, due to (17). Therefore, k is a cyclic element of L in $C(\Delta)$. □

The above characterization allows us to reformulate Theorem 12 as follows:

Theorem 15. *A multiplier M of the convolutional algebra* $(C(\Delta), \overset{\Phi}{*})$ *with* Φ *defined in (8) is a topological automorphism on the space* $C(\Delta)$ *iff the function* $m := M\{1\} \in C(\Delta)$ *is a cyclic element of the operator L.*

References

1. I. Dimovski, *Convolutional Calculus* (Kluwer Acad. Publ., Dordrecht, 1990).
2. I. Dimovski, S. Mincheva, Automorphisms of C which commutes with the integral operator, *Integral Transforms and Special Functions* **4** (1996) 69–76.
3. I. Dimovski, A. Kamiński, S. Mincheva, The commutant of an integration operator with multipoint functional, *Fractional Calculus & Applied Analysis* **4** (2001) 245–254.
4. I. Dimovski, S. Grozdev, Bernoulli operational calculus, in: *Mathematics and Education in Mathematics*, (Sofia, 1980) 30–36 (in Bulgarian).
5. R. Edwards, *Functional Analysis. Theory and Applications*. Holt, Rinehart and Winston, New York, 1965.
6. R. Larsen, *An Introduction to the Theory of Multipliers*. Springer-Verlag, Berlin, 1972.
7. L. Máté, Multiplier operators and quotiend algebra, *Bull. Acad. Polon. Sciences, Ser. Math. Astr. et Phys.* **13** (1972) 523–526.
8. J. Mikusiński, *Operational Calculus*, vol. I–II, PWN–Pergamon Press, Warszawa, 1983, 1987.
9. S. Mincheva, Automorphisms commuting with integration operators in some functional spaces, *Institute of Mathematics and Informatics, Bulgarian Academy of Sciences* (Sofia, 2000) 1–151 (Ph.D. Dissertation; in Bulgarian).

Functional Analysis and its Applications
V. Kadets and W. Żelazko

On Some Classes of Mappings Preserving of Measure

Oksana A. Ochakovskaya

Institute of Applied Mathematics and Mechanics NAS Ukraine, Donetsk, Ukraine

Abstract

Main result in the present paper contains the conditions which give that certain one-to-one mapping is a measure-preserving transformation. Some of these conditions can not be relaxed.

Key words: Measure-preserving transformations, ergodic theory
2000 MSC: 28D05

1. Introduction

Measure-preserving transformations plays an important role in ergodic theory and related questions (see, for instance, [1], [2] and the bibliography therein). In the present paper we study such transformations acting from $\mathbb{R}^n$ into $\mathbb{R}^n$. Throughout, we assume that $n \geq 2$.

Denote by $C^k(\mathbb{R}^n, \mathbb{R}^n)$ the collection of all C^k – mappings from $\mathbb{R}^n$ into $\mathbb{R}^n$. For $r > 0$, $y \in \mathbb{R}^n$ we set $B_r(y) = \{x \in \mathbb{R}^n : |x-y| \leq r\}$, $S_r(y) = \{x \in \mathbb{R}^n : |x-y| = r\}$, where $|\cdot|$ is the Euclidean norm in $\mathbb{R}^n$. We write $m(G)$ for Lebesgue measure of the set $G \subset \mathbb{R}^n$. Let $f * g$ be a convolution of functions f and g. Denote by χ_G the characteristic function of the set $G \subset \mathbb{R}^n$. For $r > 0$ let $V_r(\mathbb{R}^n)$ be a collection of all functions $f \in C(\mathbb{R}^n)$ with zero integrals over each ball with radius r. We set also $\Lambda_n = \{t > 0 : t = \frac{\lambda}{\beta}$, where $J_{\frac{n}{2}}(\alpha) = J_{\frac{n}{2}}(\beta) = 0,\ \beta \neq 0\}$, where J_ν is the Bessel function of order ν. Let $E^n = \{x \in R^n : x_n > 0\}$.

2. Main results

Theorem 1. *Assume that the injective map $F \in C^1(\mathbb{R}^n, \mathbb{R}^n)$ satisfies the following conditions:*
a) $m(F(B_1(y))) = m(B_1(y))$ for all $y \in \mathbb{R}^n$;
b) there exist a positive numbers M_q, $q = 1, 2, ...$ such that

$$\sum_{m=1}^{\infty} \left(\inf_{q \geq m} M_q^{\frac{1}{q}} \right)^{-1} = \infty \tag{1}$$

and

$$\int_{\mathbb{R}^{n-1}} |m(F(B_r(y))) - m(B_r(y))|(1 + |y_1| + ... + |y_{n-1}|)^q dy_1...dy_{n-1} \leq M_q e^{\alpha|y_n|} \tag{2}$$

for some $\alpha > 0$ and all $y_n \in R^1$, $q \in \mathbb{N}$. Then $m(F(G)) = m(G)$ for each Lebesgue measurable set $G \subset \mathbb{R}^n$.

Theorem 2. *For every $r > 0$, $\alpha > 0$ and every a sequence $\{M_q\}_{q=1}^{\infty}$ of positive numbers such that*

$$\sum_{m=1}^{\infty} \left(\inf_{q \geq m} M_q^{\frac{1}{q}} \right)^{-1} < \infty \tag{3}$$

there exists a one-to-one mapping $F \in C^1(E^n, E^n)$ with positive Jacobian such that $m(F(B_1(y))) = m(B_1(y))$ for all $B_1(y) \subset E^n$ and

$$\int_{\mathbb{R}^{n-1}} |m(F(B_r(y))) - m(B_r(y))|(1 + |y_1| + ... + |y_{n-1}|)^q dy_1...dy_{n-1} \leq M_q e^{-\alpha y_n} \tag{4}$$

for all $y_n > r$, $q \in \mathbb{N}$.
Moreover $m(F(B)) \neq m(B)$ for some ball $B \subset E^n$.

Theorem 3. *For each $r \in \Lambda_n$ there exists an one-to-one mapping $\Phi \in C^\infty(\mathbb{R}^n, \mathbb{R}^n)$ with positive Jacobian such that $m(\Phi(B_1(y))) = m(B_1(y))$ and $m(\Phi(B_r(y))) = m(B_r(y))$ for all $y \in \mathbb{R}^n$. Besides, $m(\Phi(B)) \neq m(B)$ for some ball $B \subset \mathbb{R}^n$.*

3. Proofs

To prove the main results we require the following auxiliary statements:
1) Let $f \in V_r(\mathbb{R}^n)$ and assume that there exist a positive numbers $M_q, q = 1, 2, ...$ satisfying (1) such that the following inequality holds

$$\int_{\mathbb{R}^n} |f(x_1, ..., x_n)|(1 + |x_1| + ... + |x_{n-1}|)^q dx_1...dx_{n-1} \leq M_q e^{\alpha|x_n|} \tag{5}$$

for some $\alpha > 0$ and all $x_n \in \mathbb{R}^1$, $q \in \mathbb{N}$. Then $f = 0$.

2) For every $\alpha > 0$ and every a sequence $\{M_q\}_{q=1}^{\infty}$ of positive numbers satisfying (3) there exists a real-valued function $f_\alpha \in V_r(\mathbb{R}^n) \cap C^\infty(\mathbb{R}^n)$ such that

$$\int_{\mathbb{R}^{n-1}} |f_\alpha(x_1, ..., x_n)|(1 + |x_1| + ... + |x_{n-1}|)^q dx_1 ... dx_{n-1} \leq M_q e^{-\alpha x_n}$$

for all $x_n > 0$, $q \in \mathbb{N}$. Besides, the function $\frac{\partial f_\alpha}{\partial x_1}$ is nonzero and bounded in E^n.

The proof of this statement can be obtained by repeating the arguments from the paper [3].

We proceed now to the proof of the main results.

Proof of Theorem 1. Using the formula $m(F(G)) = \int_G |J(F,x)| dx$, where $J(F,x)$ is the Jacobian of mapping F, we can rewrite the condition a) in Theorem 1 as

$$\int_{B_1(y)} (|J(F,x) - 1) \, dx = 0 \quad \text{for all} \quad y \in \mathbb{R}^n \tag{6}$$

Besides

$$m(F(B_r(y))) - m(B_r(y)) = \int_{B_r(y)} (|J(F,x)| - 1) \, dx = v * \chi_{B_r} = 0, \tag{7}$$

where $v(x) = |J(F,x)| - 1$.

Thus the function $f = v * \chi_{B_r}$ satisfies the condition (5). Hence, $f = 0$ on $\mathbb{R}^n$. Taking into account that $r \notin \Lambda_n$, by two-radii theorem (see, [4]) we obtain that $v = 0$. This means that $m(F(G)) = m(G)$ for every Lebesgue measurable set $G \subset \mathbb{R}^n$. The proof of Theorem 1 is now complete.

Proof of Theorem 2. Let $\alpha > 0$, $r > 0$ and assume that the sequence $\{M_q\}_{q=1}^{\infty}$ satisfies (3). Then the sequence $M_q' = \dfrac{M_q e^{\alpha r}}{(2+r)^q (n-1)^q}$ also satisfies (3). We set $\varphi(x) = f_\alpha(x)$ where f_α is a function in Lemma for the sequence M_q'. Then $|\dfrac{\partial \varphi}{\partial x_1}| < C$ for all $x \in E^n$. We consider the mapping $F(x) = (F_1(x), ..., F_n(x)) = (x_1 + \dfrac{\varphi(x)}{N}, x_2, ..., x_n)$, where $N > \max\left\{C, \dfrac{\pi^{n/2} r^{n-1}}{\Gamma(\frac{n}{2})}\right\}$. Then we have

$$|J(F,x)| - 1 = \frac{1}{N} \frac{\partial \varphi(x)}{\partial x_1}. \tag{8}$$

We now claim that F is one-to-one mapping. Let $x, x' \in E^n$ such that $x \neq x'$. If $(x_2, ..., x_n) \neq (x_2', ..., x_n')$ we see that $F(x) \neq F(x')$. Otherwise, $(x_2, ..., x_n) = (x_2', ..., x_n')$ and $x_1 \neq x_1'$. In this case the $F(x) \neq F(x')$ since the function $F_1(x) = x_1 + \dfrac{\varphi(x)}{N}$ is increasing with respect to variable x_1. Indeed, from the definition of N we have the inequality $\dfrac{\partial F_1}{\partial x_1} = 1 + \dfrac{1}{N} \dfrac{\partial \varphi}{\partial x_1} > 0$. This yields that F is one-to-one.

We now prove the condition b). Setting $s_q(x) = (1 + |x_1| + ... + |x_{n-1}|)^q$ for $x = (x_1, ..., x_n) \in \mathbb{R}^n$, we obtain

$$\int_{\mathbb{R}^n} |m(F(B_r(y))) - m(B_r(y))| s_q(y) dy_1 ... dy_{n-1} =$$

$$= \int_{\mathbb{R}^{n-1}} \left| \int_{B_r(y)} \frac{1}{N} \frac{\partial \varphi(x)}{\partial x_1} dx \right| s_q(y) dy_1 ... dy_{n-1}.$$

Using Green's theorem we estimate the left part of last equality by the expression

$$\frac{1}{N} \int_{S_r(0)} \left(\int_{\mathbb{R}^{n-1}} |\varphi(y_1, ..., y_{n-1}, y_n + x_n)| s_q(y - x) dy_1 ... dy_{n-1} \right) d\sigma(x) \qquad (9)$$

where $d\sigma(x)$ is the surface element of sphere $S_r(0)$.

Further, for $x \in S_r(0)$ and $y \in \mathbb{R}^{n-1}$ we have

$$s_q(y - x) \leq (n-1)^q (2+r)^q s_q(y)$$

Then (9) can be estimated by the expression

$$\frac{1}{N}(n-1)^q(2+r)^q \int_{S_r(0)} (\int_{\mathbb{R}^{n-1}} |\varphi(y_1, ..., y_{n-1}, x_n + y_n)| \cdot$$

$$\cdot s_q(y) dy_1 ... dy_{n-1}) d\sigma(x) \leq M_q e^{-\alpha|y_n + x_n|} \leq M_q e^{\alpha|x_n|} e^{-\alpha|y_n|}$$

Bearing in mind that $\varphi \in V_r(\mathbb{R}^n)$ and using (6), we obtain that F satisfies the condition a) in Theorem 1. Besides, $|J(F, x)|$ is not identically one (see (7)). This implies that there exists a ball $B \subset R^n$ such that $m(F(B)) \neq m(B)$. This completes the proof of Theorem 2.

Proof of Theorem 3. Consider the mapping $\Phi(x) = (\Phi_1(x), ..., \Phi_n(x)) = (x_1 + \varepsilon \cos \nu x_1, x_2, ..., x_n)$, where $\nu > 0$, $J_{\frac{n}{2}}(\nu) = J_{\frac{n}{2}}(\nu r) = 0$, $\varepsilon \in (0; \frac{1}{\nu})$. Then the function $|J(\Phi, x)| - 1 = -\varepsilon\nu \sin(\nu x_1)$ satisfies the conditions (6), (7) (see, for instance, [4]). This means that the mapping Φ preserves the measure of all unit balls and all balls with radius $r \in \Lambda_n$.

Repeating the arguments from the proof of Theorem 2 we conclude that Φ is one-to-one mapping. Besides, since $|J(\Phi, x)|$ is not identically unit there exists a ball $B \subset \mathbb{R}^n$ such that $m(\Phi(B)) \neq m(B)$. This completes the proof of Theorem 3.

References

1. P. Billingsley, *Ergodic Theory and Information.* John Wiley and Sons, New York, 1965.
2. I.P.Kornfeld, J.G. Sinay, S.V. Fomin, *Ergodic theory.* Nauka, Moskow, 1980. (In Russian).
3. O.A. Ochakovskaya, Functions with zero integrals over balls with fixed radius, *Mathematical Physics, Analysis, Geometry* **10** (3) (2002) 43–51.
4. V.V. Volchkov, A definite version of the local two-radii theorem, *Math. Sb.* **186** (6) (1995) 15–34; English transl. in Sbornik Math. **186** (1995).

Functional Analysis and its Applications
V. Kadets and W. Żelazko

Extreme Problems and Scales of the Operator Spaces

Igor V. Orlov

Taurida National V. Vernadsky University, Simferopol, Ukraine

Abstract

The repeated compact–normal differentiability of the Euler–Lagrange functional is proved. The sufficient extreme conditions for such a functional in the case of one and two variables are obtained.

Key words: Euler–Lagrange functional, locally convex space, inductive scale of spaces, extreme problems, nuclear space, normal derivative, compact extremum
2000 MSC: 58C20, 46T20, 46A03

Introduction

The extreme problems for integral functionals play an important role in the modern nonlinear analysis and its application. Diverse methods are employed for the investigation of these problems (see, e.g. [1–3]). One of the fundamental difficulties arising in these questions is absence, with the exception of trivial cases [16], the second strong derivative for general integral functionals. The other aspect of the problem above is that extremae of such functionals are seek, as a rule, on some compact set. These cases exclude application of the classical Fréchet extreme conditions and enclose the investigation in the framework of the (in any case generalized) variational calculus.

In the present work, on the contrary, the notion of strong differentiability is rather weakened. As a result, the modified Fréchet extreme conditions are extended to the extremae of integral functionals on compact sets. The obtained results are new both in the general case of integrand function taking values in locally convex space (LCS) and in the classical scalar or Banach case.

The work consists of two sections. In the first section the extreme theory for the normally differentiable and K-differentiable functions in LCS is constructed. The normal differentiability and the repeated K-differentiability of Euler-Lagrange functional are proved. It allowed to obtain the sufficient K-extreme conditions for such a functional.

In the second section the sufficient extreme and K-extreme conditions for the two variables functions in the products of the nuclear spaces are obtained. It allowed to obtain the sufficient K-extreme conditions for two variables Euler-Lagrange functional.

1. *K*-extremae of the integral functionals

As it was shown by I.V.Skrypnik [1], the integral functionals of the classical Euler–Lagrange functional type

$$\Phi(y) = \int_a^b f(x, y(x), y'(x))dx \tag{1}$$

are not, in general, twice Fréchet differentiable. It is shown by M.Z.Zgurovsky and V.S.Melnik that the derivatives of such functionals are T-differentiable (see [19,7]). We'll show that the functional (1) satisfies a much more strong requirement of the repeated K-differentiability, even in the case of mapping $y(\cdot)$ into LCS. It allows us to apply the classical extreme conditions in terms of K-derivatives for investigation of the compact extremae (or, K-extremae, def. 17) for such a functional.

The introduced below notion of normal and (further) compact – normal differentiability (or, K-differentiability) is jointly connected to decompositions of the operator spaces into the inductive scales of LCS.

In what follows, E, E_i, F, F^j are LCS with the corresponding defining systems of semi–norms $\{\|\cdot\|_t\}_{t\in T}$, $\{\|\cdot\|_{t_i}\}_{t_i\in T_i}$, $\{\|\cdot\|^s\}_{s\in S}$, $\{\|\cdot\|^{s_j}\}_{s_j\in S_j}$, inductively ordered in ascending order of the semi-norms; $(E;F)$ is a space of the linear continuous operators acting from E to F; $(E_1, E_2; F)$ is a space of the bilinear continuous operators acting from $E_1 \times E_2$ to F.

Definition 1. Let $A \in (E;F)$. Assume, for any $s \in S$,

$$n_A(S) = \{t \in T \ : \ \|A\|_t^s := \sup_{\|x\|_t \leqslant 1} \|Ax\|^s < +\infty\}.$$

The multivalued mapping $n_A : S \to \text{ray}(T)$, here ray$(T)$ is the set of rays in T, is called the normal index, or n-index of the operator A; the quantities $\|A\|_t^s$ are called the co-norms.

The main properties and examples of normal indices are considered in [8,9].

Definition 2. Let $\mathcal{N} = \mathcal{N}(E;F)$ be set of the normal indices of operators from $(E;F)$. Put, for every $n \in \mathcal{N}$,

$$(E;F)_n = \{A \in (E;F) \ : \ n_A \preccurlyeq n\}.$$

Provide the every $(E;F)_n$ with the projective topology τ_n relative to the defining systems of semi–norms $\{\|\cdot\|_t^s\}_{t\in n(s)}$, $s \in S$. The inductive scale of LCS

$$\overrightarrow{(E;F)} := \{((E;F)_n, \tau_n)\}_{n\in\mathcal{N}} \tag{2}$$

is called the normal decomposition of the space $(E;F)$.

An information about inductive scales of LCS see in [17,4]. Note that the scales under consideration below are more general than the classical interpolational scales and the S.G.Krein-Yu.I.Petunin scales [5,6].

By analogy with the decomposition (2), one can introduce the normal indices and decompositions into the inductive scales of LCS for the space of linear continuous operators, acting from E_1 to $\overrightarrow{(E_2;F)}$:

$$\overrightarrow{(E_1;(E_2;F))},$$

and for the space of bilinear continuous operators, acting from $E_1 \times E_2$ to F :

$$\overrightarrow{(E_1,E_2;F)}.$$

Proposition 3. (The canonical isomorphism of normal indices for the linear and bilinear operators [10]). *The canonical relation $B(h_1,h_2) = (Ah_1)h_2$ between the linear operators $A : E_1 \to (E_2;F)$ and bilinear operators $B : E_1 \times E_2 \to F$ generates an isomorphism of the corresponding normal indices. In addition, the corresponding spaces of scales $\overrightarrow{(E_1;\overrightarrow{(E_2;F)})}$ and $(\overrightarrow{(E_1,E_2;F)}$ are isometrically isomorphic; that allows to say about an isometric isomorphism of the normal decompositions:*

$$\overrightarrow{(E_1;\overrightarrow{(E_2;F)})} \cong \overrightarrow{(E_1,E_2;F)}. \tag{3}$$

Definition 4. Let a mapping $\varphi : E \to F$ be defined in some zero neighborhood of E. Say that $\varphi(h) = o(h)$ if

$$\forall s \in S \exists t \in T : (\|\varphi(h)\|^s / \|h\|_t) \to 0 \text{ as } h \to 0.$$

Definition 5. A mapping $f : E \to F$ defined in some neighborhood of $x \in E$ is called normally differentiable at the point x if $\Delta f(x,h) = A_x h + \varphi_x(h)$, here $A_x \in (E;F)$, $\varphi_x(h) = o(h)$. The operator A_x is called the normal derivative and is denoted by the usual symbol $f'(x)$. The n-index of $f'(x)$ is called the n-index of f at the point x and is denoted by $n_f(x)$.

Obviously, in the case of Banach space E we have $\varphi_x(h) = o(\|h\|)$ and we get Fréchet differentiability. In the general case of an arbitrary LCS E the approach proposed above is well known in infinite-dimensional differential calculus. This approach is closed to the names of D.H.Hyers, H.R.Fisher, S.Lang, E.Binz and a number of other mathematicians (see, e.g., survey [1]). The distinction of our approach is consideration of the first normal derived mapping as a mapping from E to the scale $\overrightarrow{(E;F)}$.As the result, beginning with the continuous differentiability the proposed approach essentially differs from the abovementioned one and leads to some new applications. At first, let formulate the "normal form" of the mean value theorem [9].

Proposition 6 (Normal form of the mean value theorem). *Let E and F are real LCS, $[a;b] \subset E, D$ be finite or countable subset of $[a;b]$, $f : [a;b] \to F$. If f is continuous on $[a;b]$ and normally differentiable on $[a;b] \backslash D$, here $n_f(x) \preccurlyeq n \in \mathcal{N}$ for $x \in [a;b] \backslash D$, then for all $s \in S$ and $t \in n(s)$:*

$$\|f(b) - f(a)\|^s \leqslant \sup_{x \in [a;b] \backslash D} \|f'(x)\|_t^s \cdot \|b - a\|_t. \tag{4}$$

A typical situation for which the upper bound (4) is applicable, is the continuous normal differentiability, that is the continuity of the mapping $f' : E \to \overrightarrow{(E;F)}$.

Definition 7. Let a mapping $f : E \to F$ be continuously normally differentiable at a point $x \in E$. Then the derived mapping $f' : E \to \overrightarrow{(E;F)}$ can be considered at some neighborhood of x as the mapping to some LCS $(E;F)_n$, $n \in \mathcal{N}$. If $f' : E \to (E;F)_n$ is normally differentiable at a point x then we'll say that f is twice normally differentiable at the point x; moreover $f''(x) := (f')'(x)$. Analogously, one can defined the normal derivatives $f^{(k)}(x), k > 2$.

Note that by virtue of the canonical isomorphism (3), $f''(x)$ can be considered as a bilinear continuous operator. So, $f'' : E \to \overrightarrow{(E,E;F)}$, in addition the corresponding co-norms coincide: $\|f''(x)\|_{t_1}^{st_2} = \|f''(x)\|_{t_1 t_2}^s$.

Definition 8. Say that a mapping $r : E \to F$ defined in some zero neighborhood in E is the small mapping of the order k ($k = 0, 1, 2, \ldots$), or $r(h) = o\big((h)^k\big)$, if

$$\forall s \in S \quad \exists t \in T : \quad \lim_{h \to 0} \Big(\|r(h)\|^s / (\|h\|_t)^k \Big) = 0.$$

Theorem 9 (Taylor's formula in the asymptotic form). *If a mapping $f : E \to F$ is m times normally differentiable at a point $x \in E$ then*

$$f(x+h) = \sum_{k=0}^{m} \frac{1}{k!} f^{(k)}(x)(h)^k + o((h)^n). \tag{5}$$

The proof see in [13]. Let introduce now a notion of the positive definite quadratic form in LCS [12].

Definition 10. Let φ be a continuous quadratic form in LCS E generated by a bilinear form g on E^2. The form φ is called nondegenerate if

$$\inf_{\|x\|_{t_1}\neq 0} \frac{\|g(x,\cdot)\|^{t_2}}{\|x\|_{t_1}} > 0 \tag{6}$$

for some $t_1, t_2 \in n_g$. If φ is nondegenerate and nonnegative then the form φ is called positive definite on E : $\varphi \gg 0$.

Note that even in the case of Banach space E the condition (6) is weaker than the classical Cartan's condition (see [3]) requiring the operator $x \mapsto g(x,\cdot)$ be an isomorphism of the spaces E and E^*.

Proposition 11. *If a quadratic form φ is nondegenerate and g is the associated bilinear form, then*

$$(\forall x_2 \in E : \ g(x_1, x_2) = 0) \quad \Rightarrow \quad (x_1 = 0).$$

Theorem 12. (Condition of positive definiteness, [12]). *If $\varphi \gg 0$ on LCS E then there exist such index $t \in T$ and constant $C_t > 0$ that*

$$\varphi(x) \geqslant C_t \cdot \|x\|_t^2$$

for all $x \in E$. The inverse statement is also valid.

On the base of these results it's not difficult to obtain the extremal conditions for the normally differentiable functions.

Proposition 13 (Fermat Lemma). *If a function $f : E \to \mathbb{R}$ has a local extremum at a point x and f is normally differentiable at this point then $f'(x) = 0$.*

Proposition 14 (Second order necessary condition). *If a function $f : E \to \mathbb{R}$ has a local minimum at a point x and f if twice normally differentiable at a point x then not only $f'(x) = 0$ but $f''(x) \geqslant 0$.*

Theorem 15. (Sufficient condition of local minimum, [11].) *If a function $f : E \to \mathbb{R}$ is twice normally differentiable at a point x, moreover $f'(x) = 0$ and $f''(x) \gg 0$ then f has the strong minimum at x.*

Now let introduce a weakened variant of normal differentiability, that is admissible for the derivatives of integral functionals.

Definition 16. A mapping $f : E \to F$ is called compact-normally differentiable (or K-differentiable) at a point $x \in E$ if for every convex compactum $C \subset E$ containing x, the restriction $f|_C$ is normally differentiable at x (in this case a value

$(f|_C)'(x)$ not depends on the choice of C). Denote by $f'_K(x)$ this value. Thus, $f'_K(x) \in (E; F)$.

Note that T-differentiability follows from K-differentiability in an obvious way.

Definition 17. Say that a function $f : E \to \mathbb{R}$ has a locally compact extremum, or K-extremum at a point $x \in E$ if a restriction $f|_C$ has a local extremum at x for every convex compactum $C \ni x$.

It is easy to see that after the replacement of the positive definiteness, normal derivatives and local extremums with the K-positive definiteness, K-derivatives and K-extremums, respectively, in the formulation of Propositions 13, 14 and Theorem 15, we obtain the corresponding conditions of the K-extremums.

Proposition 18. *If a function $f : E \to \mathbb{R}$ has a local K-extremum at a point x and f is K-normally differentiable at this point then $f'_K = 0$.*

Proof. Consider an arbitrary convex compactum C that is symmetric respective to x. Then local extremum of restriction $f|_C$ implies, obviously, equalities $\sigma f(x, h) = 0$ for any directional derivative such that $x + \lambda h \in C$ for sufficiently small λ. So, $\sigma f(x, h) = 0$ for any $h \in E$, whence $f'_K(x) = 0$.□

Proposition 19. *If a function $f : E \to \mathbb{R}$ has local K-minimum at a point x and f is twice K-normally differentiable at x then not only $f'_K(x) = 0$ but $f''_K(x) \geqslant 0$.*

Proof. It suffices to apply Proposition 14 to each restriction $f|_C$, here C is convex compactum that is symmetric respective to x.□

Theorem 20. *If a function $f : E \to \mathbb{R}$ is twice K-normally differentiable at a point x, moreover $f'_K(x) = 0$ and $f''_K(x) \gg 0$ then f has the strong K-minimum at x.*

Proof. It suffices to apply Theorem 15 to each restriction $f|_C$, here C is convex compactum that is symmetric respective to x.□

Remark 21. Actually, the condition $f''_K(x) \gg 0$ in Theorem 15 can be replaced by a much more weak condition of K-positive definiteness, namely, positive definiteness of each restriction $f''_K(x)|_{E_C}$, here $E_C = \mathrm{Span}\,(\mathrm{C} - \mathrm{x})$, C is any convex compactum in E containing x.

As a base for proof of repeated K-differentiability of Euler–Lagrange functional, it serves the following

Lemma 22. *If a mapping $g : E \to F$ is continuously normally differentiable on convex compact set $C \subset E$ then*

$$\frac{\|g(x+h) - g(x) - g'(x)h\|^s}{\|h\|_t} \rightrightarrows 0 \quad as \quad h \to 0, \quad x \in C \tag{7}$$

for some normal index $n \in \mathcal{N}(E; F)$ and for arbitrary $s \in S$, $t \in n(s)$.

Proof. As far as mapping $g' : C \to \overrightarrow{(E; F)}$ is continuous then, according to definition 7, for every $x \in C$ there exists such a neighborhood $U(x)$ and a normal index $n_x \in \mathcal{N}$ that g' can be considered as continuous mapping from U_x into $(E; F)_{n_x}$. Choosing from the covering $\{U(x)\}_{x \in C}$ a finite one $U_1(x_1) \cup U_2(x_2) \cup \ldots \cup U_k(x_k) \supset C$ and getting a normal index $n \succcurlyeq n_{x_1}, n_{x_2}, \ldots n_{x_k}$, we obtain g' as continuous mapping from C into $(E; F)_n$. Let us apply the normal form of mean value theorem (Proposition 6) to auxiliary mapping $G(h) = g(x+h) - g'(x)h$ for the arbitrary $s \in S$, $t \in n(s)$, $x \in C$:

$$\|g(x+h) - g(x) - g'(x)h\|^s = \|G(h) - G(0)\|^s \leqslant$$

$$\leqslant \sup_{k \in [0;h]} \|G'(k)\|_t^s \cdot \|h\|_t = \sup_{k \in [0;h]} \|g'(x+k) - g'(x)\|_t^s \cdot \|h\|_t.$$

From here, by virtue of uniform continuity g' on C, it follows (7).□

Generalize now the notion of Sobolev space W_2^1 for the LCS-valued mapping.

Definition 23. Let E be a complete LCS. Denote by $W_2^1([a;b]; E)$ the completion of the space $C^1([a;b]; E)$ relative to the defining system of semi-norms

$$\|y\|^t = \left(\int_a^b \|y(x)\|_t^2 dx + \int_a^b \|y'(x)\|_t^2 dx \right)^{\frac{1}{2}}, \quad t \in T.$$

Let us formulate the principal results of this section.

Theorem 24. *If a function $u = f(x, y, z)$ is continuously normally differentiable on $[a;b] \times E \times E$ then the Euler–Lagrange functional (1) is continuously normally differentiable on $W_2^1([a;b]; E)$.*

Proof. For an arbitrary $y(\cdot)$ and $h(\cdot)$ from $W_2^1([a;b]; E)$ we obtain

$$\Phi(y+h) - \Phi(y) = \int_a^b [f(x, y(x) + h(x), y'(x) + h'(x)) -$$

$$-f(x, y(x), y'(x))]\, dx = \int_a^b \left[\frac{\partial f}{\partial y}(x, y(x), y'(x))h(x) + \right.$$

$$+\frac{\partial f}{\partial z}(x, y(x), y'(x))h'(x)\Big] \, dx + \int_a^b r_x(h(x), h'(x)) \, dx;$$

here r_x is the remainder term of the increment of f at the point $(x, y(x), y'(x))$. Let us fix $\varepsilon > 0$ and choose for every $x \in [a; b]$ an index $t = t(x) \in T$ and $\delta = \delta(\varepsilon, x) > 0$ such that

$$\begin{aligned}(\|h(x)\|_t + \|h'(x)\|_t < \delta) \Rightarrow (|r_x(h(x), h'(x))| < \\ < \varepsilon \cdot (\|h(x)\|_t + \|h'(x)\|_t)).\end{aligned} \tag{8}$$

This condition holds true, by continuity, in some neighborhood $U(x)$ of x for the same t and δ.

Note that for sufficiently small $\|h\|^t$ and $\eta > 0$ one can choose such a compactum $K_\delta \subset [a; b]$ that

$$\|h(x)\|_t + \|h'(x)\|_t < \delta \quad \text{as} \quad x \in K_\delta;$$
$$\text{mes}\,([a; b] \backslash K_\delta) < \eta.$$

Choose from the covering $\{U(x) | \, x \in K_\delta\}$ a finite one

$$U(x_1) \cup U(x_2) \cup \ldots \cup U(x_n) \supset K_\delta$$

and set $\delta = \min(\delta(x_1), \ldots, \delta(x_n))$, $t \succcurlyeq t(x_1), \ldots, t(x_n)$. Then the condition (8) is fulfilled for all $x \in K_\delta$. Hence,

$$\left| \int_{K_\delta} r_x(h(x), h'(x)) \, dx \right| < \varepsilon \cdot \int_{K_\delta} [\|h(x)\|_t + \|h'(x)\|_t] \, dx <$$

$$< C\varepsilon \left(\sqrt{\int_{K_\delta} \|h(x)\|_t^2 \, dx} + \sqrt{\int_{K_\delta} \|h'(x)\|^2 \, dx} \right). \tag{9}$$

From here, setting η be sufficiently small, we get

$$\left| \int_a^b r_x(h(x), h'(x)) \, dx \right| < C\varepsilon \cdot \|h\|^t,$$

i.e. Φ is normally differentiable at every point $y(\cdot) \in W_2^1([a; b]; E)$ and its first normal derivative is

$$\Phi'(y)h = \int_a^b \left[\frac{\partial f}{\partial y}(x, y(x), y'(x))h(x) + \frac{\partial f}{\partial z}(x, y(x), y'(x))h'(x) \right] dx.$$

Let $\|\Phi'(y)\|^t < \infty$, $t = t(y) \in T$. Using $f \in C^1$, compactness of $[a; b]$ and repeating the construction above, we find $t \in T$ that does not depend on y, and such that $\|\Phi'(y)\|^t < \infty$ for all $y \in W_2^1([a; b]; E)$. Hence Φ' is continuous mapping into the scale $\overrightarrow{W_2^{-1}([a; b]; E)}$.□

Corollary 25. *Under the suppositions of Theorem 24 the condition $\Phi'(y) = 0$ holds true if and only if the Euler–Lagrange variational equation*

$$\frac{\partial f}{\partial y}(x, y, y') - \frac{d}{dx}\left[\frac{\partial f}{\partial z}(x, y, y')\right] = 0 \tag{10}$$

is valid for all $x \in [a; b]$.

Proof. As it was shown in [11], Corollary 6.1, the standard representation

$$\Phi'(y)h = \int_a^b \left(\frac{\partial f}{\partial y} - \frac{d}{dx}\left[\frac{\partial f}{\partial z}(x, y, y')\right]\right) h\, dx,$$

here $h(a) = h(b) = 0$, is extended on the case of normal derivatives. Together with $\Phi'(y)h \equiv 0$ it turns out, obviously, the validity of (10). Inverse implication is trivial.□

Theorem 26. *If a function $u = f(x, y, z)$ is twice continuously normally differentiable on $[a; b] \times E \times E$ then the Euler–Lagrange functional (1) is C^2–K-differentiable on $W_2^1([a; b]; E)$.*

Proof. Straightforward calculation gives:

$$(\Phi'(y+k) - \Phi'(y)) \cdot h = \int_a^b \left[\frac{\partial^2 f}{\partial y^2}(x, y, y')(k(x), h(x)) + \right.$$

$$+ \frac{\partial^2 f}{\partial y \partial z}(x, y, y')\left[(k'(x), h(x)) + (k(x), h'(x))\right] + \frac{\partial^2 f}{\partial z^2}(x, y, y') \cdot \tag{11}$$

$$\left. \cdot (k'(x), h'(x))\right] dx + \int_a^b \left[r_x^y(k(x), k'(x)) \cdot h(x) + r_x^z(k(x), k'(x)) \cdot h'(x)\right] dx;$$

here r_x^y and r_y^z are, respectively, remainder terms of the increments of $\partial f / \partial y$ and $\partial f / \partial z$ at the point $(x, y(x), y'(x))$.

Let C be an arbitrary convex compact set in $W_2^1([a; b]; E)$ containing $y(\cdot)$. Then, obviously, the sets

$$K = \{y \in y([a; b]) | \, y(\cdot) \in C\} \quad \text{and} \quad K_1 = \{z \in y'([a; b]) | \, y(\cdot) \in C\}$$

are convex compacts in E. By virtue of $\partial f / \partial y \in C^1$ and $\partial f / \partial z \in C^1$ on convex compact set $[a; b] \times K \times K_1$ one can apply Lemma 22 to r_x^y and r_x^z, whence

$$\frac{\|r_x^y(k(x), k'(x))\|^t}{\|k(x)\|_t + \|k'(x)\|_t} \rightrightarrows 0 \quad \text{and} \quad \frac{\|r_x^z(k(x), k'(x))\|^t}{\|k(x)\|_t + \|k'(x)\|_t} \rightrightarrows 0$$

as $x \in [a; b]$, $(k(x), k'(x)) \to 0$, for some $t \in T$. Taking into account the estimates

$$|r_x^y(k(x), k'(x)) \cdot h(x)| \leqslant \|r_x^y(k(x), k'(x))\|^t \cdot \|h(x)\|_t;$$

$$|r_x^z(k(x), k'(x)) \cdot h'(x)| \leqslant \|r_x^z(k(x), k'(x))\|^t \cdot \|h'(x)\|_t;$$

and denoting by $R_y(k)h$ the general remainder term in (11) a calculation analogous with (9) gives

$$\left(|R_y(k)h| / \|h\|^t \cdot \|k\|^t\right) \to 0$$

as $h \to 0$, $k \to 0$ in $W_2^1([a;b];E)$. It follows from here that $R_y(k) = o(k)$ and Φ' is K-differentiable at $y(\cdot)$. The continuity of $\Phi''_K = (\Phi')'_K$ is verified analogously with one of Φ' in the previous proof. □

Corollary 27. *If under the conditions of Theorem 26 the Euler–Lagrange variational equation (10) holds at a point $y(\cdot) \in W_2^1([a;b];E)$ and $\Phi''_K(y) \gg 0$, then Φ has K-minimum at this point.*

Proof. Since, by Corollary 25, $\Phi'(y) = 0$ then $\Phi'_K(y) = \Phi'(y) = 0$. It allows to apply for this case the sufficient K-extremum condition (Theorem 20).□

Corollary 28. *If under the conditions of Theorem 26 the Euler–Lagrange variational equation (10) holds at a point $y(\cdot) \in W_2^1([a;b];E)$ and for each fixed $x \in [a;b]$ the bilinear form $f''(x, y(x), y'(x))$ is positive definite at the point $(y(x), y'(x)) \in E^2$ then Φ has K-minimum at the point $y(\cdot)$.*

Proof. According to Theorem 12, for each $x \in [a;b]$ there exist $t = t(x) \in T$ and $C_t > 0$ such that

$$f''(x, y(x), y'(x)) \cdot ((h_1(x), h_2(x)), (h_1(x), h_2(x))) =$$

$$= \frac{\partial^2 f}{\partial y^2}(h_1(x), h_1(x)) + 2\frac{\partial^2 f}{\partial y \partial z}(h_1(x), h_2(x)) + \frac{\partial^2 f}{\partial z^2}(h_2(x), h_2(x)) \geqslant$$

$$\geqslant C_{t(x)} \cdot \left(\|h_1(x)\|_{t(x)}^2 + \|h_2(x)\|_{t(x)}^2\right)$$

for all $(h_1(x), h_2(x)) \in E^2$. Using compactness of $[a;b]$ by analogy with the proof of Theorem 24 it is possible to choose $t \in T$ and $C_t > 0$ in the last estimate non depending on $x \in [a;b]$. In particular,

$$\frac{\partial^2 f}{\partial y^2}(h(x), h(x)) + 2\frac{\partial^2 f}{\partial y \partial z}(h(x), h'(x)) + \frac{\partial^2 f}{\partial z^2}(h'(x), h'(x)) \geqslant$$

$$\geqslant C_t \cdot \left(\|h(x)\|_t^2 + \|h'(x)\|_t^2\right)$$

for all $x \in [a;b]$ and $h(\cdot) \in W_2^1([a;b];E)$. From here, using (11), we get

$$\Phi''_K(y)(h,h) \geqslant C_t \cdot \left[\int_a^b \|h(x)\|_t^2\, dx + \int_a^b \|h'(x)\|_t^2\, dx\right] = C_t \cdot (\|h\|^t)^2,$$

i.e., according to Theorem 12, $\Phi''_K(y) \gg 0$. It remains to apply Corollary 27. □

Note at the conclusion that the last results are transferred to the general Euler–Lagrange functional on a compact domain $\Omega \subset \mathbb{R}^n$.

2. Extremae of functionals in the products of the nuclear spaces

In this section the classical sufficient conditions for extremum of two variables function are generalized for the case of twice normally and K-normally differentiable function defined in the product of the nuclear spaces. The obtained results are applied to a wide class of integral functionals.

As an initial point it serves a generalization of the Young theorem on symmetry of second derivative considered as bilinear operator for the case of the second order normal derivatives.

Theorem 29 ([13].). *If a mapping $f: E \to F$ is twice normally differentiable at a point $x \in E$ then $f''(x)$ is a symmetric bilinear operator from $E \times E$ to F.*

Proof. Following by standard scheme for proof of the Young theorem in Banach spaces [3]), introduce auxiliary mapping

$$A(h,k) = f(x+h+k) - f(x+h) - f(x+k) + f(x); \quad A(h,k) = A(k,h).$$

Let $n_2^1 = (n_1, n_{12})$ be a second order normal index of $f''(x)$; thus we consider $f''(x)$ as a linear operator acting from E to $\overrightarrow{(E;F)}$. In addition, according to definition 7, $f''(x) : E \to (E;F)_{n_1}$ and n_{12} is normal index of $f''(x)$ in this paire of spaces. For every $s \in S$ we have

$$\begin{aligned} \|A(h,k) - (f''(x)k)h\|^s \leqslant \|A(h,k) - f'(x+k)h + f'(x)h\|^s + \\ + \|f'(x+k)h - f'(x)h - (f''(x)k)h\|^s. \end{aligned} \tag{12}$$

Fix $t_1 \in n_1(s)$ and estimate each term from the right in (12).

a). Since, by virtue of continuity f' at x, $f' : E \to (E;F)_{n_1}$ in some neighborhood of x then the normal indices of operators $f'(x+k)$, $f'(x)$ and $f''(x)k$ don't exceed n_1 for sufficiently small k. Hence

$$\begin{aligned} &\|f'(x+k)h - f'(x)h - (f''(x)k)h\|^s \leqslant \\ &\leqslant \|f'(x+k) - f'(x) - f''(x)k\|^s_{t_1} \cdot \|h\|_{t_1}. \end{aligned} \tag{13}$$

Next, by definition of the second normal derivative,

$$f'(x+k) - f'(x) - f''(x)k = o(k). \tag{14}$$

Let ν_{12} be corresponding o-index. Then for $t_1 \in \nu_{12}(s)$ it follows from (14)

$$(\|f'(x+k) - f'(x) - f''(x)k\|^s / \|k\|_{t_1}) \to 0 \quad \text{as} \quad k \to 0. \tag{15}$$

Hence, for $t_1 \in (n_1 \vee \nu_{12})(s)$ it follows from (13) and (15)

$$\|f'(x+k)h - f'(x)h - (f''(x)k)h\|^s / \|k\|_{t_1} < \varepsilon \tag{16}$$

under $\|k\|_{t_1} < \delta_1(\varepsilon)$.

b). Consider auxiliary mapping

$$B(h) = f(x+k+h) - f(x+h) - f'(x+k)h + f'(x)h.$$

Applying the normal form of mean value theorem (Prop. 6) to $B(\cdot)$ on $[0;h]$ for $t_1 \in n_1(s)$ we get

$$\begin{aligned}
&\|A(h,k) - f'(x+k)h + f'(x)h\|^s = \|B(h) - B(0)\|^s \leqslant \\
&\leqslant \sup_{0\leqslant\lambda\leqslant 1} \|B'(\lambda h)\|^s_{t_1} \cdot \|h\|_{t_1} = \sup_{0\leqslant\lambda\leqslant 1} \|f'(x+k+\lambda h) - f'(x+\lambda h) - \\
&-f'(x+k) + f'(x)\|^s_{t_1} \cdot \|h\|_{t_1} = \sup_{0\leqslant\lambda\leqslant 1} \|[f'(x+k+\lambda h) - \\
&-f'(x) - f''(x)(k+\lambda h)] - [f'(x+\lambda h) - f'(x) - f''(x)(\lambda h)] - \\
&-[f'(x+k) - f'(x) - f''(x)k]\|^s_{t_1} \cdot \|h\|_{t_1}.
\end{aligned} \tag{17}$$

Since

$$\begin{cases} f'(x+k+\lambda h) - f'(x) - f''(x)(k+\lambda h) = o(k+\lambda h); \\ f'(x+\lambda h) - f'(x) - f''(x)(\lambda h) = o(\lambda h); \\ f'(x+k) - f'(x) - f''(x)k = o(k); \end{cases} \tag{18}$$

moreover, it can assume ν_{12} as o-index for each of these equalities then for $t_1 \in \nu_{12}(s)$ it follows from (17) and (18)

$$(\|A(h,k) - f'(x+k)h + f'(x)h\|^s / (\|h\|_{t_1} + \|k\|_{t_1})) < \varepsilon \cdot \|h\|_{t_1} \tag{19}$$

under $\|h\|_{t_1} < \delta_2(\varepsilon)$, $\|k\|_{t_1} < \delta_2(\varepsilon)$. It follows from (12), (16) and (19)

$$(\|A(h,k) - (f''(x)k)h\|^s / (\|h\|_{t_1} + \|k\|_{t_1})) < \varepsilon\|h\|_{t_1} < \varepsilon(\|h\|_{t_1} + \|k\|_{t_1})$$

under $t_1 \in (n_1 \vee \nu_{12})(s))$, $\|h\|_{t_1} < \delta_2$, $\|k\|_{t_1} < \min(\delta_1, \delta_2)$. The last estimate means that

$$A(h,k) - (f''(x)k)h = o((h,k)^2). \tag{20}$$

By analogous way we get

$$A(k,h) - (f''(x)h)k = o((k,h)^2). \tag{21}$$

Finally, it follows from (20) and (21).

$$(f''(x)h)k - (f''(x)k)h = o((h,k)^2).$$

Replacing with $h \mapsto \lambda h$, $k \mapsto \lambda k$ in the last equality we receive

$$(f''(x)h)k - (f''(x)k)h = (o(\lambda^2 \cdot (h,k)^2)/\lambda^2) \to 0$$

as $\lambda \to 0$, whence $f''(x)h)k = f''(x)k)h$. □

Corollary 30. (Schwartz theorem.) *If a mapping $f: E = E_1 \times E_2 \to F$ is twice normally differentiable at a point $x \in E$ then the equality*

$$(\partial^2 f/\partial x_i \partial x_j)(x)(h_i, h_j) = (\partial^2 f/\partial x_j \partial x_i)(x)(h_j, h_i) \quad (i, j = 1, 2)$$

holds, i.e. the mixed second order partial derivatives coincide up to the symmetry of the arguments.

Proof. On the strength of [9], (16), the equality

$$f'(\cdot)(h_1, h_2) = \frac{\partial f}{\partial x_1}(\cdot)h_1 + \frac{\partial f}{\partial x_2}(\cdot)h_2$$

is valid in some neighborhood of x. It follows

$$f''(x)(k_1, k_2) = \frac{\partial f'}{\partial x_1}(x)k_1 + \frac{\partial f'}{\partial x_2}(x)k_2,$$

and hence

$$(f''(x) \cdot (k_1, k_2)) \cdot (h_1, h_2) = \left(\frac{\partial f'}{\partial x_1}(x)k_1 + \frac{\partial f'}{\partial x_2}(x)k_2 \right) \cdot (h_1, h_2). \tag{22}$$

Differentiating (22) in x_i ($i = 1, 2$) and using continuity of the evaluation $\overrightarrow{E^*} \times E \to \mathbb{R}$ (see [14]) we get

$$\left(\frac{\partial f'}{\partial x_i}(x)k_i \right) \cdot (h_1, h_2) = \left(\frac{\partial^2 f}{\partial x_i \partial x_1}(x)k_i \right) h_1 + \left(\frac{\partial^2 f}{\partial x_i \partial x_2}(x)k_i \right) h_2. \tag{23}$$

Note that, by virtue of canonical isomorphism (3) for normal and binormal indices,

$$\left(\frac{\partial^2 f}{\partial x_i \partial x_j} \right)(x) \in \overrightarrow{(E_i; \overrightarrow{(E_j; F)})} \cong \overrightarrow{(E_i, E_j; F)}. \quad (i, j = 1, 2)$$

Substituting (23) into (22), we get

$$(f''(x) \cdot (k_1, k_2)) \cdot (h_1, h_2) = \sum_{i,j=2}^{2} \left(\frac{\partial^2 f}{\partial x_i \partial x_j}(x)k_i \right) h_j. \tag{24}$$

Then, applying Theorem 29, it follows from (24)

$$\sum_{i,j=1}^{2} \left(\frac{\partial^2 f}{\partial x_i \partial x_j}(x)k_i \right) h_j = \sum_{i,j=1}^{2} \left(\frac{\partial^2 f}{\partial x_j \partial x_i}(x)h_j \right) k_i,$$

from here we receive that the equalities

$$\frac{\partial^2 f}{\partial x_i \partial x_j}(x)(k_i, h_j) = \frac{\partial^2 f}{\partial x_j \partial x_i}(x)(h_j, k_i)$$

are valid for any $i, j = 1, 2$, $k_i \in E_i$, $h_j \in E_j$.□

As the following important initial point it serves a condition of positive definiteness in the product of the nuclear spaces. Since any complete nuclear LCS is isomorphic to a projective limit of the Hilbert spaces [15] then we need at first to obtain such a condition in the product of the Hilbert spaces.

Proposition 31 ([13]). *Let H_i $(i = 1, 2)$ are the real Hilbert spaces, $H_1 \cong H_2$, $B = (B_{ij})$ be a self-adjoint linear continuous operator in $H_1 \times H_2$, here $B_{ij} : H_j \to H_i$ $(i, j = 1, 2)$. Hence, $B_{11} = B_{11}^*$, $B_{22} = B_{22}^*$, $B_{12} = B_{21}^*$. If*

1) $B_{11} \gg 0$ and B_{11} commutes with B_{12}, B_{21} and B_{22};

2) $B_{11}B_{22} - B_{12}B_{21} \gg 0$;

then B is positive definite on $H_1 \times H_2$.

Proof. Let

$$A_{11} = \begin{pmatrix} B_{11} & 0 \\ 0 & B_{11} \end{pmatrix}; \quad \text{then} \quad A_{11}^{-1} = \begin{pmatrix} B_{11}^{-1} & 0 \\ 0 & B_{11}^{-1} \end{pmatrix}.$$

Consider operator

$$A_{11}B = \begin{pmatrix} B_{11}^2 & B_{11}B_{12} \\ B_{11}B_{21} & B_{11}B_{22} \end{pmatrix}.$$

For $h = (h_1, h_2) \in H_1 \times H_2$ we get

$$\begin{aligned} \langle A_{11}Bh, h\rangle &= \langle B_{11}^2 h_1, h_1\rangle + 2\langle B_{11}B_{12}h_2, h_1\rangle + \langle B_{11}B_{22}h_2, h_2\rangle = \\ &= (\langle B_{11}h_1, B_{11}h_1\rangle + 2\langle B_{12}h_2, B_{11}h_1\rangle + \langle B_{12}h_2, B_{12}h_2\rangle) + \\ &+ \langle (B_{11}B_{22} - B_{21}B_{12})h_2, h_2\rangle = \|B_{11}h_1 + B_{12}h_2\|^2 + \\ &+ \langle (B_{11}B_{22} - B_{21}B_{12})h_2, h_2\rangle \geqslant C \cdot \|h_2\|^2, \quad \text{here} \quad C > 0. \end{aligned}$$

Analogously, for $h' = (h_2, h_1)$ we obtain

$$\langle A_{11}Bh', h'\rangle = \langle A_{11}Bh, h\rangle \geqslant C\|h_1\|^2,$$

whence

$$\langle A_{11}Bh, h\rangle \geqslant C \cdot \max\left(\|h_1\|^2, \|h_2\|^2\right) \geqslant \frac{C}{2} \cdot \|h\|^2,$$

i.e. $A_{11}B \gg 0$.

Note that the operator A_{11}^{-1} is positive definite and self-adjoint together with $A_{11}B$. In addition A_{11}^{-1} and $A_{11}B$ commute by virtue of the equalities

$$\left(B_{11}B_{ij}B_{11}^{-1} = B_{ij}\right) \Longleftrightarrow \left(B_{11}B_{ij} = B_{ij}B_{11}\right).$$

As it is well known these conditions guarantee positive definiteness of the product

$$A_{11}^{-1}(A_{11}B) = B \gg 0 \qquad \square.$$

Basing on the condition of positive definiteness (Theorem 12) and the canonical isomorphism (3), it is not difficult to generalize Proposition 31 for the case of the product of nuclear spaces.

Theorem 32. *Let E_i $(i = 1, 2)$ are the real complete nuclear LCS, $E_1 \cong E_2$, $E = E_1 \times E_2$, φ be a continuous quadratic form on E, generated by a symmetric bilinear form g on E^2, B be a linear operator from E to E^* associated with g (i.e., $g(h, k) = \langle h, Bh\rangle$), $B = (B_{ij})$, here $B_{ij} : E_j \to E_i$ $i, j = 1, 2$. If*

1) $B_{11} \gg 0$ *and* B_{11} *commutes with* B_{12}*,* B_{21} *and* B_{22}*;*
2) $B_{11}B_{22} - B_{12}B_{21} \gg 0$*;*
then φ *is positive definite on* E*.*

Proof. Since any complete nuclear LCS is isomorphic to a projective limit of the Hilbert spaces [15] then, without loss of generality, it can suppose that

$$E_1 = E_2 \cong \operatorname{proj}\lim_{t \in T} H_t,$$

here H_t are Hilbert spaces. Further, by canonical isomorphism theorem (Proposition 3) one can consider B as a continuous linear operator from E to $\overrightarrow{E}^*$. So, one can consider B as continuous linear operator from $H_{t_1} \times H_{t_2}$ to $H^*_{t'_1} \times H^*_{t'_2} \cong H_{t_1} \times H_{t_2}$ for some $t'_1, t'_2 \in T$; moreover B is self-adjoint by virtue of symmetry of g. Analogously, the conditions 1) and 2) of the Theorem are realized on some $H_{t''_1} \times H_{t''_2}$ and $H_{t'''_1} \times H_{t'''_2}$, respectively. Choosing $t_i \succcurlyeq t'_i, t''_i, t'''_i$ $(i = 1, 2)$, we arrive to realization of the conditions of Proposition 31 on $H_{t_1} \times H_{t_2}$. Hence B is positive definite on $H_{t_1} \times H_{t_2}$, whence

$$\varphi(h) = \langle h, Bh \rangle \geqslant C_t \cdot \|h\|_t^2; \quad (C_t > 0)$$

for $h \in E^t = H_{t_1} \times H_{t_2}$, $t = (t_1, t_2)$. Finally, applying Theorem 12 we get positive definiteness of the form φ everywhere on E.□

Together with Corollary 28, the last Theorem give us the sufficient extreme conditions for Euler–Lagrange functional (1) in terms of second order partial derivatives of the integrand function.

Corollary 33. *Let* E *be a real complete nuclear LCS, function* $f(x, y, z)$ *be twice contituously normally differentiable on* $[a; b] \times E \times E$*. If the Euler–Lagrange variational equation (10) holds at a point* $y(\cdot) \in W_2^1([a; b]; E)$ *and for each fixed* $x \in [a; b]$*:*
1) $\frac{\partial^2 f}{\partial y^2}(x, y(x), y'(x)) \gg 0$ *in* E^2 *and commutes with* $\partial^2 f/\partial y \partial z(x, y(x), y'(x))$ *and* $\partial^2 f/\partial z^2(x, y(x), y'(x))$*;*
2) $\frac{\partial^2 f}{\partial y^2}(x, y(x), y'(x)) \cdot \frac{\partial^2 f}{\partial z^2}(x, y(x), y'(x)) - \left[\frac{\partial^2 f}{\partial y \partial z}(x, y(x), y'(x))\right]^2 \gg 0$ *in* E^2*;*
then Φ *has* K*-minimum at the point* $y(\cdot)$*.*

Proof. Applying Theorem 32 to $f''(x, y(x), y'(x))$ we get positive definiteness of the form above at $(y(x), y'(x))$ for each $x \in [a; b]$. It remains to apply Corollary 28.
□

Next, combining the result of Theorem 32 with the sufficient extreme condition (Theorem 15) it can easily obtain a sufficient extreme condition for two variables function in the product of the nuclear spaces.

Theorem 34. *Let* E_i $(i = 1, 2)$ *are the real complete nuclear LCS,* $E_1 \cong E_2$*, real function* $f(x_1, x_2)$ *be twice continuously differentiable at a point* $(x_1, x_2) \in E_1 \times E_2$*, moreover* $\partial_1 f(x_1, x_2) = \partial_2 f(x_1, x_2) = 0$*. If*
1) $\partial_{11} f \gg 0$ *and* $\partial_{11} f$ *commutes with* $\partial_{12} f$*,* $\partial_{21} f$*,* $\partial_{22} f$*;*

2) $\partial_{11}f \cdot \partial_{22}f - \partial_{12}f \cdot \partial_{21}f \gg 0$;
at (x_1, x_2) *then* f *has a strong minimum at this point.*

Proof. Since, according to [13], Th.2.1, $f''(x_1, x_2)$ is symmetric bilinear form on $(E_1 \times E_2)^2$ and for associated Hessian $B = (B_{ij})$, here $B_{ij} = \partial_{ij}f(x_1, x_2)$, the conditions of Theorem 32 are realized then the form $f''(x_1, x_2)$ is positive definite. By virtue of Theorem 15 the conditions $f'(x_1, x_2) = 0$ and $f''(x_1, x_2) \gg 0$ provide minimum for f at this point.□

Example 35. Let $f(x_1, x_2)$ be a twice continuously differentiable function in $\mathbb{R}^2$, Ω be a compact domain in $\mathbb{R}^n$, $S(\Omega)$ be the corresponding L.Schwartz space of the quickly decreasing functions (see [2]). Consider the integral functional

$$F(x_1, x_2) = \int_{\Omega} f(x_1(t), x_2(t))dt; \quad (x_i(\cdot) \in S(\Omega), \quad i = 1, 2). \tag{25}$$

A straightforward calculation shows that

$$\partial_{ij}F(x_1, x_2)(h_j, h_i) = \int_{\Omega^2} \partial_{ij}f(x_1(t_1), x_2(t_2))h_j(t_j)h_i(t_i)dt_1dt_2; \tag{26}$$

whence by Fubini theorem it follows that

$$(\partial_{ij}F \cdot \partial_{kl}F)(h_l, h_i) = \int_{\Omega^3} \partial_{ij}f(t_1, t_2)\partial_{kl}f(t_1, u_2)h_i(t_1)h_l(u_2)dt_1dt_2du_2. \tag{27}$$

Hence the partial normal derivatives $\partial_{ij}F$ commute. At last, it's clear from (26)–(27) taking into account nuclearity of $S(\Omega)$ [2] that the inequalities

$$\partial_{11}f > 0; \quad \partial_{11}f \cdot \partial_{22}f - \partial_{12}f \cdot \partial_{21}f > 0 \tag{28}$$

fulfilled on the set $\{(x_1(t), x_2(t)) | t \in \Omega\}$ provide the conditions 1)–2) of Theorem 34. Thus, together with the necessary extreme conditions

$$\partial_1 f(x_1(t), x_2(t)) = \partial_2 f(x_1(t), x_2(t)) = 0 \quad (t \in \Omega)$$

the inequalities (28) guarantee minimum of the integral functional (25) at the point $(x_1(\cdot), x_2(\cdot)) \in S(\Omega) \times S(\Omega)$.

Remark 36. In fact, we have used in the reasonings above only the fact that the spaces under consideration are projective limits of the Hilbert spaces, without taking account of nuclearity of the embeddings. Thus, Theorems 32–34 are transferred to the projective limits of the Hilbert spaces.

Remark 37. By analogy with the results of Section 1, Theorems 29 and 34 are transferred to the twice continuously K-differentiable two variables functions and K-extremums.

Theorem 38. *If a mapping $f : E \to F$ is twice K-normally differentiable at a point $x \in E$ then f''_K is a symmetric bilinear operator from $E \times E$ to F.*

Proof. It is sufficient to note that for each convex compact set $C \ni x$ the equality $\left((f'_K)|_C\right)' = (f|_C)''$ holds true. Because $(f|_C)''(x)$ is symmetric by Theorem 29 then $f''_K(x)$ is symmetric bilinear operator, too.□

Theorem 39. *Let E_i $(i = 1,2)$ are the real complete nuclear LCS, $E_1 \cong E_2$, real function $f(x_1, x_2)$ be twice continuously K-differentiable at a point $(x_1, x_2) \in E_1 \times E_2$, moreover $\partial_{1,K} f(x_1, x_2) = \partial_{2,K} f(x_1, x_2) = 0$. If*
(i) $\partial_{11,K} f \gg 0$ and $\partial_{11,K} f$ commutes with $\partial_{12,K} f$, $\partial_{21,K} f$, $\partial_{22,K} f$;
(ii) $\partial_{11,K} f \cdot \partial_{22,K} f - \partial_{12,K} f \cdot \partial_{21,K} f \gg 0$;
at (x_1, x_2) then f has a strong K-minimum at this point.

Proof. Since a product $C_1 \times C_2$ of two convex compact sets $C_i \subset E_i$ $(i = 1,2)$ is also convex compactum in $E_1 \times E_2$ it follows that all partial K-derivatives $\partial_{i,K} f$ and $\partial_{ij,K} f$ can be considered as usual partial normal derivatives for each restriction $f|_C$, here C is an arbitrary convex compactum in E. Therefore Theorem 34 provides a strong minimum of $f|_C$ at (x_1, x_2), i.e. a strong K-minimum of f at this point.□

Applying Theorem 39 to the Euler–Lagrange functional

$$\Phi(y_1, y_2) = \int_a^b f(x, y_1(x), y_2(x), y'_1(x), y'_2(x))dx \tag{29}$$

we obtain the following

Theorem 40. *If a function $u = f(x, y_1, y_2, z_1, z_2)$ is twice continuously normally differentiable on $[a;b] \times E \times E \times E \times E$, here E is a projective limit of the Hilbert spaces, Euler–Lagrange variational equations*

$$\frac{\partial f}{\partial y_1} - \frac{d}{dx}\left(\frac{\partial f}{\partial z_1}\right) = 0; \quad \frac{\partial f}{\partial y_2} - \frac{d}{dx}\left(\frac{\partial f}{\partial z_2}\right) = 0; \tag{30}$$

are fulfilled at a point $(y_1(\cdot), y_2(\cdot)) \in W_2^1([a,b]; E \times E)$, $\partial^2\Phi_K/\partial y_1^2$ is K-positive definite and commutes with $\partial^2\Phi_K/\partial y_i \partial y_j$ $(i,j = 1,2)$ and the determinant

$$\partial_{11,K}\Phi \cdot \partial_{22,K}\Phi - \partial_{12,K}\Phi \cdot \partial_{21,K}\Phi \tag{31}$$

is K-positive definite at $(y_1(\cdot), y_2(\cdot))$ then Φ has a strong K-minimum at this point.

Proof. Set $\overline{y}(\cdot) = (y_1(\cdot), y_2(\cdot)) : [a;b] \to E^2$, $\overline{z} = (z_1, z_2) \in E^2$, $\overline{f}(x, \overline{y}, \overline{z}) = f(x, y_1, y_2, z_1, z_2)$. Then the functional (29) can be considered as

$$\overline{\Phi}(\overline{y}) = \int_a^b \overline{f}(x, \overline{y}(x), \overline{y}'(x))\, dx, \tag{32}$$

here $\overline{f} \in C^2([a;b] \times E^2 \times E^2)$. According to Theorem 24 the functional (329) is continuously normally differentiable. The conditions (30) are equivalent to

$$\frac{\partial \overline{f}}{\partial \overline{y}} - \frac{d}{dx}\left(\frac{\partial \overline{f}}{\partial \overline{z}}\right) = 0,$$

that, in turn, is equivalent (see [11], Corollary 6.1) to $\overline{\Phi}'(\overline{y}) = 0$, or

$$\partial_1 \Phi(y_1, y_2) = 0, \quad \partial_2 \Phi(y_1, y_2) = 0. \tag{33}$$

Further, by virtue of Theorem 26 the functional (32) and hence the functional (29) is twice continuously K-differentiable. It allows us, taking into account (33) and (31), to apply to this case Theorem 39.□

References

1. V.I. Averbuch, O.G. Smolyanov, The various definitions of derivative in linear topological spaces, *Uspekhi matematicheskih nauk* **23** (4) (1968) 67–116. (Russian).
2. Yu.M. Berezansky, Z.G. Sheftel, G.F. Us, *Functional Analysis. Vol.2.* Birkhauser Verlag, Basel–Boston–Berlin, 1996.
3. H. Cartan, *Calcul differentiel. Formes differentiel.* Hermann, Paris, 1967.
4. S.G. Gindikin, L.R. Volevich, *Distributions and Convolution Equations.* Gordon and Breach Sc.Publ., 1992.
5. S.G. Krein, Yu.I. Petunin, Scales of the Banach spaces, *Uspekhi matematicheskih nauk* **21** (2) (1966) 89–169. (Russian).
6. S.G. Krein, Yu.I. Petunin, E.M. Semenov, *Interpolation of linear operators.* Nauka, Moscow, 1978. (Russian).
7. V.S. Melnik, About $T-$ differentiable functionals, *Doklady RAN* **324** (5) (1992) 928–932. (Russian).
8. I.V. Orlov, Normal indices of linear and nonlinear mappings in the locally convex spaces and scales of the spaces, *Spectral and Evolution Problems* **11** (2001), TNU Publ., Simferopol, 18–29.
9. I.V. Orlov, Normal indices and normal differentiation in the locally convex spaces, *Ucheniye zapiski TNU* **15** (1) (2002)112–121. (Russian).
10. I.V. Orlov, Normal decompositions of Operator Spaces over Locally Convex Spaces, *Funktsional'niy Analiz i Ego Prilozheniya* **36** (4) (2002) 78–80. (Russian).
11. I.V. Orlov, Normal differentiability and functional extremums in locally convex space, *Cybernetics and System Analysis,* (4) (2002) 23-34. (Russian).
12. I.V. Orlov, Quadratic forms in locally convex spaces, *Dinamicheskie systemi* **17** (2001), KFT Publ., Simferopol, 196-205. (Russian).
13. I.V. Orlov, Young theorem for normal derivatives and functional extremums in the products of the nuclear spaces, *Uchenye Zapiski TNU,* **15** (2) (2002). (Russian).

14. I.V. Orlov, Normal functional indices and normal duality, *Methods of Functional Analysis and Topology* **8** (3) (2002).
15. H. Shaefer, *Topological Vector spaces*. MacMillan Company, New York – London, 1966.
16. I.V. Skrypnik, *Nonlinear elliptic high order problems*. Naukova Dumka, Kiev, 1973. (Russian).
17. H. Tribel, *Theory of the function spaces*. Mir, Moscow, 1986, (Russian).
18. A.V. Uglanov, Calculus of variations in the Banach spaces, *Matematicheskii Sbornik* **191** (10) (2000) 105–118. (Russian).
19. M.Z. Zgurovsky, Melnik, V.S., *Nonlinear analysis and infinitive–dimensional system control*. Naukova Dumka, Kiev, 1999. (Russian).

Functional Analysis and its Applications
V. Kadets and W. Żelazko

On some Problem of the Convolution of Bounded Functions

Krzysztof Piejko[a] **Janusz Sokół**[a] **Jan Stankiewicz**[b]

[a]Department of Mathematics, Rzeszów University of Technology, Poland
[b]Institute of Mathematics, University of Rzeszów, Poland

Abstract

We consider convolution properties of regular functions subordinated to the homography $\frac{1+Az}{1-Bz}$. *We give some extension of result from [11].*

Key words: convolution, subordination, bounded functions
2000 MSC: 30C45

1. Introduction

Let $\mathcal{H}$ be the class of regular functions in the open unit disc $\Delta = \{z \in \mathbb{C} : |z| < 1\}$ and let $\mathcal{N}$ be the class of functions $f \in \mathcal{H}$ normalized by $f(0) = 1$. By S we denote the class of functions $f \in \mathcal{H}$ which are univalent in Δ and normalized by $f(0) = f'(0) - 1 = 0$.

Let S^* and S^c are the classes of starlike and convex functions:

$$S^* = \{f \in S : f(\Delta) \text{ is starlike with respect to the origin}\},$$

$$S^c = \{f \in S : f(\Delta) \text{ is convex}\}.$$

The class of Schwarz functions, denoted by Ω, is the class of functions $\omega \in \mathcal{H}$, such that $\omega(0) = 0$ and $|\omega(z)| < 1$ for $z \in \Delta$.

Email addresses: `piejko@prz.rzeszow.pl` (Krzysztof Piejko), `jsokol@prz.rzeszow.pl` (Janusz Sokół), `jan.stankiewicz@prz.rzeszow.pl` (Jan Stankiewicz).

We say that a function f is subordinate to a function g in Δ (and write $f \prec g$ or $f(z) \prec g(z)$) if there exists a function $\omega \in \Omega$ such that $f(z) = g(\omega(z))$; $z \in \Delta$. If the function g is univalent in Δ, then

$$f \prec g \iff f(0) = g(0) \quad \wedge \quad f(\Delta) \subset g(\Delta).$$

In this case we have $\omega(z) = g^{-1}(f(z))$.

Let the functions f and g are of the form

$$f(z) = \sum_{n=0}^{\infty} a_n z^n, \quad g(z) = \sum_{n=0}^{\infty} b_n z^n, \quad z \in \Delta.$$

We say that Hadamard product of f and g is the function $f \star g$ if

$$(f \star g)(z) = f(z) \star g(z) = \sum_{n=0}^{\infty} a_n b_n z^n.$$

J. Hadamard [3] proved, that the radius of convergence of $f \star g$ is greater or equal to the product of radii of convergence of the corresponding series f and g. In particular, if $f, g \in \mathcal{H}$, then also $f \star g \in \mathcal{H}$. Because for $f, g \in \mathcal{H}$

$$(f \star g)(z) = \frac{1}{2\pi} \int_0^{2\pi} f\left(\frac{z}{\rho} e^{it}\right) g\left(\rho e^{-it}\right) dt,$$

where $|z| < \rho < 1$, then the function $f \star g$ is also called the *convolution* of the functions f and g.

Very often it was investigated the problem of connections between functions f, g and their convolution $f \star g$. Many conjectures was given. A lot of them are still not verified. Many years was considered a conjecture given in 1958 by G. Pólya and I. J. Schoenberg [7]:

Conjecture I. *The Hadamard product of two convex mappings is a convex mapping.*

In 1961 H. S. Wilf [12] gave more generally supposition:

Conjecture II.

If F and G are convex mappings in Δ and f is subordinate to F, then the convolution $f \star G$ is subordinate to $F \star G$.

In 1973 S. Rusheweyh i T. Sheil-Small [9] proved both conjectures and more results of this type. Their very important results we may write as:

Theorem A. *If $f \in S^c$ and $g \in S^c$, then $f \star g \in S^c$.*

Theorem B. *If $F \in S^c$, $G \in S^c$ and $f \prec F$, then $f \star G \prec F \star G$.*

In 1985 S. Ruscheweyh and J. Stankiewicz [10] proved some generalization of Theorems A and B:

Theorem C. *If the functions F and G are convex in Δ, then for all functions f and g we have*

$$f \prec F \;\wedge\; g \prec G \implies f \star g \prec F \star G.$$

For the classes $Q_1 \subset \mathcal{H}$ and $Q_2 \subset \mathcal{H}$ the convolution $Q_1 \star Q_2$ is defined as

$$Q_1 \star Q_2 = \{h \in \mathcal{H};\; h = f \star g,\; f \in Q_1,\; g \in Q_2\}.$$

For the given complex numbers A, B such that $A + B \neq 0$ and $|B| \leq 1$ let us denote

$$P(A,B) = \left\{ f \in \mathcal{N} :\; f(z) \prec \frac{1 + Az}{1 - Bz} \right\}.$$

W. Janowski introduced the class $P(A,B)$ in [4] and considered it for $-1 \leq B < A \leq 1$. If $A = B = 1$ then the class $P(1,1)$ is the class of functions with positive real part (Carathéodory functions). In 1988 J. Stankiewicz and Z. Stankiewicz [11] investigated convolution properties of the class $P(A,B)$ and proved the following theorem:

Theorem D. *If A, B, C, D are some complex numbers such that $A + B \neq 0$, $C + D \neq 0$, $|B| \leq 1$, $|D| \leq 1$, then*

$$P(A,B) \star P(C,D) \subset P(AC + AD + BC, BD),$$

moreover, if $|B| = 1$ or $|D| = 1$, then

$$P(A,B) \star P(C,D) = P(AC + AD + BC, BD).$$

The equality between the class $P(A,B) \star P(C,D)$ and the class $P(AC + AD + BC, BD)$ for $|B| < 1$ and $|D| < 1$ was an open problem. In this paper we prove that the above mentioned classes are different.

For the proof we need the following two theorems.

Theorem E. (G. Eneström [1], S. Kakeya [5], see also A. W. Goodman [2]). If $a_0 > a_1 > a_2 > \cdots > a_n > 0$; $n \in \mathbb{N}$, then the polynomial

$$p(z) = a_0 + a_1 z + a_2 z^2 + \cdots + a_n z^n$$

has no roots in $\overline{\Delta} = \{z \in \mathbb{C} :\; |z| \leq 1\}$.

Theorem F. (W. Rogosinski [8]). If the function $f(z) = \sum\limits_{n=0}^{\infty} \alpha_n z^n$ is subordinate to the function $F(z) = \sum\limits_{n=0}^{\infty} \beta_n z^n$ then for $N = 0, 1, 2, 3, \ldots$ we have

$$\sum_{n=0}^{N} |\alpha_n|^2 \leq \sum_{n=0}^{N} |\beta_n|^2 .$$

2. Main result

The following theorem is a fulfillment of Theorem D.

Theorem 1. *Let A, B, C, D are some complex numbers such that $B + A \neq 0$, $D + C \neq 0$. If $|B| < 1$ and $|D| < 1$, then*

$$P(A, B) \star P(C, D) \neq P(AD + AC + BC, BD). \tag{1}$$

Proof. The proof will be divided into three steps. First we give a family of bounded functions ω^ν; $\nu = 1, 2, 3\ldots$ having special property of coefficients. Afterwards we construct, using ω^ν, a function belonging to the class $P(AD + AC + BC, BD)$ and finally we will show that such function is not in the class $P(A, B) \star P(C, D)$.

Now we use a method of E. Landau [6] and we find some functions ω^ν, $\nu = 1, 2, \ldots$, which are basic in this proof. We observe that

$$\frac{1}{1-z} = \left(\frac{1}{\sqrt{1-z}}\right)^2 = \left(\sum_{k=0}^{\infty} p_k z^k\right)^2 = 1 + z + z^2 + z^3 + \cdots,$$

where

$$p_0 = 1 \quad \text{and} \quad p_k = \frac{1 \cdot 3 \cdot \ldots \cdot (2k-1)}{2 \cdot 4 \cdot \ldots \cdot 2k}; \quad k = 1, 2, 3, \ldots \tag{2}$$

For some $\nu \in \mathbb{N}$ we set

$$K_\nu(z) = \sum_{k=0}^{\nu} p_k z^k = 1 + p_1 z + p_2 z^2 + p_3 z^3 + \cdots + p_\nu z^\nu$$

and note that

$$K_\nu^2(z) = 1 + z + z^2 + \cdots + z^\nu + b_{\nu+1} z^{\nu+1} + \cdots + b_{2\nu} z^{2\nu},$$

where $b_{\nu+1}, \ldots, b_{2\nu} \in \mathbb{C}$.

Let ω^ν be given by

$$\omega^\nu(z) = \frac{z^{\nu+1} K_\nu\left(\frac{1}{z}\right)}{K_\nu(z)} = z \frac{z^\nu + p_1 z^{\nu-1} + p_2 z^{\nu-2} + \cdots + p_\nu}{1 + p_1 z + p_2 z^2 + \cdots + p_\nu z^\nu}. \tag{3}$$

Since

$$p_k > p_k \frac{2k+1}{2k+2} = p_{k+1} \quad \text{where} \quad k = 0, 1, 2, 3, \ldots,$$

then for $\nu \in \mathbb{N}$ we have

$$1 > p_1 > p_2 > p_3 > \ldots > p_\nu > 0.$$

Applying Theorem E to the polynomial K_ν we obtain $K_\nu(z) \neq 0$ for $|z| \leq 1$ hence the function ω^ν is regular in Δ. Moreover $\omega^\nu(0) = 0$ and on the circle $|z| = 1$ we have

$$\left|\omega^{\nu}\left(e^{it}\right)\right| = \left|\frac{e^{(\nu+1)it}K_{\nu}\left(e^{-it}\right)}{K_{\nu}\left(e^{it}\right)}\right| = \frac{\left|\overline{K_{\nu}\left(e^{it}\right)}\right|}{\left|K_{\nu}\left(e^{it}\right)\right|} = 1; \quad t \in \mathbb{R}.$$

In this way we conclude that for $\nu \in \{1, 2, 3, \ldots\}$

$$\omega^{\nu} \in \Omega. \tag{4}$$

Let for a certain $\nu \in \mathbb{N}$ the function ω^{ν} is represented by following power series expansions:

$$\omega^{\nu}(z) = \gamma_1^{\nu} z + \gamma_2^{\nu} z^2 + \gamma_3^{\nu} z^3 + \cdots,$$

and let $s_n^{\nu}(z)$ denote partial sum

$$s_n^{\nu}(z) = \sum_{k=1}^{n} \gamma_k^{\nu} z^k = \gamma_1^{\nu} z + \gamma_2^{\nu} z^2 + \gamma_3^{\nu} z^3 + \cdots + \gamma_n^{\nu} z^n$$

for all $n \in \{1, 2, 3, \ldots, \nu + 1\}$.

Now we will estimate $s_n^{\nu}(1) = \gamma_1^{\nu} + \gamma_2^{\nu} + \gamma_3^{\nu} + \cdots + \gamma_n^{\nu}$. If we integrate on a circle $C: \ |z| = r$ in a counterclockwise direction with $0 < r < 1$ then we obtain

$$\int_C a z^m dz = 0 \quad \text{and} \quad \int_C \frac{a}{z} dz = 2\pi a i, \tag{5}$$

for all integer $m \neq -1$ and $a \in \mathbb{C}$. Hence

$$\begin{aligned} s_n^{\nu}(1) &= \gamma_1^{\nu} + \gamma_2^{\nu} + \gamma_3^{\nu} + \cdots + \gamma_n^{\nu} = \\ &= \frac{1}{2\pi i} \int_C \omega^{\nu}(z) \left(\frac{1}{z^2} + \frac{1}{z^3} + \frac{1}{z^4} + \cdots + \frac{1}{z^{n+1}}\right) dz = \\ &= \frac{1}{2\pi i} \int_C \frac{\omega^{\nu}(z)}{z^{n+1}} \left(1 + z + \cdots + z^{n-1}\right) dz = \frac{1}{2\pi i} \int_C \frac{\omega^{\nu}(z)}{z^{n+1}} Q(z) dz, \end{aligned}$$

where

$$Q(z) = 1 + z + \cdots + z^{n-1} + d_n z^n + d_{n+1} z^{n+1} + \cdots + d_s z^s,$$

is any polynomial whose first n terms are equal to $1 + z + z^2 + \cdots + z^{n-1}$.

If for $n \leq \nu + 1$ we set

$$Q(z) = K_{\nu}^2(z) = 1 + z + \cdots + z^n + \cdots + z^{\nu} + b_{\nu+1} z^{\nu+1} + \cdots + b_{2\nu} z^{2\nu},$$

then we obtain

$$s_n^{\nu}(1) = \frac{1}{2\pi i} \int_C \frac{\omega^{\nu}(z)}{z^{n+1}} K_{\nu}^2(z) dz.$$

From (3) it follows that

$$s_n^\nu(1) = \frac{1}{2\pi i}\int_C \frac{\frac{z^{\nu+1}K_\nu(\frac{1}{z})}{K_\nu(z)}}{z^{n+1}} K_\nu^2(z)dz = \frac{1}{2\pi i}\int_C z^{\nu-n}K_\nu\left(\frac{1}{z}\right)K_\nu(z)dz =$$

$$= \frac{1}{2\pi i}\int_C z^{\nu-n}\left(1+p_1\frac{1}{z}+\cdots+p_\nu\frac{1}{z^\nu}\right)(1+p_1z+\cdots+p_\nu z^\nu)\,dz.$$

Using (5) we get

$$s_n^\nu(1) = p_{\nu-n+1}+p_{\nu-n+2}p_1+p_{\nu-n+3}p_2+\cdots+p_\nu p_{n-1},$$

where (p_k) are given by (2), and so

$$s_n^\nu(1) = \sum_{k=1}^{n}\gamma_k^\nu = \sum_{k=1}^{n} p_{\nu-n+k}p_{k-1}. \tag{6}$$

By the above we have

$$\gamma_1^\nu = s_1^\nu(1) = p_\nu$$

and for all $\nu \in \{1,2,3,\ldots\}$ and $n \in \{2,3,4,\ldots,\nu+1\}$

$$\gamma_n^\nu = s_n^\nu(1) - s_{n-1}^\nu(1) = \sum_{k=1}^{n} p_{\nu-n+k}p_{k-1} - \sum_{k=1}^{n-1} p_{\nu-n+1+k}p_{k-1}$$

and therefore

$$\gamma_n^\nu = \sum_{k=1}^{n-1}(p_{\nu-n+k}-p_{\nu-n+1+k})p_{k-1} + p_\nu p_{n-1}. \tag{7}$$

The sequence (p_n) is positive and decreasing, then from (7) it follows that

$$\gamma_n^\nu > 0 \quad \text{for all} \quad \nu\in\mathbb{N} \quad \text{and} \quad n\in\{1,2,3,\ldots,\nu+1\}. \tag{8}$$

We conclude from (6) that for $n=\nu+1$

$$s_{\nu+1}^\nu(1) = \gamma_1^\nu+\gamma_2^\nu+\cdots+\gamma_\nu^\nu+\gamma_{\nu+1}^\nu = \sum_{k=1}^{\nu+1} p_{k-1}p_{k-1} = 1+\sum_{k=1}^{\nu} p_k^2.$$

Since $\sum_{k=1}^{\infty} p_k^2 = \infty$ then

$$\lim_{\nu\to\infty} s_{\nu+1}^\nu(1) = \lim_{\nu\to\infty}\left(1+\sum_{k=1}^{\nu}p_k^2\right) = +\infty. \tag{9}$$

This means that ω^ν, $\nu\in\mathbb{N}$, is the Schwarz function, for which the initial $\nu+1$ coefficients are positive and their sum increase to infinity when $\nu\to\infty$.

Now using the properties of ω^ν we construct the function belonging to the class $P(AC+AD+BC, BD)$ which is not in the class $P(A,B)\star P(C,D)$, where $|B|<1$ and $|D|<1$. Since for all $x \in \mathbb{C}$; $|x|=1$ we have

$$P(A,B) = P(Ax, Bx).$$

Thus we can assume without loss of generality that $B \in [0;1)$ and $D \in [0;1)$.

For fixed $\nu \in \mathbb{N}$ let h^ν be given by

$$h^\nu(z) = \frac{1+(AC+AD+BC)\omega^\nu(z)}{1-BD\omega^\nu(z)}, \tag{10}$$

where $\omega^\nu \in \Omega$ is given by (3).

It is clearly that

$$h^\nu(z) \prec \frac{1+(AC+AD+BC)z}{1-BDz}$$

and so

$$h^\nu \in P(AC+AD+BC, BD).$$

Suppose that there exist the functions $f \in P(A,B)$, $g \in P(C,D)$ such that

$$f(z)\star g(z) = h^\nu(z). \tag{11}$$

The functions f and g have the following form:

$$f(z) = \frac{1+A\omega_1(z)}{1-B\omega_1(z)} = 1+(A+B)\frac{\omega_1(z)}{1-B\omega_1(z)}$$

and

$$g(z) = \frac{1+C\omega_2(z)}{1-D\omega_2(z)} = 1+(C+D)\frac{\omega_2(z)}{1-D\omega_2(z)},$$

where $\omega_1, \omega_2 \in \Omega$. For simplicity of notations we write

$$f(z) = 1+(A+B)\tilde{f}(z), \qquad g(z) = 1+(C+D)\tilde{g}(z),$$

where

$$\tilde{f}(z) = \frac{\omega_1(z)}{1-B\omega_1(z)} \quad \text{and} \quad \tilde{g}(z) = \frac{\omega_2(z)}{1-D\omega_2(z)}. \tag{12}$$

Using these notation we can rewrite (11) as

$$\left[1+(A+B)\tilde{f}(z)\right]\star[1+(C+D)\tilde{g}(z)] = \frac{1+(AC+AD+BC)\omega^\nu(z)}{1-BD\omega^\nu(z)}$$

and so

$$1+(A+B)(C+D)\tilde{f}(z)\star\tilde{g}(z) = 1+(A+B)(C+D)\frac{\omega^\nu(z)}{1-BD\omega^\nu(z)}.$$

This means that

$$\tilde{f}(z) \star \tilde{g}(z) = \tilde{h}^{\nu}(z), \tag{13}$$

where

$$\tilde{h}^{\nu}(z) = \frac{\omega^{\nu}(z)}{1 - BD\omega^{\nu}(z)}.$$

Let the functions $\tilde{h}^{\nu}$, $\tilde{f}$ and $\tilde{g}$ have the following expansions in Δ:

$$\tilde{h}^{\nu}(z) = \sum_{n=1}^{\infty} c_n^{\nu} z^n, \quad \tilde{f}(z) = \sum_{n=1}^{\infty} a_n z^n, \quad \tilde{g}(z) = \sum_{n=1}^{\infty} b_n z^n. \tag{14}$$

From (12) it follows, that

$$\tilde{f}(z) \prec \frac{z}{1 - Bz} \quad \text{and} \quad \tilde{g}(z) \prec \frac{z}{1 - Dz},$$

and hence by Theorem F we obtain

$$\sum_{n=1}^{N} |a_n|^2 \leq \frac{1 - |B|^{2N}}{1 - |B|^2} \quad \text{and} \quad \sum_{n=1}^{N} |b_n|^2 \leq \frac{1 - |D|^{2N}}{1 - |D|^2}; \quad N = 1, 2, 3 \ldots$$

Letting $N \to \infty$ we get

$$\sum_{n=1}^{\infty} |a_n|^2 + \sum_{n=1}^{\infty} |b_n|^2 \leq \frac{1}{1 - |B|^2} + \frac{1}{1 - |D|^2}$$

hence

$$\sum_{n=1}^{\infty} \left(|a_n|^2 + |b_n|^2\right) = \sum_{n=1}^{\infty} \left[\left(|a_n| - |b_n|\right)^2 + 2|a_n||b_n|\right] \leq \tag{15}$$

$$\leq \frac{1}{1 - |B|^2} + \frac{1}{1 - |D|^2}.$$

From (13) and (14) we obtain

$$a_n b_n = c_n^{\nu}, \quad \text{for} \quad n = 1, 2, 3, \ldots, \tag{16}$$

therefore (15) and (16) give

$$\sum_{n=1}^{\infty} \left(|a_n| - |b_n|\right)^2 \leq \frac{1}{1 - |B|^2} + \frac{1}{1 - |D|^2} - 2\sum_{n=1}^{\infty} |c_n^{\nu}|. \tag{17}$$

Now we observe that

$$\sum_{n=1}^{\infty} c_n^{\nu} z^n = \tilde{h}^{\nu}(z) = \frac{\omega^{\nu}(z)}{1 - BD\omega^{\nu}(z)}$$

$$= \omega^\nu(z) + BD\left[\omega^\nu(z)\right]^2 + B^2D^2\left[\omega^\nu(z)\right]^3 + \cdots$$
$$= \left(\gamma_1^\nu z + \gamma_2^\nu z^2 + \cdots\right) + BD\left(\gamma_1^\nu z + \gamma_2^\nu z^2 + \cdots\right)^2 +$$
$$+B^2D^2\left(\gamma_1^\nu z + \gamma_2^\nu z^2 + \cdots\right)^3 + \cdots = \gamma_1^\nu z + \left(\gamma_2^\nu + BD(\gamma_1^\nu)^2\right)z^2 +$$
$$\left(\gamma_3^\nu + 2BD\gamma_1^\nu\gamma_2^\nu + B^2D^2(\gamma_1^\nu)^3\right)z^3 + \cdots$$

Since $B \in [0;1)$ and $D \in [0;1)$ and since (8) we have

$$\sum_{n=1}^{\infty} |c_n^\nu| = |\gamma_1^\nu| + \left|\gamma_2^\nu + BD(\gamma_1^\nu)^2\right| + \left|\gamma_3^\nu + 2BD\gamma_1^\nu\gamma_2^\nu + B^2D^2(\gamma_1^\nu)^3\right| + \cdots$$
$$\geq \sum_{n=1}^{\nu+1} |c_n^\nu| \geq \gamma_1^\nu + \gamma_2^\nu + \gamma_3^\nu + \ldots + \gamma_{\nu+1}^\nu = s_{\nu+1}^\nu(1).$$

From the above we have for all $\nu \in \{1,2,3,\ldots\}$

$$\sum_{n=1}^{\infty} |c_n^\nu| \geq s_{\nu+1}^\nu(1). \tag{18}$$

Combining (17) and (18) we obtain

$$0 \leq \sum_{n=1}^{\infty} \left(|a_n| - |b_n|\right)^2 \leq \frac{1}{1-|B|^2} + \frac{1}{1-|D|^2} - 2s_{\nu+1}^\nu(1). \tag{19}$$

From (9) it follows that we be able to choose a suitable ν such that the right side of (19) is negative. In this way (19) follows the contradiction and the proof is complete.

From Theorem 1 and Theorem D we immediately have that this two classes $P(A,B) \star P(C,D)$ and $P(AD + AC + BC, BD)$ are equal if and only if $|B| = 1$ or $|D| = 1$.

References

1. G. Eneström, Härledning af en allmä formel för antalet pensionärer som vid en godtycklig tidpunkt förefinnas inom en sluten pensionskassa, *Öfversigt af Kongl. Vetenskaps – Akademiens Förhandlingar* **50** (1893) 405-415.
2. A.W. Goodman, *Univalent functions*.vol. 2, Mariner Publishing Co., Tampa, Florida, 1983.
3. J. Hadamard, Théoréme sur les séries entiéres, *Acta Mathematica* **22** (1898) 55-63.
4. W. Janowski, Some extremal problems for certain families of analytic functions, *Ann. Polon. Math.* **28** (1957) 297-326.
5. S. Kakeya, On the limits of the roots of an algebraic equation with positive coefficients, *Tôhoku Math. Jour.* **2**(1912) 140-142.
6. E. Landau, *Darstellung und Begrundung einiger neurer Ergebnise der Funktionenentheorie*. Chelsea Publishing Co., New York, 1946.

7. G. Pólya, I. J. Schoenberg, Remarks on de la Vallée Pousin means and convex conformal maps of the circle, *Pacific J. Math.* **8** (1958) 295-334.
8. W. Rogosinski, On the coefficients of subordinate functions, *Proc. London Math. Soc.* **48** (2) (1943) 48-82.
9. S. Ruscheweyh, T. Sheil-Small, Hadamard product of schlicht functions and the Pólya-Schoenberg conjecture, *Comtnt. Math. Helv.* **48** (1973) 119-135.
10. S. Ruscheweyh, J. Stankiewicz, Subordination under convex univalent functions, *Bull. Polon. Acad. Sci. Math* **33** (1985) 499-502.
11. J. Stankiewicz, Z. Stankiewicz, Convolution of some classes of function, *Folia Sci. Univ. Techn. Resoviensis* **48**(1988) 93-101.
12. H. S. Wilf, Subordintating factor sequences for convex maps of the unit circle, *Proc. Amer. Math. Soc.* **12** (1961) 689-693.

Functional Analysis and its Applications
V. Kadets and W. Żelazko

Superstrictly Singular and Superstrictly Cosingular Operators ⋆

Anatolij Plichko [a]

[a]Institute of Mathematics, Cracow University of Technology, Cracow, Poland

Abstract

We prove some known and new results on superstrictly singular operators and establish full duality between superstrictly singular and superstrictly cosingular operators.

Key words: Singular operators, superstrictly singular operators, Banach spaces
2000 MSC: 46B07, 46B28, 47B06

1. Introduction

Through in this paper the *space* means a Banach space and the *operator* means a bounded linear operator.

Definition 1. [6]. An operator T from a space X into a space Y is called *strictly singular* if there does not exist a number $c > 0$ and an infinite dimensional subspace $E \subset X$ such that $\|Tx\| \geqslant c\|x\|$ for all x in E.

The following local version of this definition is natural.

Definition 2. An operator T is called *superstrictly singular* (*SSS* for short) if there does not exist a number $c > 0$ and a sequence of subspaces $E_n \subset X$, $\dim E_n = n$, such that

$$\|Tx\| \geqslant c\|x\| \quad \text{for all } x \text{ in } \cup_n E_n . \tag{1}$$

⋆ Partially supported by DAAD foundation
Email address: aplichko@kspu.kr.ua (Anatolij Plichko).

Put for an operator T

$$b_n(T) = \sup \inf_{x \in S(E_n)} \|Tx\| \,, \tag{2}$$

where supremum is taken over all n-dimensional subspaces $E_n \subset X$ and $S(E_n)$ is the unit sphere of E_n.

Obviously

$$\|T\| = b_1(T) \geq b_2(T) \geq \cdots \geq 0 \,,$$

T is SSS if and only if

$$b_n(T) \to 0 \quad \text{as} \quad n \to \infty \tag{3}$$

and the greatest constant c for which (1) is satisfied, is equal to $\lim_{n\to\infty} b_n(T)$.

The numbers $b_n(T)$, which are called the *Bernštein numbers*, were considered in Approximation Theory (see [21] and references therein). They are among the classical widths. However those widths are considered mostly for compact sets. Evidently, every compact operator is SSS. We almost do not see noncompact SSS operators in [21]. For operators, defined on a Hilbert space, the Bernštein numbers are connected with so-called AMD numbers (see [2]).

SSS operators were introduced implicitly in [14] and explicitly, under the name "$\mathfrak{S}_0^*$-operators" or "operators of class C_0^*", in [10]-[12]. Probably, the difference can be explained by absence of good letter $\mathfrak{S}$ in Kharkiv. Implicitly these operators were considered also in [23]. The term "superstrictly singular operator" was introduced in [4]; where this class was investigated by technique of superideals. Our note is inspired by the paper [4]. We do not try to calculate the exact values of $b_n(T)$, but we shall determine whether they converge to zero or not.

Obviously, every SSS operator is strictly singular. It is easy to see that T has the finite rank if and only if $b_n(T) = 0$ begining with some number N. Observe also, that if T is infinite dimensional then the supremum in (2) can be taken over subspaces E_n such that $T|_{E_n}$ are injective, only. Then the formula (2) turns into following

$$b_n(T) = \sup \frac{1}{\|(T|_{E_n})^{-1}\|} \tag{2'}$$

where the supremum is taken over all n-dimensional subspaces E_n such that the restrictions $T|_{E_n}$ are injective.

2. Some properties of SSS operators

Using the spectrum of continuous function and Dvoretzky's theorem, V. Milman [11] has shown that SSS operators form a closed ideal, i.e. that a sum of SSS operators is SSS, a composition (right or left) of SSS operator with the bounded operator is SSS and $L(X,Y)$-norm limit of SSS operators is SSS. Using superideal technique and the Dvoretzky theorem, this result was proved in [4] too. We shall deduce Milman's result from the following Theorem 1. This theorem was proved in [11] too (using the spectrum and Dvoretzky theorem). Our proof of Theorem 1 is elementary and does not use Dvoretzky's theorem.

Theorem 1. *The operator $T : X \to Y$ is SSS if and only if for every sequence of subspaces $E_n \subset X$, $\dim E_n = n$, there exist subspaces $F_n \subset E_n$ such that $\dim F_n \to \infty$ and*

$$\|T|_{F_n}\| \to 0 \text{ as } n \to \infty .$$

The "if" part of this statement is evident. To prove the "only if" part we need:

Lemma 1. *For every $\varepsilon > 0$ there exists a sequence $k(n) \to \infty$ as $n \to \infty$ such that every normed space E_n, $\dim E_n = n$, contains subspaces $G_1, \dots, G_{k(n)}$, each are of the dimension $k(n)$, whose finite dimensional decomposition constant is less than $1 + \varepsilon$.*

The proof of this lemma is easily can deduced from Mazur's lemma on the basic sequences ([8], p.4). Lemma 1 also follows immediately from the Dvoretzky's theorem, but we prefer not to use the Dvoretzky theorem here. □

Proof of Theorem 1. Let $\varepsilon > 0$. Construct, using Lemma 1, subspaces $(G_i^n)_{i=1}^{k(n)}$ of E_n so that $\dim G_i^n = k(n) \to \infty$ and $(G_i^n)_{i=1}^{k(n)}$ have, for fixed n, the finite dimensional decomposition constant less than $1 + \varepsilon$.

Using (2), choose elements x_i^n in $S(G_i^n)$, $i = 1, \dots, k(n)$ such that $\|Tx_i^n\| \leq b_{k(n)}(T)$ (because of finite dimensional nature of E_n, infimums in (2) are attained).

Now choose a sequence $1 \leq l(n) \leq k(n)$ so that $l(n) \to \infty$ and $l(n) \cdot b_{k(n)}(T) \to 0$ as $n \to \infty$. Put $F_n = \text{lin}(x_i^n)_{i=1}^{l(n)}$. Then for $x \in S(F_n)$, $x = \sum_{i=1}^{l(n)} a_i x_i^n$, we have according to finite dimensional decomposition nature

$$|a_1| < 1 + \varepsilon \ , \ |a_2| < 2(1 + \varepsilon) \ , \dots , \ |a_{l(n)}| < 2(1 + \varepsilon).$$

Hence

$$\sum_{i=1}^{l(n)} |a_i| \|Tx_i^n\| \leq (\sum |a_i|) b_{k(n)}(T) < 2(1 + \varepsilon) l(n) b_{k(n)}(T) \to 0 \text{ as } n \to \infty . \ \square$$

Remark 1. Theorem 1 is, to a certain extent, a development of the well known Kato's result [6]: *Let X and Y be infinite dimensional spaces. Assume that $T : X \to Y$ is an operator such that the restriction of T to any subspace of finite codimension is not an isomorphism. Then for every $\varepsilon > 0$ there is an infinite dimensional subspace E of X so that $T|_E$ is compact and $\|T|_E\| \leq \varepsilon$.*

Corollary 1. *An operator T is SSS if and only if for every sequence of finite dimensional subspaces $E_n \subset X$, $\dim E_n \to \infty$, there are subspaces $F_n \subset E_n$ such that $\dim F_n \to \infty$ and*

$$\|T|_{F_n}\| \to 0 \quad as \quad n \to \infty .$$

Recall that for given $\varepsilon > 0$, a space E is called $(1 + \varepsilon)$-*Euclidean* if there exists an operator U from E onto some Euclidean space with

$$(1 - \varepsilon)\|x\| \leq \|Ux\| \leq (1 + \varepsilon)\|x\| \quad \forall x \in E .$$

Application of Dvoretzky's theorem gives immediately

Corollary 2 *The following conditions are equivalent:*
(i) T is SSS;
(ii) for every $\varepsilon > 0$ and every sequence of $(1+\varepsilon)$-Euclidean subspaces $E_n \subset X$ with $\dim E_n \to \infty$, we have

$$\inf_{x \in S(E_n)} \|Tx\| \to 0 \quad as \quad n \to \infty \,;$$

(iii) for every $\varepsilon > 0$ and every sequence of $(1+\varepsilon)$-Euclidean subspaces $E_n \subset X$ with $\dim E_n \to \infty$, there are subspaces $F_n \subset E_n$, $\dim F_n \to \infty$ such that

$$\|T|_{F_n}\| \to 0 \quad as \quad n \to \infty \,.$$

Corollary 3. *A sum $T + U$ of two SSS operators is SSS.*

Proof. Let $E_n \subset X$ be an arbitrary sequence of subspaces, $\dim E_n \to \infty$. There are, by Theorem 1, subspaces $F_n \subset E_n$ so that $\dim F_n \to \infty$ and $\|T|_{F_n}\| \to 0$. By definition, there exists a sequence $x_n \in S(F_n)$, such that $\|Ux_n\| \to 0$. Then

$$\|(T+U)x_n\| \to 0 \text{ as } n \to \infty \,. \square$$

The fact that a composition of SSS and bounded operator is SSS, can be proved very simply.

Corollary 4 *Let $X = X_1 \oplus X_2$, $T_1 : X_1 \to Y$ and $T_2 : X_2 \to Y$ be SSS. Then the operator $T : X \to Y$ is defined by the formula $T(x_1 + x_2) = T_1x_1 + T_2x_2$, $x_1 \in X_1$, $x_2 \in X_2$ is SSS.*

Proposition 1. *An $L(X,Y)$-norm limit of SSS operators (T_k) is SSS.*

Proof. If we assume conversely that there exists $c > 0$ and a sequence of subspaces $E_n \subset X$, $\dim E_n \to \infty$, with the property (1) for T. Then choose k such that $\|T - T_k\| < c/2$. Since T_k is SSS, $\inf\{\|T_kx\| : x \in S(E_n)\} < c/2$ for some n. Hence $\inf\{\|Tx\| : x \in S(E)\} < c$. This contradicts (1). $\square$

The following statement is contained in [11] and [4] but we give a more transparent proof.

Theorem 2. *If T is not SSS then for every sequence $\varepsilon_n > 0$ there are $(1+\varepsilon_n)$-Euclidean subspaces $E_n \subset X$, $\dim E_n \to \infty$, and a constant $d > 0$ so that for every n*

$$(d - \varepsilon_n)\|x\| \le \|Tx\| \le (d + \varepsilon_n)\|x\| \quad \forall x \in E_n \,.$$

Proof. Of course, we can not do without the Dvoretzky theorem here. Applying it twice, we obtain the existence of $(1+\varepsilon_n)$-Euclidean subspaces $F_n \subset X$, $\dim F_n \to \infty$, and a constant $c > 0$ such that

$$c\|x\| \leq \|Tx\| \quad \forall x \in \cup_n F_n ,$$

and the subspaces TF_n are $(1+\varepsilon_n)$-Euclidean too.

To continue the proof we use the easily-proved

Lemma 2. *For any segment $[a,b]$, $\varepsilon > 0$, and a natural number l there is a natural number $k = k(l,a,b,\varepsilon)$ such that any l-element subset of $[a,b]$ contains a k-element subset of diameter $< \varepsilon$. Moreover $k(l,a,b,\varepsilon)$ can be chosen so that $k(l,a,b,\varepsilon) \to \infty$ as $l \to \infty$.*

And also

Lemma 3. *Let $0 < a \leq b$, $0 < \varepsilon < a$ and $l \in \mathbb{N}$. Then there exists a natural number $k = k(l,a,b,\varepsilon)$ such that $k \to \infty$ as $l \to \infty$ and for any operator T from l-dimensional Euclidean space F onto l-dimensional Euclidean space G satisfying the condition*

$$a\|x\| \leq \|Tx\| \leq b\|x\| \quad \forall\, x \in F$$

there is a subspace $E \subset F$ with $\dim E = k$, and a number $d \in [a,b]$ satisfying the condition

$$(d-\varepsilon)\|x\| \leq \|Tx\| \leq (d+\varepsilon)\|x\| \quad \forall\, x \in E . \tag{4}$$

Proof of Lemma 3. Let us choose in F an orthonormal basis $(e_i)_1^l$ such that $(Te_i)_1^l$ is orthogonal basis in G (see f.e. [3], p.175). According to Lemma 2, there exists $d \in [a,b]$ and natural numbers $i_1, \ldots, i_k$ for which

$$|\|Te_{i_1}\| - d| < \varepsilon, \ldots, |\|Te_{i_k}\| - d| < \varepsilon .$$

Then for any numbers $(a_j)_1^k$, we have

$$\|T(\sum_{j=1}^{k} a_j e_{i_j})\|^2 = \sum_{j=1}^{k} |a_j|^2 \|Te_{i_j}\|^2 \leq \sum |a_j|^2 (d+\varepsilon)^2 = (d+\varepsilon)^2 \|\sum a_j e_{i_j}\|^2 . \tag{5}$$

Put $E = \mathrm{lin}(e_{i_j})_{j=1}^k$. The right inequality in (4) follows from (5). The left one is verified in the same way. □

Now it is clear how to finish the proof of Theorem 2. □

Remark 2 An infinite dimensional version of Theorem 2 is also true: *If T is an isomorphism from one Hilbert space into another, then there exists an infinite dimensional subspace $E \subset H$ such that the restriction $T|_E$ is proportional to an isometry up to ε.* The same fact is true for l_p spaces, $1 \leq p < \infty$ and c_0 [11].

Note two another properties of SSS operators.

Proposition 2. *Let $T : X \to Y$ be an operator and let $\overline{T}$ be the operator T from X into $\overline{TX} \subset Y$. Then $\overline{T}$ is SSS if and only if T is SSS, moreover for every n*

$$b_n(T) = b_n(\overline{T}).$$

Proposition 3. *Let $T : X \to Y$ be an operator and $\widehat{T} : X/\ker T \to Y$ be the corresponding quotient operator. If $\widehat{T}$ is SSS, then T is SSS too, moreover for every n*

$$b_n(T) \le b_n(\widehat{T}) .$$

Remark 3. An example of quotient map $T : l_1 \to l_1/Z \simeq l_2$ shows that the converse statement is not true (we shall show further that T is SSS; of course $\widehat{T}$ is an isometry).

3. Examples

a) Strictly singular operators which are non SSS

As it is well known, every strictly singular operator on l_p , $1 \le p < \infty$ or on c_0 is compact ([8], p.76). Every operator from l_p into l_q , $p \ne q$, $1 \le p, q < \infty$, and from l_p into c_0 and from c_0 into l_p, is strictly singular ([8], p.75). Moreover, every operator from l_p into l_q for $p > q$ and from c_0 into l_p, is compact (Pitt's theorem ([8], p.76). Let us give

Example 1. A (strictly singular) operator from l_p into l_q , $1 < p < q < \infty$, which is not SSS.

Denote by E_n a subspace of l_p which consists of sequences with supports contained between 2^n and 2^{n+1}. Let $F_n \subset E_n$ be a subspace spanned by 2^n-dimensional version of Rademacher functions and G_n be its complement spanned by the remaining 2^n-dimensional version of Walsh functions. From Khinchine's inequality ([8], p.66) it follows that there are constants $a, b > 0$ and $d_n > 0$ such that

$$ad_n\|x\|_p \le \|x\|_q \le bd_n\|x\|_p \quad \text{for every } x \text{ in } F_n ,$$

and also that the projections in E_n onto F_n along G_n both in norm l_p and in norm l_q are uniformly bounded.

Define an operator T in the following way: if $x \in F_n$, put $Tx = d_n x$; if $x \in G_n$, put $Tx = x$. Then we extend T to a linear and bounded operator from l_p into l_q . □

Remark 4. For $p = 2$, $q > 2$ a similar example was given in [10] and [11].

Remark 5. It will follow from the next Proposition 6 that each operator from l_1 into l_p , $1 < p < \infty$, is SSS. Soon we shall consider c_0-valued operators .

Corollary. *Let $1 < p < q < \infty$. For every infinite dimensional $\mathcal{L}_p$-space X and $\mathcal{L}_q$-space Y (in the sense of [7]) there exists a strictly singular operator from X into Y which is non SSS.*

Indeed, any infinite dimensional $\mathcal{L}_p$-space contains a complemented subspace isomorphic to l_p [7].

Let us consider c_0-valued operators .

Proposition 4. *Let X be a Banach space with a finite dimensional decomposition (E_n). Then there exists an injective operator from X into c_0 which is non SSS. If X has no subspaces isomorphic to c_0 then this operator is strictly singular. There exists an operator from an arbitrary separable Banach space into c_0 which is non SSS.*

Proof. Of course, we can assume that $\dim E_n \to \infty$. Let us take $\varepsilon > 0$. Using the local universality of c_0, choose subspaces $F_n \subset c_0$ which are $(1+\varepsilon)$-isometric to E_n. So we have disjoint supports and let $T : E_n \to F_n$ be these $(1+\varepsilon)$-isometries. It remains to extend T to an injective operator from X into c_0. To prove the last statement of the proposition it is sufficient to use the well known fact that any infinite dimensional separable space has a quotient with basis ([8], p.10). □

b) **SSS operators which are non compact**

A natural example of noncompact SSS operator is the identity inclusion $l_p \hookrightarrow l_q$ for $1 \leq p < q \leq \infty$ [12]. Probably, the widths $b_n(\hookrightarrow)$ for these inclusions were calculated or estimated in the Approximation Theory. In [21] the inclusion $l_1 \hookrightarrow l_2$ (a width of octahedron) is considered. We shall deduce the result of [12] from a rather general statement on an inclusion into c_0. However let us first recall

Lemma 4. [9],[12]. *Let X be some linear space of functions of natural argument vanishing at infinity. Then every subspace $E_n \subset X$, $\dim E_n = n$, contains a function $x(i)$ such that $\max\{|x(i)| : i \in \mathbb{N}\}$ is attained in at least n points.*

Proof. Obviously, every function $|x(i)|$, $x \in X$ attains maximum in at least one point. Let for n the lemma be proved. Consider an arbitrary subspace $E_{n+1} \subset X$, $\dim E_{n+1} = n+1$. By the inductive hypothesis, there exists a function $y \in E_{n+1}$ such that $\max |v(i)|$ is attained at least in n points $(i_k)_1^n$. Without loss of generality we can assume that the maximum is attained exactly in n points. The subspace $E^n = \{x \in X : x(i_k) = 0 \text{ for } k = 1, \ldots, n\}$ has codimension n, hence has a non-zero intersection with E_{n+1}. Let $0 \neq z \in E^n \cap E_{n+1}$ and $\max |z(i)| = |z(j)|$. Since $z(i_k) = 0$, then $j \neq i_k$, $k = 1, \ldots, n$. Then the family $\{\lambda y + (1-\lambda)z : 0 \leq \lambda \leq 1\}$ contains a function, which attains the maximum of modulus in $n+1$ points $(i_k)_{k=1}^{n+1}$ (i_{n+1} may be not equal to j). □

Proposition 5. *Let X be a space with a symmetric basis (e_n) which is not equivalent to the standard basis of c_0. Then the natural inclusion $X \hookrightarrow c_0$ is SSS.*

Proof. Let $E_n \subset X$, $\dim E_n = n$. Using Lemma 4, choose an element $x_n(i) \in E_n$, whose maximum modulus is attained at least in n points and is equal to one. Obviously, $\|x_n\|_X \to \infty$, which proves that our operator is SSS. □

Remark 6. From the proof of Proposition 5 it follows that for the inclusion $\hookrightarrow_p$ of l_p to c_0, we have

$$b_n(\hookrightarrow_p) = n^{-1/p} .$$

Corollary 1. *The natural inclusion $l_p \hookrightarrow l_q$ for $1 \le p < q < \infty$ is SSS.*

Proof. Indeed,

$$\|x\|_q^q = \sum |x(i)|^q = \sum |x(i)|^p |x(i)|^{q-p}$$
$$\le (\sum |x(i)|^p) \max |x(i)|^{q-p} = \|x\|_p^p \|x\|_0^{q-p} .$$

If $\|x\|_p = 1$ then $\|x\|_q \le \|x\|_0^{(q-p)/q}$. This and Proposition 5 implies SSS of our inclusion. □

Remark 7. From the proof of Corollary 1 it follows that for the inclusion $\hookrightarrow_{pq}$ of l_p to l_q, $q < p$

$$b_n(\hookrightarrow_{pq}) = n^{(p-q)/pq} .$$

Corollary 2. *Let $1 < p < q < \infty$, X be an infinite dimensional $\mathcal{L}_p$-space and Y be an infinite dimensional $\mathcal{L}_q$-space. Then there exists an SSS and non compact operator from X into Y.*

To prove this it is sufficient to recall the arguments of corollary of Example 1.

Corollary 3. *There exists an SSS and non compact operator in the space $L_p(0,1)$, $1 \le p < \infty$, $p \ne 2$.*

Proof. Let X be a subspace of the space L_p spanned by Rademacher functions, Y be its complement (for $p \ne 1$), U be a complemented subspace of L_p isometric to l_p , V be its complement.

If $p < 2$ then there exists an SSS and non compact operator from U into X (Corollary 1). If $p > 2$ then there exists an SSS and non compact operator from X into U (Corollary 1, too).

In the both case let us consider an arbitrary compact operator from V (resp. Y) into L_p and extend our operator onto the whole space (see Corollary 4 to Theorem 1). □

Remark 8. For $2 < p < \infty$ Corollary 3 was proved in [12]. Of course, Corollary 3 remains true for every space containing complemented $L_p(0,1)$, for example, for $L_p(\mathbb{R})$, $1 \le p < \infty$, $p \ne 2$.

c) **Spaces, where** $SSS(X,Y) = L(X,Y)$

Definition 7. We say that a space X *contains no uniformly complemented almost Euclidean subspaces* if there exists $\varepsilon > 0$ such that for every sequence of $(1+\varepsilon)$-Euclidean subspaces $E_n \subset X$, $\dim E_n \to \infty$, the relative projection constants $\lambda(E_n, X) \to \infty$.

Definition 9. [16]. A space X is called *locally π-Euclidean* if there is a $d \ge 1$ so that for all n and $\varepsilon > 0$ there is an N such that every N-dimensional subspace of

X contains a $(1+\varepsilon)$-isomorph of l_2^n which is d-complemented in X.

Every $\mathcal{L}_1$-space or $\mathcal{L}_\infty$-space does not contain uniformly complemented almost Euclidean subspaces ([7], Corollaries 6,2). The well known Pisier space in which the norms of all n-dimensional projections grow with n [18] does not contain such subspaces. Locally π-Euclidean spaces coincide with B-*convex* spaces, i.e. spaces which contain no l_1^n uniformly [17]. We shall use both terms B-convex and locally π-Euclidean.

Proposition 6. *Let X contain no uniformly complemented almost Euclidean subspaces and Y be B-convex. Then every operator T from X into Y is SSS.*

Proof. Suppose contrary to our claim, that there is a sequence of $(1+\varepsilon)$-Euclidean subspaces $E_n \subset X$, $\dim E_n \to \infty$, such that

$$\inf_{x\in S(E_n)} \|Tx\| \geq c > 0 \text{ for every } n .$$

Passing, if necessary, to subspaces, we may suppose that $F_n := TE_n$ are $(1+\varepsilon)$-Euclidean and d-complemented. Let P_n be a projection of norm d of the space Y onto F_n. Put $Q_n = (T^{-1}|_{F_n})P_nT$. Evidently, Q_n is a projection of X onto E_n and

$$\|Q_n\| \leq \|T^{-1}|_{F_n}\| \cdot \|P_n\| \cdot \|T\| \leq \frac{d}{c}\|T\| .$$

Contradiction. □

Remark 9. Proposition 6 is inspired by the following

Proposition MP [14]. *Let T be an operator from the space $C(S)$ of all continuous functions on a compactum S into l_2. Then*

$$b_n(T) \leq 1/\sqrt{p_n} ,$$

where p_n is the projection constant of n-dimensional Euclidean space.

Proof. Probably, we have the assume in Proposition MP $\|T\| \leq 1$. Let E_n be an n-dimensional subspace of $C(S)$ such that $T|_{E_n}$ is one to one and $F_n = TE_n$. Then there is a projection $Q_n : C(S) \to E_n$ with $\|Q_n\| \leq \|T^{-1}|_{F_n}\|$. As it is well known, the projection constant q_n of subspace E_n satisfies the inequality $p_n \leq q_n \cdot \|T\|\|T^{-1}|_{F_n}\|$. Hence,

$$p_n \leq q_n \cdot \|T^{-1}|_{F_n}\| \leq \|Q_n\| \cdot \|T^{-1}|_{F_n}\| \leq \|T^{-1}|_{F_n}\|^2 \leq b_n(T)^{-2}.$$

It implies Proposition MP. □

d) Absolutely summing and narrow operators

An operator $T : X \to Y$ is called *p-absolutely summing* if there is a constant K such that, for every choice of an integer n and vectors $\{x_i\}_{i=1}^n$ in X we have

$$(\sum_{i=1}^n \|Tx_i\|^p)^{1/p} \leq K \sup_{\|f\|=1} (\sum_{i=1}^n |f(x_i)|^p)^{1/p} .$$

The following proposition was known to the authors of [14].

Proposition 7. *Every p-absolutely summing operator T, $1 \le p < \infty$, is SSS.*

Proof. Since any p-absolutely summing operator is q-absolutely summing if $p < q$ ([8], p.65), we can assume $p > 2$. Suppose T is not SSS. Take subspaces E_n from Theorem 2. Obviously, for "almost orthogonal" bases $\{x_i\}_{i=1}^n$ in these spaces we can not find a common K in the definition of p-absolutely summing operator. □

Note that not every p-absolutely summing operator is compact.

An operator T from $L_p(0,1)$ to a space Y is *narrow* if for each measurable subset $A \subset (0,1)$ and each $\varepsilon > 1$ there is $x \in L_p$ such that $x^2 = \chi_A$ and $\|Tx\| < \varepsilon$. For properties of narrow operators see [19]. In particular, every compact operator is narrow. There is a narrow operator which is not strictly singular.

Question. Let $1 < p < \infty$. Is every SSS operator $T : L_p \to Y$ narrow, for any space Y?

4. Connection between SSS of an operator and its duals

Let us consider first the connection between STS of an operator and its dual. A simple example such as embedding $L_\infty \hookrightarrow L_2$ shows that an operator can be SSS [1] even if its dual is not strictly singular. So, as in the case of strictly singular operators, to find some connections between SSS of operator and its dual, we have to impose some additional conditions. It is found, that for SSS operators such conditions are more natural than those for strictly singular one.

Probably, the following corollary of the principle of local reflexivity is well known.

Lemma 5. *Let X be a locally π-Euclidean space. Then X^* is "locally π^*-Euclidean"; more exactly, there is a $d^* \ge 1$ so that for all n and ε there is an N such that every N-dimensional subspace of X^* contains an n-dimensional $(1+\varepsilon)$-Euclidean and d-complemented in X^* subspace , moreover this complement is weakly* closed.*

Proof. Since X is locally π-Euclidean it is B-convex [17]; hence X^* is B-convex (This is well known and it is easy to prove). So X^* is locally π-Euclidean.

Given N-dimensional subspace of X^*, let E_n be a subspace of X^* , $\dim E_n = n$, the existence of which guarantee Definition 4. Let F^n be a d-complement of E_n. Then the annihilator in the dual space $(F^n)^\perp \subset X^{**}$ has dimension n. Let $\varepsilon > 0$. By the principle of local reflexivity there exists a $(1+\varepsilon)$-isometry T from $(F^n)^\perp$ onto some subspace $X_n \subset X$ such that $\langle T\varphi, e\rangle = \langle e, \varphi\rangle$ for every $\varphi \in (F^n)^\perp$ and

[1] A simple proof of SSS of this embedding is contained, in fact, in ([20], Theorem 5.2).

$e \in E_n$. Hence, we can take $X_n^{\perp} \subset X^*$ as the desired weakly* closed complement for E_n . $\square$

Theorem 3. *Let T be an SSS operator from a B-convex space X into arbitrary space Y. Then T^* is SSS.*

Proof. Suppose T^* is not SSS. By Lemma 5, X^* is locally π^*-Euclidean. Hence there are subspaces $E_n \subset Y^*$ such that $\dim E_n \to \infty$, for some $c > 0$

$$\|T^* f\| \geq c\|f\| \quad \forall f \in \cup_n E_n$$

and subspaces $F_n = T^* E_n$ are uniformly complemented in X^*, moreover these complements G^n are weakly* closed. Put $X_n = (G^n)^{\top} \subset X$, where $\top$ denotes an annihilator in the predual space. Then $\dim X_n = n$ and for $x \in S(X_n)$

$$\begin{aligned} \|Tx\| &= \sup_{f \in S(Y^*)} \langle Tx, f\rangle = \sup_{f \in S(Y^*)} \langle x, T^* f\rangle \\ &\geq \sup_{f \in S(E_n)} \langle x, T^* f\rangle \geq c \sup_{g \in S(F_n)} \langle x, g\rangle \geq \frac{c}{d}\|x\|. \end{aligned}$$ $\square$

Remark 10. Let us recall that a space is called *c-convex* if it does not contain uniformly l_∞^n. There exists an SSS operator T from a c-convex space X into a Hilbert space H such that T^* is not even strictly singular. As an example, we can take a quotient map $l_1 \to l_1/Z \simeq H$. The SSS of this quotient follows from Proposition 6. Theorem 3 is false for strictly singular operators ([5], p.47).

The following definition was introduced in [12]: A closed subspace E of a space X is said to be *partially complemented* if there exists an infinite dimensional closed subspace $F \subset X$, $E \cap F = 0$, for which $E + F = \overline{E + F}$. In [12] it is stated

Theorem M. *If every closed subspace $E \subset X$, $\operatorname{codim} E = \infty$, is partially complemented, then the strict singularity of an operator T implies the strict singularity of T^*.*

Prof. Milman informed us that the proof of Theorem M contains a gap and that his proof works only if X^* is norm separable. So, the following question is natural; Let X be separable and suppose that every infinite codimensional subspace of X is partially complemented. Must X^* be separable? Consider the James tree JT.

Let us now consider the connection between the SSS of operator and its second dual. The following (probably, well known) fact plays a decisive role there.

Corollary of the principle of local reflexivity. *Let Φ be a finite dimensional subspace of X^{**} and suppose that for some $c > 0$*

$$\|T^{**}\varphi\| > c\|\varphi\| \quad \forall \varphi \in \Phi .$$

Then there exists a subspace $E \subset X$, $\dim E = \dim \Phi$, such that

$$\|Tx\| > c\|x\| \quad \forall x \in E. \tag{6}$$

Proof. Let F be a finite dimensional subspace of Y^* which $(1-\varepsilon)$-norms $T^{**}\Phi$, i.e. $\forall\varphi \in T^{**}\Phi$ there is $f \in S(F)$, for which $(1-\varepsilon)\|\varphi\| \le \langle f, \varphi\rangle$. By the principle of local reflexivity, there exists a subspace $E \subset X$ and $(1+\varepsilon)$-isometry $U : E \to \Phi$ such that

$$\langle x, g\rangle = \langle g, Ux\rangle \quad \forall x \in E \ , \ g \in T^*F.$$

Then for arbitrary $x \in E$ and $f \in F$

$$\langle Tx, f\rangle = \langle x, T^*f\rangle = \langle Ux, T^*f\rangle = \langle T^{**}Ux, f\rangle \ .$$

Therefore, for any $x \in E$ there is an element $f \in S(F)$ such that

$$\begin{aligned}\|Tx\| &\ge |\langle Tx, f\rangle| = |\langle T^{**}Ux, f\rangle| \\ &\ge (1-\varepsilon)\|T^{**}Ux\| > (1-\varepsilon)c\|Ux\| > \frac{(1-\varepsilon)}{(1+\varepsilon)}c\|x\| \ .\end{aligned}$$

So, (6) is satisfied for sufficiently small ε . □

Corollary. *If T is SSS then T^{**} is SSS too.*

Remark 11. The strict singularity of T does not imply the strict singularity of T^{**}. We shall give examples in the next section.

5. Superstrictly cosingular operators

Definition 5. [15]. An operator $T : X \to Y$ is called *strictly cosingular* if for every closed subspace $E \subset Y$ of infinite codimension, the map QT (where $Q : Y \to Y/E$ is a quotient map) has non-closed range.

It is easy to see ([15]) that if T^* is strictly singular, then T is strictly cosingular and if T^* is strictly cosingular then T is strictly singular.

The following quantitative characteristic of the operator T was introduced in [14]:

$$a_n(T) = \sup_{E^n} \inf\{\|\widehat{Tx}\|_{Y/E^n} : x \in X \ , \ \|\hat{x}\|_{X/T^{-1}E^n} = 1\} \ ,$$

where supremum is taken over all closed subspaces $E^n \subset Y$ of codimension n and caps denote the corresponding quotient classes.

Remark 12. More exactly, in [14]

$$a_n(T) = \sup_{E^n} \inf_{\|x\|=1} \|Tx\|_{Y/E^n} \ .$$

Note that with this definition $a_n(T) = 0$ for every operator with infinite dimensional kernel.

Definition 6. We call an operator T *superstrictly cosingular* (*SSCS* for short), if $a_n(T) \to 0$ as $n \to \infty$.

Evidently, every SSCS operator is strictly cosingular.

We do not know any papers except for [14] where the numbers $a_n(T)$ are considered. We do not know also, how the numbers $a_n(T)$ are connected with other widths. The principle of local reflexivity gives the complete duality between SSS and SSCS operators. Namely, the following is true

Theorem 4. *An operator T is SSS (SSCS) if and only if T^* is SSCS (respectively SSS).*

Proof. 1. *T^* is SSS $\Rightarrow$ T is SSCS.* Let E^n be a closed subspace of Y of codimension n and $F_n = (E^n)^\perp$. Then $\dim F_n = n$. Let $f_0 \in S(F_n)$ be an element such that

$$\|T^* f_0\| = \inf_{f \in S(F_n)} \|T^* f\| \, .$$

Put $\widehat{E}_n = X/T^{-1}E^n$. Then $(\widehat{E}_n)^* = T^* F_n$. Let $\hat{x} \in S(\widehat{E}_n)$ be an element such that

$$\langle \hat{x}, T^* f_0 \rangle = \sup_{f \in S(F_n)} \langle \hat{x}, T^* f \rangle \, .$$

Then for every representative $x \in \hat{x}$

$$\|\widehat{Tx}\| = \sup_{f \in S(F_n)} \langle \widehat{Tx}, f \rangle = \sup_{f \in S(F_n)} \langle Tx, f \rangle =$$
$$\sup_{f \in S(F_n)} \langle x, T^* f \rangle = \langle x, T^* f_0 \rangle \le \|x\| \|T^* f_0\| \, .$$

Hence, $\|\widehat{Tx}\| \leqslant \|T^* f_0\|$ and implication 1 is proved.

2. *T is SSS$\Rightarrow$ T^* is SSCS.* By the Corollary of the principle of local reflexivity, T^{**} is SSS. The T^* is SSCS by item 1.

3. *T is SSCS $\Rightarrow$ T^* is SSS.* Let F_n be an n-dimensional subspace of Y^*. We can suppose, by (2′), that the restriction $T^*|_{F_n}$ is injective. Put $\widehat{E}_n = X/(T^* F_n)^\top$. Then $\dim \widehat{E}_n = n$. Take an element $\hat{x}_0 \in S(\widehat{E}_n)$ such that (the norm $\|\widehat{Tx}\|$ is taken in the quotient space $Y/F_n^\top$)

$$\|\widehat{Tx_0}\| = \inf_{\|\hat{x}\|=1} \|\widehat{Tx}\|.$$

Take an element $f \in S(F_n)$ such that

$$\langle Tx_0, f \rangle = \sup_{\|\hat{x}\|=1} \langle \widehat{Tx}, f \rangle \, .$$

Then

$$\|T^* f\| = \sup_{\|\hat{x}\|=1} \langle \hat{x}, T^* f \rangle = \sup_{x \in \hat{x}\ ,\ \|\hat{x}\|=1} \langle x, T^* f \rangle = \sup_{x \in \hat{x}\ ,\ \|\hat{x}\|=1} \langle Tx, f \rangle =$$
$$\sup_{\|\hat{x}\|=1} \langle Tx, f \rangle = \langle \widehat{Tx_0}, f \rangle \le \|\widehat{Tx_0}\| \, .$$

4. *T^* is SSCS $\Rightarrow$ T is SSS.* By item 3, T^{**} is SSS and hence T is SSS too. □

Theorem 4 allows to transform results, known for SSS operators, into SSCS operators. Let us present some of them.

Corollary 1. *SSCS operators form a closed ideal.*

Proof. See Corollary 3 of Theorem 1 and Proposition 1.

Corollary 2. *If an operator T is SSCS then there exists a sequence of closed subspaces $E^n \subset Y$ of codimension n such that $X/T^{-1}E^n$ are $(1+\frac{1}{n})$-Euclidean and*

$$\inf\{\|\widehat{Tx}\|_{Y/E^n} : \|\hat{x}\|_{X/T^{-1}E^n} = 1\} \to 0 \ \text{as} \ s \to \infty \, .$$

Proof. See Corollary 2 of Theorem 1.

Corollary 3. *The natural embedding $l_p \hookrightarrow l_q$, where $1 < p < q < \infty$, is SSCS.*
Proof. See Corollary 1 of Proposition 5.

Corollary 4. *Let X be B-convex and suppose that Y does not contain uniformly complemented almost Euclidean subspaces. Then every operator T from X into Y is SSCS.*

Proof. See Proposition 6.
Corollary 5. *Let T be an SSCS operator from an arbitrary space X into a B-convex space Y. Then T^* is SSCS too.*

Proof. See Theorem 6.
Theorem 4 allows us also to prove non SSS or non SSCS of some particular operators. Let us present examples.

Corollary 6. *An operator $T : L_1(0, 2\pi) \to c_0$ which assigns to a function from L_1 the sequence of its Fourier coefficients, is neither SSS nor SSCS, although it is strictly singular.*

Proof. As it is known ([22]) this operator is strictly singular but its adjoint is neither strictly singular nor strictly cosingular. It remains to use Theorem 2. □

Corollary 7. *The quotient map from l_1 onto $l_1/Z \simeq l_p$, $1 < p < \infty$, is SSS, by Proposition* 6. *The quotient map Q from l_1 onto $l_1/Y \simeq c_0$ is strictly singular but the adjoint map Q^* is not strictly cosingular* ([15]; *Example* 1). *So, Q is not SSS.*

6. SSS and uniformly strictly singular operators

In [13] the notion of uniformly strictly singular operator was introduced and it was shown that in well known Gowers-Maurey space every operator is a sum of scalar and uniformly strictly singular operators. After some small correction, the definition of [13] reads as follows.

Definition 3. Let X be a space with a basis (e_k) and Y be an arbitrary space. An element in X is called a *block* if it has finite support, that is , if it is a finite linear combination of elements of the basis. The blocks v and w are said to be *consecutive* if the support of v (the set of elements of the basis that form v as a linear combination) ends before the support of w begins. One says that an operator $T : X \to Y$ is *uniformly strictly singular* with respect to the basis (e_k) if for every $\varepsilon > 0$ there exists a number N such that for arbitrary consecutive blocks $(x_n)_1^N$ there exists an element $x \in S(\text{lin}(x_n)_1^N)$ such that $\|Tx\| < \varepsilon$.

Let us recall that a sequence (x_n, f_n), $x_n \in X$, $f_n \in X^*$, is called an *M-basis* if $f_n(x_m) = \delta_{nm}$, $[x_n]_1^\infty = X$ and for any $x \in X \backslash \{0\}$ there is n such that $f_n(x) \neq 0$. The M-basis is called 1-*norming* if

$$\|x\| = \sup\{|f(x)| : f \in \text{lin}(f_n)_1^\infty \ , \ \|f\| = 1\}$$

for every $x \in X$. Of course we can carry Definition 6 to M-basis. Evidently, every SSS operator is uniformly strictly singular with respect to any basis (and with respect to any M-basis).

Proposition 8. *Suppose that an operator $T : X \to Y$, where X is separable, is not SSS. Then T is not uniformly strictly singular with respect to some 1-norming M-basis.*

Proof. Let (x_n, f_n) be a 1-norming M-basis of X. Then for some $c > 0$ there is a sequence $n_k \uparrow \infty$ such that the subspaces $E_k = \text{lin}(x_n : n_k \leq n < n_{k+1})$ contain subspaces F_k so that $\inf\{\|Tx\| : x \in S(F_k)\} > c$ and $\dim F_k \to \infty$. Hence, we can construct a basis $e_{n_k+1}, \ldots, e_{n_{k+1}}$ in every E_k so that F_k is a span of some of the first elements of this basis and the biorthogonal functionals span $\text{lin}(f_n : n_k < n \leq n_{k+1})$. Obviously, T is not uniformly strictly singular with respect to (e_n) . □

Question. Does exist a non SSS operator which is uniformly strictly singular with respect to each basis ?

Proposition 9. *Every operator T from l_p into a space Y of type q, $1 \leq p < q \leq 2$, is uniformly strictly singular with respect to the standard basis of l_p.*

Proof. Since Y has type q, there exists a constant M such that for an arbitrary collection $(y_n)_1^N$ in Y there are signs (θ_n) for which

$$\|\sum_1^N \theta_n y_n\| \leq M(\sum_1^N \|y_n\|^q)^{1/q} \ .$$

Let $y_n = Tx_n$ where $(x_n)_1^N \subset S(l_p)$ are consecutive blocks. Then

$$\|T(\sum \theta_n x_n)\| \leq M(\sum \|Tx_n\|^q)^{1/q} \leq M\|T\|N^{1/q} \ .$$

Since $\| \sum_{n=1}^N \theta_n x_n\| = N^{1/p}$, this proves Proposition 9. □

Let us give an

Example. of an operator which is uniformly strictly singular with respect to one basis but is not uniformly strictly singular with respect to another basis, and hence is non SSS.

Let $T : l_p \to l_q$ be the operator from Example 1, where $1 < p < q \leq 2$. By Proposition 9, this operator is uniformly strictly singular with respect to the standard basis. Obviously, it is not uniformly strictly singular with respect to basis consisting of blocks $w_i^n : 1 \leq i \leq 2^n$, $n = 1, 2, \dots$ of Walsh systems.

Remark 13. Prof. R. Wagner informs us, that from results of [1] it follows the existence of strictly singular operator which is not uniformly strictly singular with respect to any basis.

His arguments are following: In [1] it was constructed a strictly singular operator T from a space X with a basis $e_k \to 0$ weakly to a space Y with the property: Whenever (e_{k_i}) are elements of the basis with $n < k_1 < \cdots < k_n$ then $T|_{\text{lin}(e_{k_i})_1^n}$ is isometry.

Now take any other basis (x_k) in X. Fix an n and let $k_1 = n + 1$. Approximate e_{k_1} by a finite combination of x_k. Let x_{m_1} be the last vector used to approximate e_{k_1}.

There exists a number k_2 such that none of $x_1, \dots, x_{m_1}$ has a significant role in the expression of e_{k_2} as a combination of vectors x_k (because $e_k \to 0$ weakly). Approximate e_{k_2} by a finite combination of x_k, $k > m_1$. Let x_{m_2} be the last vector used to approximate e_{n_2}.

Continue until n is reached. The sequence $e_{k_1}, \dots, e_{k_n}$ is almost a block sequence of (x_k). It is also satisfies $n < k_1 < \cdots < k_n$ so T is an isometry on $\text{lin}(e_{k_i})_1^n$. Hence T is not uniformly strictly singular with respect to (x_k).

Obviously, we can take an M-basis instead of basis x_k in this argumentation.

Acknowledgement. The author wishes to express his thanks to prof. A.Pietsch for suggesting the problems and for many stimulating conversations.

References

1. S. Argyros and I. Deliyanni, Examples of asymptotic-l_1 Banach spaces, *Trans. Amer. Math. Soc.* **349** (1997) 973-995.
2. A. Castejón, E. Corbacho and V. Tarieladze, AMD numbers, compactness, strict singularity and essential spectrum of operators *Georgian Math. J.* **9** (2002) 227-270.
3. I.M. Glazman and Yu.I. Lyubich, *Finite dimensional linear analysis.* Nauka, Moscow, 1969 (Russian). English transl.: The M.I.T. Press, Cambridge, Mass-London, 1974.
4. A. Hinrichs and A. Pietsch, *Closed ideals and Dvoretzky's theorem for operators.* (unpublished).
5. N.J. Kalton and N.T. Peck, Twisted sums of sequence spaces and the three space problem, *Trans. Amer. Math. Soc.* **255** (1979) 1-30.
6. T. Kato, Perturbation theory for nullity deficiency and other quantities of linear operators, *J. Analyse Math.* **6** (1958) 273-222.

7. J. Lindenstrauss and A. Pełczyński, Absolutely summing operators on $\mathcal{L}_p$-spaces and their applications, *Studia Math.* **29** (1986) 275-328.
8. J. Lindenstrauss and L. Tzafriri, *Classical Banach spaces.* I. Springer, Berlin e.a., 1977.
9. G.G. Lorentz, Lower bounds for the degree of approximation, *Trans. Amer. Math. Soc.* **97** (1960) 25-34.
10. V.D. Milman, Some properties of strictly singular operators, *Funktsional. Anal. i Prilozhen.* **3** (1) (1969) 93-94. (Russian).
11. V.D. Milman, Spectrum of bounded continuous functions specified on a unit sphere in Banach space, *Funksional. Anal. i Prilozhen.* **3** (2) (1969) 69-79. (Russian).
12. V.D. Milman, Operators of class C_0 and C_0^*, *Teor. Funktsii Funktsional. Anal. i Prilozen.* **10** (1970) 15-18. (Russian).
13. V.D. Milman and R. Wagner, Asymptotic versions of operators and operator ideals *Convex geometric analysis,* Cambridge Univ. Press, Math. Sci. Res. Inst. Publ., **34** (1999) 165-179.
14. B.S. Mitiagin and A. Pełczyński, *Nuclear operators and approximative dimensions.* Proceedings of International Congress of Mathematicians. Moscow, 1966, 366-372.
15. A. Pełczyński, Strictly singular and strictly cosingular operators. I,II. *Bull. Acad. Polon. Sci. Sér. Sci. Math. Astronom. Phys.* **13** (1965) 34-41.
16. A. Pełczyński and H.P. Rosenthal, Localization techniques in L^p-spaces. *Studia Math.* **52** (1974/75) 263-289.
17. G. Pisier, Holomorphic semi-groups and the geometry of Banach spaces, *Ann. of Math.* **115** (1982) 375-392.
18. G. Pisier, Counterexamples to a conjecture of Grothendieck. *Acta Math.* **151** (1983) 181-208.
19. A.M. Plichko and M.M. Popov, Symmetric function spaces on atomless probability spaces, *Dissertationes Math.* **306** (1990).
20. W. Rudin, *Functional analysis.* McGraw-Hill Book Co. New York e.a. 1973.
21. V.M. Tikhomirov, *Some questions in approximation theory.* Izdat. Moskov. Univ. Moscow. 1976 (Russian).
22. L. Weis, Perturbation classes of semi-Fredholm operators. *Math. Z.* **178** (1981) 429-442.
23. R. Whitley, Markov and Bernstein's inequalities and compact and strictly singular operators, *J. Approx. Theory* **34** (1982) 277-285.

V. Kadets and W. Żelazko

m-Adic Multivariate Linear Splines and their Applications to Approximation Theory

Aliaksandr Radyna

Mechanical and Mathematical Dept., Belarusian State University, 220050, 4 F. Skaryna avenue, Minsk, Belarus'

Abstract

Let m be an integer greater then 1 and $\mathbb{Z}_m$ be a ring of m-adic integers. Let f be a continuous function acting from $\mathbb{Z}_m^r$ to real numbers. By the function we introduce an m-adic linear spline which is interpolate f in given points. It is shown that the spline approximates f uniformly.

Key words: m-adic linear spline, approximation, interpolation, m-adic numbers
2000 MSC: 11D88, 11J17

1. Introduction

During the last 10 years, p-adic analysis were used intensively in quantum physics, some aspects of biology, social sciences et.c., see, for example, [12], [5], [8]. Thus, it is natural to develop p-adic analysis and approximation theory in such a direction. There were many papers about p-adic interpolation and approximation of a continuous function but most of them were devoted to some generalizations of p-adic interpolation theorems of Mahler and Dieudonne, see [6], [7], [2]. Works such as [1], [4], [9], [11] contain such results. We also point out the article of Grintsyavichyus and Markshaĭtis [3] as more related to the topic of our work. Above articles except

Email address: Radyna@bsu.by (Aliaksandr Radyna).

last one dealt with interpolation and approximation of a $\mathbb{Q}_p$-valued function. The results of our article refer to approximation for a real-valued continuous function.

In this work we introduce m-adic linear spline for an r-dimensional case. Interpolation principles for construction of a sequence of splines which will approximate a given continuous real-valued function are applied. In fact, it will be proved that m-adic linear splines of r variables is dense in space $C(\mathbb{Z}_m^r;\mathbb{R})$.

In section 2 we will define an m-adic linear spline of r variables for a function $f \in C(\mathbb{Z}_m^r;\mathbb{R})$ using an interpolation problem. In section 3 we will prove that such a spline will approximate f.

Let us remind some basic notations. Let $\mathbb{N}$ be a set of positive integers. Denote by $|\cdot|$ a standard absolute value. Let $\mathbb{Q}_m$ is a set of a series $a = \sum_{k=s}^{\infty} a_k m^k$, $a_k \in \{0,\dots,m-1\}$, $a_s \neq 0$ (m-adic numbers) with m-adic absolute value $|a|_m = m^{-s}$. Note, that $|\cdot|_m$ has properties of a norm, moreover a property of ultra-norm: $|a+b|_m \leq \max(|a|_m,\ |b|_m)$. Let $B_\gamma(y) = \{x \in \mathbb{Q}_m : |x-y|_m \leq m^\gamma\}$. Denote a characteristic function of a set B by $I_B(x)$. Let $\{e_i\}_{i=1}^d$ be a standard basis of vectors $e_i = (0,\dots,0,1,0,\dots,0) \in \mathbb{R}^d$, where $d = m^{rn}$.

2. A definition of an m-adic linear spline in $\mathbb{Z}_m^r$

For any $n \in \mathbb{N}$ we have the canonical partition $\mathbb{Z}_m = \bigsqcup_{i=0}^{m^n-1} B_{-n}[i]$ (see [10] for details). That implies the canonical partition of $\mathbb{Z}_m^r$:

$$\mathbb{Z}_m^r = \bigsqcup_{i_1=0}^{m^n-1} \bigsqcup_{i_2=0}^{m^n-1} \cdots \bigsqcup_{i_r=0}^{m^n-1} B_{-n}[i_1] \times \cdots \times B_{-n}[i_r]. \tag{1}$$

Denote by $\overrightarrow{x}$ an element of $\mathbb{Z}_m^r$. Let
us consider a set of vectors $\mathcal{S}_n \subset \mathbb{Z}_m^r$

$$\mathcal{S}_n = \{(i_1,\dots,i_r)\ :\ i_k = 0,1,\dots,m^n-1, \quad k=1,2,\dots,r\}\ .$$

We consider the ultra-norm $\|\overrightarrow{x}\| = \max_{k=1,\dots,r} |x_k|_m$ on $\mathbb{Z}_m^r$. Denote $B_{-n}[\overrightarrow{i}] = \{\overrightarrow{x} \in \mathbb{Z}_m^r\ :\ \|\overrightarrow{x} - \overrightarrow{i}\| \leq m^{-n}\}$. Therefore, formula (1) can be rewritten as

$$\mathbb{Z}_m^r = \bigsqcup_{\overrightarrow{i} \in \mathcal{S}_n} B_{-n}[\overrightarrow{i}]. \tag{2}$$

Definition. Let $f \in C(\mathbb{Z}_m^r;\mathbb{R})$. By the function f we introduce the spline:

$$L_n(\overrightarrow{x}) = \sum_{\overrightarrow{i} \in \mathcal{S}_n} \lambda(\overrightarrow{i})\ \|\overrightarrow{x} - \overrightarrow{i}\|\ , \tag{3}$$

where $\lambda(\overrightarrow{i}) \in \mathbb{R}$, $\overrightarrow{i} \in \mathcal{S}_n$ can be found from the system of linear equations

$$\sum_{\overrightarrow{i} \in \mathcal{S}_n} \lambda(\overrightarrow{i})\ \|\overrightarrow{j} - \overrightarrow{i}\| = f(\overrightarrow{j}), \quad \overrightarrow{j} \in \mathcal{S}_n\ . \tag{4}$$

The matrix of the system (4) will depend on enumeration of $\mathcal{S}_n$. In this section we will propose a scheme of constructing of the matrix and inverting one. Firstly, let us enumerate $\mathcal{S}_n$. Every number $i_k = 0, 1, \dots, m^n - 1$ can be represented by $i_k = \sum_{\nu=0}^{n-1} i_k(\nu) m^\nu$ or as

$$i_k = (i_k(0), i_k(1), \dots, i_k(n-1)), \text{ where } i_k(\nu) \in \{0, \dots, m-1\} .$$

So, $\overrightarrow{i} \in \mathcal{S}_n$ can be represented as

$$\begin{aligned} \overrightarrow{i} &= (i_1, i_2, \dots, i_r) = (i_1(0), i_1(1), \dots, i_1(n-1); \\ &i_2(0), i_2(1), \dots, i_2(n-1); \dots, i_r(0), i_r(1), \dots, i_r(n-1)) \end{aligned} \tag{5}$$

Now, we can order representations (5) by the lexicographical law. That means

$$\begin{aligned} \overrightarrow{i} \prec \overrightarrow{j} \Leftrightarrow i_k(\nu) &= j_k(\nu),\ k = 1, \dots, s\ , \nu = 0, \dots, t, \\ &i_s(t+1) < j_s(t+1), \text{ if } t = n-2 \text{ then } i_{s+1}(0) < j_{s+1}(0). \end{aligned} \tag{6}$$

Therefore, $\mathcal{S}_n$ can be enumerated as $\overrightarrow{i}_1 \prec \overrightarrow{i}_2 \prec \dots \prec \overrightarrow{i}_{m^{rn}}$. In accordance with such enumeration, close vectors by a subscript are close by the distance $\| \cdot - \cdot \|$. We rewrite (4) as

$$\sum_{k=1}^{m^{rn}} \lambda_k \, \| \overrightarrow{i}_l - \overrightarrow{i}_k \| = f(\overrightarrow{i}_l), \quad l = 1, 2, \dots, m^{rn} . \tag{7}$$

The matrix of the system is $K_n = \{\| \overrightarrow{i}_l - \overrightarrow{i}_k \|\}_{l,k=1}^{m^{rn}}$. To solve (7) and find numbers λ_k, $k = 1, \dots m^{rn}$ we have to invert K_n. To do this we have to introduce a finite sequence of $m^{rn} \times m^{rn}$ block-diagonal matrices

$$I_k(n) = \left(\delta_{\left[\frac{\alpha-1}{m^{rk}}\right], \left[\frac{\beta-1}{m^{rk}}\right]} \right)_{\alpha,\beta=1}^{m^{rn}}, \quad k = 0, 1, \dots, n, \tag{8}$$

where $\delta_{x,y}$ equals one if $x = y$ or zero if $x \neq y$ and $[\cdot]$ denotes an integer part of a number. These matrices have two properties:

$$\mathrm{Diag}\{\overbrace{I_k(n), \dots, I_k(n)}^{m^r}\} = I_k(n+1), \tag{9}$$

$$I_i(n) \cdot I_j(n) = I_j(n) \cdot I_i(n) = m^{ri} \cdot I_j(n), \text{ when } 0 \le i \le j \le n . \tag{10}$$

It is not hard to see that matrices (8) forms a linear independent system.

Lemma 1. *The following representation of K_n is valid*

$$K_n = -\frac{1}{m^{n-1}} I_0(n) - \sum_{k=1}^{n-1} \left(\frac{m-1}{m^{n-k}} \, I_k(n) \right) + I_n(n) . \tag{11}$$

Proof. Here we use self-similarity of $\mathbb{Z}_m^r$ and property (9) of matrices (8). We will prove the representation inductively.

For $n = 1$ we have

$$K_1 = -I_0(1) + I_1(1).$$

For $n = 2$,

$$K_2 = I_2(2) - I_1(2) + m^{-1}\mathrm{Diag}\{\overbrace{K_1, \dots, K_1}^{m^r}\} =$$
$$= I_2(2) - I_1(2) + m^{-1}(-I_0(2) + I_1(2)) = -m^{-1}I_0(2) - \frac{m-1}{m}I_1(2) + I_2(2)$$

Suppose that (11) is true for K_{n-1} (inductive assumption). When the subscript of the matrix equals n we get

$$K_n = I_n(n) - I_{n-1}(n) + m^{-1}\mathrm{Diag}\{\overbrace{K_{n-1}, \dots, K_{n-1}}^{m^r}\} =$$
$$= I_n(n) - I_{n-1}(n) + m^{-1}(-\frac{1}{m^{n-2}}I_0(n) - \sum_{k=1}^{n-2} \frac{m-1}{m^{n-k-1}}I_k(n) + I_{n-1}(n)),$$

which equals (11). ■

Further we will omit (n) in formulas with $I_k(n)$.

Lemma 2. (Criterion of an invertibility of a matrix from span$\langle I_0, I_1, \dots I_n\rangle$). *A matrix $A = \sum_{k=0}^{n} a_k I_k$ is an invertible iff $\sum_{i=0}^{k} a_i m^{ri} \neq 0$ for all $k = 0, 1, \dots n$ and $A^{-1} = \sum_{k=0}^{n} b_k I_k$, where,*

$$b_0 = \frac{1}{a_0},\ b_k = -\frac{a_k}{\sum\limits_{i=0}^{k-1} a_i m^{ri} \sum\limits_{i=0}^{k} a_i m^{ri}}\ ,\ k = 1, 2, \dots, n. \tag{12}$$

Proof. Property (10) implies that span$\langle I_0, I_1, \dots I_n\rangle$ is an algebra. Therefore, if A^{-1} exists, then, it must has the same expansion as A. Namely, $A^{-1} = \sum_{j=0}^{n} b_j I_j$, where b_j, $j = 0, 1, \dots n$ are unknown coefficients which can be derived from the identity

$$\sum_{k=0}^{n} a_k I_k \cdot \sum_{j=0}^{n} b_j I_j = I_0.$$

Let us multiple the left-hand side of the last equation taking into account (10) and equate coefficients with I_k on both sides. We have

$$a_0 \cdot b_0 = 1,\ a_k \sum_{i=0}^{k-1} m^{ri} b_i + \left(\sum_{i=0}^{k} a_i m^{ri}\right) b_k = 0,\quad k = 1, 2, \dots, n.$$

This is the system of linear equations with respect to b_k, $k = 0, 1, \dots n$ which has a solution iff $\sum_{i=0}^{k} a_i m^{ri} \neq 0$, $k = 0, 1, \dots n$ and next recurrent relations hold:

$$b_0 = \frac{1}{a_0},\ b_k = \frac{-a_k \left(\sum\limits_{i=0}^{k-1} b_i m^{ri}\right)}{\sum\limits_{i=0}^{k} a_i m^{ri}},\quad k = 1, 2, \dots, n. \tag{13}$$

Remark that if $a_k = 0$ for some indexes $k \in \{1, \dots, n\}$ then $b_k = 0$ and formula (12) is true. So, we eliminate from considerations all $a_k = 0$. Let $a_{k_l} \neq 0$, $k_l \in \Lambda \subset \{0, 1, 2, \dots, n\}$ for all $l = 0, 1, 2, \dots, L$. We set $a_{k_0} := a_0$, $b_{k_0} := b_0$. As a result, (13) are written as

$$b_{k_0} = \frac{1}{a_{k_0}}, \; b_{k_l} = \frac{-a_{k_l}\left(\sum\limits_{i=0}^{k_{l-1}} b_i m^{ri}\right)}{\sum\limits_{i=0}^{k_l} a_i m^{ri}}, \quad l = 1, 2, \dots, L. \tag{14}$$

One can rewrite (14) as

$$b_{k_0} = \frac{1}{a_{k_0}}, \quad b_{k_1} = \frac{-a_{k_1} \cdot b_{k_0}}{\sum\limits_{i=0}^{k_1} a_i m^{ri}},$$

$$b_{k_l} = \frac{a_{k_l}}{a_{k_{l-1}}} \frac{\sum\limits_{i=0}^{k_{l-2}} a_i m^{ri}}{\sum\limits_{i=0}^{k_l} a_i m^{ri}} b_{k_{l-1}}, l = 2, 3 \dots, L. \tag{15}$$

Let us multiple $j+1$ first equations in (15) and reduce coefficients $b_{k_0}, b_{k_1}, \dots, b_{k_{j-1}}$ from both sides. We have

$$b_{k_j} = -\frac{1}{a_{k_0}} \frac{a_{k_1}}{\sum\limits_{i=0}^{k_1} a_i m^{ri}} \prod_{l=2}^{j} \left(\frac{a_{k_l}}{a_{k_{l-1}}} \frac{\sum\limits_{i=0}^{k_{l-2}} a_i m^{ri}}{\sum\limits_{i=0}^{k_l} a_i m^{ri}} \right) =$$

$$= -\frac{a_{k_j}}{\sum\limits_{i=0}^{k_{l-1}} a_i m^{ri} \sum\limits_{i=0}^{k_l} a_i m^{ri}}, \; j = 2, 3, \dots, L. \tag{16}$$

¿From (15) one can conclude that last equality in (16) holds for $j = 1$ too. Since $k_{l-1} \leq k_l - 1$ then $a_i = 0$ for all $i \in (k_{l-1}, k_l - 1)$. Therefore, $\sum\limits_{i=0}^{k_{l-1}} a_i m^{ri} = \sum\limits_{i=0}^{k_l-1} a_i m^{ri}$. Finally, we can rewrite (16) as

$$b_{k_0} = \frac{1}{a_{k_0}}, \; b_{k_j} = -\frac{a_{k_j}}{\sum\limits_{i=0}^{k_l-1} a_i m^{ri} \sum\limits_{i=0}^{k_l} a_i m^{ri}}, \; j = 1, 2, \dots, L. \tag{17}$$

Comparing (17) and (12) we see that both cases $a_k = 0$ and $a_k \neq 0$ can be described by (12). ∎

Theorem 1. *Let f be a function acting from $\mathbb{Z}_m^r$ to $\mathbb{R}$. Let $\{\overrightarrow{i}_k\}_{k=1}^{m^{rn}}$ be a set $\mathcal{S}_n$ (see (5)) ordered by the lexicographical law (see (6)). Then there exists the unique function $L_n(\overrightarrow{x}) = \sum_{k=1}^{m^{rn}} \lambda_k \|\overrightarrow{x} - \overrightarrow{i}_k\|$ such that $L_n(\overrightarrow{i}_l) = f(\overrightarrow{i}_l)$ for all $l = 1, 2, \dots m^{rn}$.*

Proof. Applying lemma 2 to K_n (see (11)) we find $K_n^{-1} = \sum_{k=0}^n b_k I_k$ where

$$b_0 = -m^{n-1}$$

$$b_k = \frac{\left((m^{r+1}-1)^2/(m-1)\right)\cdot m^{n+k-2}}{\left(m^{(r+1)k-1}+\frac{m^r-1}{m-1}\right)\left(m^{(r+1)k+r}+\frac{m^r-1}{m-1}\right)}, \quad 1 \le k \le n-1 \tag{18}$$

$$b_n = \frac{(m^{r+1}-1)^2\cdot m^{2n-2}}{(m-1)(m^r-1)\left(m^{(r+1)n-1}+\frac{m^r-1}{m-1}\right)\left(m^{(r+1)n}-1\right)}.$$

Now we have to find λ_i, $i = 1, \ldots, m^{rn}$. The above considerations imply that

$$\sum_{i=1}^{m^{rn}} \lambda_i e_i = K_n^{-1}\cdot\sum_{i=1}^{m^{rn}} y_i e_i = \sum_{k=0}^{n} b_k I_k \sum_{i=1}^{m^{rn}} y_i e_i = \sum_{k=0}^{n}\sum_{i=1}^{m^{rn}} y_i b_k I_k e_i, \tag{19}$$

where $y_l = f(\overrightarrow{i}_l)$, $l = 1, 2, \ldots, m^{rn}$. The last formula in (19) contains the expression $I_k e_i$. It is not hard to see that $I_0 e_i = e_i$ and $I_k e_i = \sum_{j=1+[\frac{i-1}{m^{rk}}]m^{rk}}^{(1+[\frac{i-1}{m^{rk}}])m^{rk}} e_j$ for all $k = 1, 2, \ldots, n$ and $i = 1, 2, \ldots, m^{rn}$. Substituting the value of $I_k e_i$ into (19) we will get

$$\sum_{i=1}^{m^{rn}} \lambda_i e_i = b_0 \sum_{i=1}^{m^{rn}} y_i e_i + \sum_{k=1}^{n} b_k \sum_{i=1}^{m^{rn}} y_i \sum_{j=1+[\frac{i-1}{m^{rk}}]m^{rk}}^{(1+[\frac{i-1}{m^{rk}}])m^{rk}} e_j \tag{20}$$

(here and below $[\cdot]$ denotes an integer part of a number). Since the conditions

$$1 \le i \le m^{rn} \text{ and } 1 + \left[\frac{i-1}{m^{rk}}\right] m^{rk} \le j \le \left(1 + \left[\frac{i-1}{m^{rk}}\right]\right) m^{rk}$$

are equivalent with

$$1 \le i \le m^{rn} \text{ and } \left[\frac{i-1}{m^{rk}}\right] = \left[\frac{j-1}{m^{rk}}\right]$$

we obtain

$$\sum_{i=1}^{m^{rn}} \lambda_i e_i = b_0 \sum_{i=1}^{m^{rn}} y_i e_i + \sum_{k=1}^{n} b_k \sum_{i=1}^{m^{rn}} e_i \sum_{j=1+[\frac{i-1}{m^{rk}}]m^{rk}}^{(1+[\frac{i-1}{m^{rk}}])m^{rk}} y_j ,$$

which yields

$$\lambda_i = \sum_{k=0}^{n} b_k \sum_{j=1+[\frac{i-1}{m^{rk}}]m^{rk}}^{(1+[\frac{i-1}{m^{rk}}])m^{rk}} y_j , i = 1, 2, \ldots, m^{rn} . \tag{21}$$

Since (7) has the unique solution (21), the function L_n exists and is unique. ■

3. Statement and solution of the approximation problem

In this section we will show, that any continuous function from $C(\mathbb{Z}_m^r;\mathbb{R})$ can be approximated by the spline of type (3). But first we have to prove the lemma.

Lemma 3. *The following formula holds:*

$$\max_{\vec{y}\in\mathbb{Z}_m^r} |L_n(\vec{y}) - \sum_{k=1}^{m^{rn}} y_k I_{B_{-n}[\vec{i}_k]}(\vec{y})| = \max_{l=1,\dots,m^{rn}} |\lambda_l| \cdot \frac{1}{m^{rn}} . \qquad (22)$$

Proof. By the construction of L_n we have that $L_n(\vec{i}_k) = y_k$, $k = 1, 2, \dots, m^{rn}$. Let $\vec{y}$ be an arbitrary point in $\mathbb{Z}_m^r$ such that $\vec{y} \neq \vec{i}_k$ $\forall\, k = 1, 2, \dots, m^{rn}$. Since (2), then $\mathbb{Z}_m^r = \bigsqcup_{k=1}^{m^{rn}} B_{-n}[\vec{i}_k]$. Then there is a ball $B_{-n}[\vec{i}_\nu]$, such that $\vec{y} \in B_{-n}[\vec{i}_\nu] \setminus \{\vec{i}_\nu\}$. As a result of this we can write

$$L_n(\vec{y}) = \sum_{k=1}^{\nu-1} \lambda_k \|\vec{y} - \vec{i}_k\| + \lambda_\nu \|\vec{y} - \vec{i}_\nu\| + \sum_{k=\nu+1}^{m^{rn}} \lambda_k \|\vec{y} - \vec{i}_k\| .$$

Since a distance between points of different balls is equal to the distance between the centers, we can replace $\vec{y}$ in the first and last sums with $\vec{i}_\nu$. After that, we get the equality $L_n(\vec{y}) = L_n(\vec{i}_\nu) + \lambda_\nu \|\vec{y} - \vec{i}_\nu\|$, $\vec{y} \in B_{-n}[\vec{i}_\nu]$ which can be performed to $L_n(\vec{y}) = y_\nu + \lambda_\nu \|\vec{y} - \vec{i}_\nu\|$, $\vec{y} \in B_{-n}[\vec{i}_\nu]$. Resuming last two equalities one can conclude that

$$L_n(\vec{y}) = \sum_{\nu=1}^{m^{rn}} y_\nu I_{B_{-n}[\vec{i}_\nu]}(\vec{y}) + \sum_{\nu=1}^{m^{rn}} \lambda_\nu \|\vec{y} - \vec{i}_\nu\| I_{B_{-n}[\vec{i}_\nu]}(\vec{y}) .$$

This obviously gives us

$$\sup_{\vec{y}\in\mathbb{Z}_m^r} \|L_n(\vec{y}) - \sum_{\nu=1}^{m^{rn}} y_\nu I_{B_{-n}[\vec{i}_\nu]}(\vec{y})\| = \sup_{\vec{y}\in\mathbb{Z}_m^r} |\sum_{\nu=1}^{m^{rn}} \lambda_\nu \|\vec{y} - \vec{i}_\nu\| I_{B_{-n}[\vec{i}_\nu]}(\vec{y})| .$$

The right-hand side of the equality is a maximum of a continuous function over the compact set $\mathbb{Z}_m^r$. Therefore, it is attained at the point $\vec{y}_0$. So,

$$\sup_{\vec{y}\in\mathbb{Z}_m^r} |L_n(\vec{y}) - \sum_{\nu=1}^{m^{rn}} y_\nu I_{B_{-n}[\vec{i}_\nu]}(\vec{y})| = |\lambda_l| \cdot \|\vec{y}_0 - \vec{i}_l\|, \quad \vec{y}_0 \in B_{-n}[\vec{i}_l] .$$

In this case the point $\vec{y}_0$ must belong to $B_{-n}[\vec{i}_l] \setminus B_{-n-1}[\vec{i}_l]$ and $|\lambda_l|$ must be maximal. Otherwise, we will get a contradiction with maximality of

$$\left| \sum_{\nu=1}^{m^{rn}} \lambda_\nu \|\vec{y} - \vec{i}_\nu\| I_{B_{-n}[\vec{i}_\nu]}(\vec{y}) \right| .$$

Therefore, we obtain formula (22). ■

Theorem 2. *Let $f \in C(\mathbb{Z}_m^r;\mathbb{R})$. Then for any $\varepsilon > 0$ there exists a positive integer N_ε such that for all $n > N_\varepsilon$, $|L_n(\overrightarrow{x}) - f(\overrightarrow{x})| < \varepsilon$, $\forall \overrightarrow{x} \in \mathbb{Z}_m^r$.*

Proof. It is sufficient to prove the statement when f is a characteristic function of an arbitrary ball in $\mathbb{Z}_m^r$, because a linear span of such functions is dense in $C(\mathbb{Z}_m^r;\mathbb{R})$. For instance, the proof of this fact would be easily derived from Th.26.2, see [10]. Therefore, we shall approximate a characteristic function $I_{B_{-u}[\overrightarrow{i}_l]}$, where $B_{-u}[\overrightarrow{i}_l]$ is an arbitrary ball in $\mathbb{Z}_m^r$. Here without loss of generality we can assume, that $\overrightarrow{i}_l \in \mathcal{S}_u$. It is true, because in the ultrametric space $(\mathbb{Z}_m^r, \|\cdot\|)$ every point of the ball is a center of the ball, and $\mathbb{Z}_m^r = \bigsqcup_{\overrightarrow{i} \in \mathcal{S}_u} B_{-u}[\overrightarrow{i}]$. Therefore, one can assume that $\overrightarrow{i}_l$ is a center of $B_{-u}[\overrightarrow{i}_l]$ and $\overrightarrow{i}_l$ is a point from the partition $\overrightarrow{i}_1, \overrightarrow{i}_2, \ldots, \overrightarrow{i}_{m^{ru}}$, $\mathbb{Z}_m^r = \bigsqcup_{k=1}^{m^{ru}} B_{-u}[\overrightarrow{i}_k]$. Let us divide every ball $B_{-u}[\overrightarrow{i}_k]$ into m^{rn} balls. As a result we obtain the partition $\overrightarrow{j}_1, \overrightarrow{j}_2, \ldots, \overrightarrow{j}_{m^{r(n+u)}}$, where $\overrightarrow{j}_{m^{rn}(l-1)+1}, \ldots, \overrightarrow{j}_{m^{rn}l} \in B_{-u}[\overrightarrow{i}_l]$. Then $y_k = 1$ if $k = m^{rn}(l-1)+1, \ldots, m^{rn}l$ and $y_k = 0$ otherwise. We can construct the function $L_{n+u}(\overrightarrow{x}) = \sum_{\nu=1}^{m^{r(n+u)}} \lambda_\nu \|\overrightarrow{x} - \overrightarrow{j}_\nu\|$, where λ_ν can be found from the formula

$$\lambda_\nu = \sum_{k=0}^{n+u} b_k \sum_{\mu=1+[\frac{\nu-1}{m^{rk}}]m^{rk}}^{(1+[\frac{\nu-1}{m^{rk}}])m^{rk}} y_\mu \ , \ \nu = 1, 2, \ldots, p^{n+u} \ . \tag{23}$$

Taking into account the values of y_μ we can rewrite (23) as

$$\lambda_\nu = \sum_{k=0}^{n} b_k m^{rk} + \sum_{k=n+1}^{n+u} b_k m^{rn}, \quad \nu \in \{m^{rn}(l-1)+1, \ldots, m^{rn}l\} \ , \tag{24}$$

$$\lambda_\nu = m^{rn} \sum_{k=n+N}^{n+m} b_k, \ \nu \in \{1, \ldots, m^{rn}(l-1)\} \cup \{m^{rn}l+1, \ldots, m^{r(n+u)}\} \ , \tag{25}$$

where $N = N(\nu, l)$ is a number from $\{1, \ldots, u\}$,

$$b_0 = -m^{n+u-1}$$

$$b_k = \frac{\left(\left(m^{r+1}-1\right)^2 / (m-1)\right) \cdot m^{n+u+k-2}}{\left(m^{(r+1)k-1} + \frac{m^r-1}{m-1}\right)\left(m^{(r+1)k+r} + \frac{m^r-1}{m-1}\right)} \ , \quad 1 \le k \le n+u-1 \tag{26}$$

$$b_{n+u} = \frac{\left(m^{r+1}-1\right)^2 \cdot m^{2n+2u-2}}{(m-1)(m^r-1)\left(m^{(r+1)(n+u)-1} + \frac{m^r-1}{m-1}\right)\left(m^{(r+1)(n+u)} - 1\right)} \ .$$

Remark that if $u = 0$, then $l := 1$ and we will obtain only the first sum in (24). Denote by $B_n := m^{rn} \sum_{k=n+1}^{n+u} b_k$. Then (24) and (25) are rewritten as

$$\lambda_\nu = \sum_{k=0}^{n} b_k m^{rk} + B_n \ , \quad \text{for } m^{rn}(l-1)+1 \le \nu \le m^{rn}l \ , \tag{27}$$

$$0 < \lambda_\nu \le B_n \ , \quad \text{for } \nu \in \{1, \ldots, m^{rn}(l-1)\} \cup \{m^{rn}l+1, \ldots m^{r(n+u)}\} \ . \tag{28}$$

Let us substitute b_k from (26) into (27), (28) and find an asymptotic for $|\lambda_i|$ when n goes to infinity. Using (26), it is easy to see that

$$b_k = O(m^{-(2r+1)k+r+n+u}),\ k = 1, 2, \ldots, n+u\ . \tag{29}$$

By definition of B_n this imply

$$B_n = O(m^{-rn-r+u-1})\ , n \to \infty. \tag{30}$$

Formulas (28), (30) and the condition $B_n > 0$ gives us an asymptotic for $|\lambda_\nu|$, $\nu \in \{1, \ldots, m^{rn}(l-1)\} \cup \{m^{rn}l+1, \ldots m^{r(n+u)}\}$:

$$|\lambda_\nu| = O(m^{-rn-r+u-1}),\ n \to \infty\ . \tag{31}$$

Let us find an asymptotic for $|\lambda_\nu|$ from (27). Denote by $a := \frac{m^r-1}{m-1}$. Applying

$$\frac{m^{(r+1)k-1}(m^{r+1}-1)}{(m^{(r+1)k-1}+a)(m^{(r+1)(k+1)-1}+a)} = \frac{1}{m^{(r+1)k-1}+a} - \frac{1}{m^{(r+1)(k+1)-1}+a}$$

to $\sum\limits_{k=0}^{n} b_k m^{rk}$ and using values of b_k from (26), we obtain

$$\sum_{k=0}^{n} b_k m^{rk} = -\frac{(m^{r+1}-1)m^{n+u-1}}{(m-1)(m^{(r+1)n+r}+a)} = O(m^{-rn+u-1})\ . \tag{32}$$

Applying (30), (32) to (27), we get the asymptotic

$$|\lambda_\nu| = O(m^{-rn+u-1}),\ n \to \infty, \quad \nu \in \{m^{rn}(l-1)+1, \ldots, m^{rn}l\}. \tag{33}$$

Comparing (31) and (33), we obtain that $\max\limits_{\nu=1,\ldots,m^{r(n+u)}} |\lambda_\nu| = O(m^{-rn+u-1})$ as $n \to \infty$. By (22) we have that for a big n, $\max\limits_{\vec{x} \in \mathbb{Z}_m^r} |L_{n+u}(\vec{x}) - I_{B_{-u}[\vec{i}_l]}(\vec{x})| = O(p^{-2rn+(1-r)u-1})$ which is equivalent of the statement of the theorem. ■

References

1. Y. Amice, Interpolation p-adique, *Bull. Soc. Math. France* **92** (1964) 117–180.
2. J. Dieudonne, Sur les fonctions continues p-adiques, *Bull. Sci. Math.* **68** (2) (1944) 79–95.
3. P.K. Grintsyavichyus, G.N. Markshaĭtis, Expansion of characteristic functions of open sets of a p-adic integer ring in certain polynomials of a p-adic norm, and a special case of the Weierstrass–Stone theorem, *Lituanian Math. J.* **23** (1) (1983) 40–43.
4. Hà Huy Khoái, p-Adic interpolation, *Mat. Zam.* **26** (1) (1979) 101–112.
5. A. Khrennikov, *Non-Archimedean analysis: quantum paradoxes, dynamical systems and biological models*. Kluwer Academic Publishers, Dordrecht, 1997.
6. K. Mahler, An interpolation series for continuous functions of a p-adic variable, *J. Reine Angew. Math.* **199** (1958) 23–34.
7. K. Mahler, *Introduction to p-adic numbers and their functions*. Cambridge Univ. Press, 1973.

8. G. Parisi, N. Sourlas, p-Adic numbers and replica symmetry breaking, *Eur. Phys. J. B.* **14** (2000) 535–542.
9. L. van Hamme, Jackson's interpolation formula in p-adic analysis, *Proceedings of the Conference on p-adic Analysis* (Nijmegen,1978) 119–125, Report 7806, Katholieke Univ., Nijmegen, 1978.
10. W.H. Schikhof, *Ultrametric calculus*. Cambridge Univ. Press, Cambridge 1984.
11. V.K. Srinivasan, Cam Van Tran, On an approximation theorem of Walsh in the p-adic field, *J. Approx. Theory* **35** (2) (1982) 191–193.
12. V.S. Vladimirov, I.V. Volovich, and E.I. Zelenov, *p-Adic analysis and mathematical physics.* World Scientific Publ., Singapore 1993.

V. Kadets and W. Żelazko

Change of Variable in Integrals over p-adic and Adelic Domains

Yauhen Radyna

Belarusian State University, Minsk, Belarus

Abstract

Here we consider integration of complex-valued functions over p-adic and adelic domains. Formulas of change of variable are obtained resembling one for integral over real domain

$$\int_{\varphi(A)} f(y)dy = \int_A f(\varphi(x))|\varphi'(x)|dx.$$

A general approach involving absolutely continuous mappings and theorem of Radon and Nikodim is considered in p-adic case. A zero-measure feature of adelic analytic mappings is described.

Key words: p-adic integrals, adelic integrals, continuously differentiable mappings, absolutely continuous mappings, analytic mappings
2000 MSC: 11S80, 11S31

1. Change of variable and continuous differentiation

Let $\mathbb{Q}_p$ denote the field of p-adic numbers with standard valuation $|\ |_p$ and Haar measure λ; $B[a,\delta]$ and $B(a,\delta)$ denote closed and open balls respectively with center at a and radius δ; $\mathbb{Z}_p = B[0,1]$ is the ring of p-adic integers.

The continuity of usual derivative does not imply even local injectivity of p-adic mapping.

Email address: yauhen_radyna@tut.by (Yauhen Radyna).

Example 1. Let us consider non-intersecting balls $A_n = B[p^n, p^{-2n}]$, $n = 1, 2, \ldots$. Define $\varphi : \mathbb{Z}_p \to \mathbb{Z}_p$, $\varphi(x) = x - p^{2n}$ if x belongs to some of the A_n and x otherwise. The mapping has non-vanishing constant derivative which equals 1. Yet $\varphi(p^n) = p^n - p^{2n} = \varphi(p^n - p^{2n})$ and local injectivity fails at 0.

We can define continuous differentiability slightly in other way [1].

Definition 2. Let $A \subset \mathbb{Q}_p$ be open. Mapping $f : A \to \mathbb{Q}_p$ is called continuously differentiable on A if there exists continuous $F : A^2 \to \mathbb{Q}_p$ such that

$$f(x) - f(y) = (x - y)F(x, y) \quad \forall x, y \in A.$$

We assume $f'(x) = F(x, x)$ and write $f \in C^1(A)$.

Every continuously differentiable mapping is differentiable in ordinary sense. In case of real domain definition 2 yields the usual notion of continuous differentiability.

The following theorems can be proved for C^1 mappings.

Theorem 3. (local reversibility of C^1 mappings, [1]) *Let $f \in C^1(A)$ and let $a \in A$ satisfy $f'(a) \neq 0$. There exists open ball $B(a, \delta)$ contained in A such that restriction $f : B(a, \delta) \to B(f(a), \delta|f'(a)|_p)$ is a bijection. It inverse g is also continuously differentiable and $g'(f(a)) = f'(a)^{-1}$*

Theorem 4. (formula of change of variable, [1]) *Let $A \subset \mathbb{Q}_p$ be open set. Let homeomorphism $\varphi \in C^1(A)$ satisfy $\varphi'(x) \neq 0$ $\forall x \in A$. Then for any complex-valued function $f \in L_1(\varphi(A))$ integrable with respect to Haar measure the following equality holds*

$$\int_{\varphi(A)} f(y) d\lambda(y) = \int_A f(\varphi(x)) |\varphi'(x)| d\lambda(x).$$

Note that in [2] only analytic mappings on $\mathbb{Q}_p$ were used for change of variable. They all are continuously differentiable on their domains of convergence.

Example 5. Consider $\alpha : \mathbb{Z}_p \to \mathbb{Z}_p : \sum_{i=0}^{\infty} x_i p^i \mapsto \sum_{i=0}^{\infty} x_i p^{2i+1}$. Mapping $\varphi : \mathbb{Z}_p \to \mathbb{Z}_p : x \mapsto x + \alpha(x)$ is continuously differentiable isometry and $\varphi' \equiv 1$. Yet φ is not analytic on $\mathbb{Z}_p$ because α is not. Indeed, derivative $\alpha' \equiv 0$. The only analytic mappings with zero derivative are constants, but α is not a constant.

2. Change of variable and Radon-Nikodim derivative

There exists another approach based on notion of Radon-Nikodim derivative.

Let $A \subset \mathbb{Q}_p$ be open and $\varphi : A \to \varphi(A)$ be continuous homeomorphism. For any subset $X \subset A$ measurable with respect to Haar measure λ we define measure μ by formula $\mu(X) = \lambda(\varphi(X))$. For any integrable functions $f \in L_1(\varphi(A))$ consider equalities

$$\int\limits_{\varphi(A)} f(y)d\lambda(y) = \int\limits_{A} f(\varphi(x))d\mu(x) = \int\limits_{A} f(\varphi(x))\rho(x)d\lambda(x).$$

The first equality is always true due to definition of measure μ, the second is true if and only if measure μ is *absolutely continuous* with respect to Haar measure λ. In this case the theorem of Radon and Nikodim guarantees the existence of function ρ called the Radon-Nikodim derivative $\rho = d\mu/d\lambda$.

By $d(X)$ denote the diameter of set $X \subset \mathbb{Q}_p$, $d(X) = \sup_{x,\,y \in X} |x-y|_p$.

Definition 6. Let $A \subset \mathbb{Q}_p$ be open. We say mapping $\varphi : A \to \mathbb{Q}_p$ is absolutely continuous if for any $\varepsilon > 0$ there exists $\delta > 0$ such that for arbitrary closed disjoint balls $B_1[x_1, \delta_1], ..., B_n[x_n, \delta_n]$, $\cup_{i=1}^n B_i \subset A$ we have: $\sum_{i=1}^n d(B_i) < \delta$ implies $\sum_{i=1}^n d(\varphi(B_i)) < \varepsilon$.

Theorem 7. *Absolutely continuous mapping φ defines measure μ which is absolutely continuous with respect to λ.*

Proof. Let $X \subset A$ have zero Haar measure. For any $\varepsilon > 0$ let us take δ from the definition of absolute continuity of φ. Consider (countable) cover of X with disjoint balls $X \subset \bigsqcup_{i=1}^{\infty} B_i[x_i, \delta_i]$, $\sum_{i=1}^{\infty} \delta_i < \delta$. Denote $\varepsilon_i = d(\varphi(B_i))$. Then $\varphi(X) \subset \bigcup_{i=1}^{\infty} B[\varphi(x_i), \varepsilon_i]$. Absolute continuity of φ yields $\sum_{i=1}^{n} \varepsilon_i < \varepsilon$. Taking the limit as $n \to \infty$ we obtain $\sum_{i=1}^{\infty} \varepsilon_i \leq \varepsilon$. The latter gives $\mu(X) = \lambda(\varphi(X)) = 0$ because ε is arbitrary. □

Lemma 8. *Let $\sigma : \{0, 1, 2, 3, ..\} \to \{0, 1, 2, 3, ...\}$ be permutation. Consider mapping $\varphi : \mathbb{Z}_p \to \mathbb{Z}_p : \sum_{k=0}^{\infty} x_k p^k \mapsto \sum_{k=0}^{\infty} x_{\sigma(k)} p^k$ interchanging digits of canonical representation of p-adic argument. Mapping φ is continuous, bijective, and its Radon-Nikodim derivative equals 1.*

Proof. Bijectivity of φ is obvious. For any $\varepsilon > 0$ let us take n satisfying $\varepsilon < p^{-n}$ and define $m = \max\{\sigma(0), \sigma(1), ...\sigma(n)\}$. Then $|x-a|_p < \delta = p^{-m}$ implies $|\varphi(x) - \varphi(a)|_p < \varepsilon$ and thus φ is continuous.

Radon-Nikodim derivative equals 1 because we have $\mu(X) = \lambda(\varphi(X)) = \lambda(X)$ for any measurable $X \subset \mathbb{Z}_p$. Indeed, Haar measure of a set of p-adic numbers with n digits fixed is p^{-n}. It does not change under φ. Any closed ball $B[x, p^{-n}] \subset \mathbb{Z}_p$ consists of p-adic numbers with first n digits fixed. Thus $\mu(B) = \lambda(B)$ for any ball B. But the balls form a semi-ring of measurable sets and their measures uniquely define measures of all other sets. □

Example 9. Mapping $\varphi : \mathbb{Z}_p \to \mathbb{Z}_p$ transposing digits at places 4^k and $2 \cdot 4^k$, $k = 0, 1, 2, ...$ in canonical representation of argument is continuous but not absolutely continuous.

Proof. Continuity follows from lemma 8. Consider balls $B_{mj} = B[j, p^{-2 \cdot 4^m}]$, $m = 1, 2, 3, ...$, $j = 0, 1, ...p^{4^m} - 1$. Number $j + p^{2 \cdot 4^m}$ belongs to B_{mj}. Sum of diameters

$$\sum_j d(B_{mj}) = p^{4^m} p^{-2 \cdot 4^m} = p^{-4^m} \to 0.$$

Yet

$$\sum_j d(\varphi(B_{mj})) \geq \sum_j |\varphi(j) - \varphi(j + p^{4^m})|_p = p^{4^m} p^{-4^m} = 1 \not\to 0.$$

□

This example also shows that absolute continuity is not necessary for existence of Radon-Nikodim derivative.

In contrast with the case of real variable there exist absolutely continuous but nowhere differentiable mappings.

Example 10. Mapping $\varphi : \mathbb{Z}_p \to \mathbb{Z}_p$ transposing digits at places $2k$ and $2k + 1$, $k = 0, 1, 2, ...$ in canonical representation of argument is absolutely continuous and nowhere differentiable in usual sense.

Proof. Absolute continuity follows from the Lipshitz inequality $|\varphi(x) - \varphi(y)|_p \leq p \cdot |x - y|_p$.

For any $a \in \mathbb{Z}_p$ consider sequence $a_n \to a$, $n = 0, 1, ...$, where a_n differs from a only at n-th digit (no matter how). The valuation of ratio $|(\varphi(a_n) - \varphi(a))/(a_n - a)|_p$ is $1/p$ for n even and p for odd. Therefore the ratio cannot have a limit as $n \to \infty$. □

3. Change of variable in integrals over adelic domains

Consider set of adeles $\mathbb{A}$ [3], i.e., set of sequences $(x_\infty, x_2, x_3, x_5, ..., x_p, ...)$, where $x_\infty \in \mathbb{Q}_\infty = \mathbb{R}$, $x_p \in \mathbb{Q}_p$, and $x_p \in \mathbb{Z}_p$ beginning from some p. Adeles form a ring with respect to pointwise addition and multiplication. There is a topology for adeles such that $\mathbb{A}$ is locally compact group with respect to addition. The basis of neighborhoods of zero consists of the sets $V_S = \prod_{p \in S} V_p \times \prod_{p \notin S} \mathbb{Z}_p$, where S is finite subset of $P = \{\infty, 2, 3, ..., p, ...\}$ containing ∞, V_p is neighborhood of zero in $\mathbb{Q}_p$. Cylindrical neighborhood of point $a \in \mathbb{A}$ is any set $a + V_S$. By λ we define Haar measure on $\mathbb{A}$.

Define function $\nu : \mathbb{A} \to \mathbb{R} : (x_\infty, x_2, \ldots, x_p, ...) \mapsto |x_\infty|_\infty \cdot |x_2|_2 \cdot ... \cdot |x_p|_p \cdot ...$ (in general product can diverge to zero).

Definition 11. Let $A \subset \mathbb{A}$ be open. We say mapping $f : A \to \mathbb{A}$ is continuously differentiable on A if there exists continuous mapping $F : A^2 \to \mathbb{A}$ such that $f(x) - f(y) = (x - y)F(x, y)\ \forall x, y \in A$. We assume $f'(x) = F(x, x)$ and write $f \in C^1(A)$.

Mapping F is unique because set of pairs $(x, y) \in A^2$ such that $x - y$ has inverse $(x - y)^{-1}$ is dense in A^2.

Definition 12. Let $A \subset \mathbb{A}$ be open. We say mapping $\varphi : A \to \mathbb{A}$ is locally cylindrical on A if any point $a \in A$ has cylindrical neighborhood $U \subset A$ such that for all $p = \infty, 2, 3, ...$ equality $\iota_p(x) = \iota_p(y)$, $x, y \in U$ implies $\iota_p(\varphi(x)) = \iota_p(\varphi(y))$.

Theorem 13. (formula of change of variable) *Let $A \subset \mathbb{A}$ be open. Let homeomorphism $\varphi : A \to \varphi(A)$ satisfy the following conditions: 1) $\varphi \in C^1(A)$; 2) φ is locally cylindrical on A; 3) $\iota_p(\varphi'(x)) \neq 0$ for $p = \infty, 2, 3, 5, \ldots$. Then for all integrable $f \in L_1(\varphi(A))$ the following equality holds*

$$\int_{\varphi(A)} f(y)d\lambda(y) = \int_{A} f(\varphi(x))\nu(\varphi'(x))d\lambda(x).$$

Proof. For any $a \in A$ consider cylindrical neighbourhood U of a as in definition 12. Group $\mathbb{A}$ is locally compact and σ-compact; so we can choose locally finite at most countable cover $A = \bigcup_i U_i$ with such neighborhoods. It is enough to prove the formula of change of variable for each $U = U_i$.

We can write $U = U_\infty \times U_2 \times ... \times U_p \times ...$, $U_p \subset \mathbb{Q}_p$ and

$$\varphi(x_\infty, x_2, ..., x_p, ...) = (\varphi_\infty(x_\infty), \varphi_2(x_2), ..., \varphi_p(x_p), ...);$$

here mappings $\varphi_p : U_p \to \mathbb{Q}_p$ are continuously differentiable and derivative φ_p' does not vanish on U_p. Derivative φ' can also be written as

$$\varphi'(x_\infty, x_2, ..., x_p, ...) = (\varphi_\infty'(x_\infty), \varphi_2'(x_2), ..., \varphi_p'(x_p), ...).$$

Applying formulas of change of variable for $\mathbb{Q}_\infty = \mathbb{R}$ and $\mathbb{Q}_p$ and using properties of Haar measure on $\mathbb{A}$ we obtain

$$\int_{\varphi(U)} d\lambda(y) = \prod_p \int_{\varphi_p(U_p)} d\lambda_p(y) = \prod_p \int_{U_p} |\varphi_p'(x)|_p d\lambda_p(x) =$$

$$= \int_U \left(\prod_p |\varphi_p(x_p)|_p\right) (d\lambda_\infty \cdot d\lambda_2 \cdot ... \cdot d\lambda_p \cdot ...) = \int_U \nu(\varphi'(x))d\lambda(x). \quad \square$$

Adeles form topological ring, so we can give the following definition.

Definition 14. Adelic analytic mapping is any formal power series with adelic coefficients converging at some neighbourhood.

Properties of such analytical mappings are very different from the properties of p-adic and real analytic mappings. The whole set of adeles can have zero measure image under non-constant analytic mapping. The following proposition demonstrates it.

Proposition 15. *Let $n \geq 2$ be an integer. Image of ring of adeles $\mathbb{A}$ under $\pi_n : x \mapsto x^n$ has zero masure.*

Proof. Theorem of Dirichlet states that there exist infinitely many primes $p \equiv 1 (\text{mod } n)$ (i.e., $(p-1)$ is divisable by n). Let α be primitive element modulo such p (i.e., among $\alpha^1, \alpha^2, ..., \alpha^{p-1} (\text{mod } p)$ we can find all of numbers $1, 2, ..., p-1$). Among powers $1^n, 2^n, ..., (p-1)^n$ (mod p) we can find only $(p-1)/n$ different numbers, exactly $\alpha^1, \alpha^2, ...\alpha^{(p-1)/n}$.

Therefore powering elements of $\mathbb{Z}_p$ we shall obtain only elements with the first nonzero digit equal to $\alpha^1, \alpha^2, ...\alpha^{(p-1)/n}$. Measure of set $\{x^n : x \in \mathbb{Z}_p\}$ is obviously not greater than $1/n$ (in fact, it is $1/n \cdot (p^n - p^{n-1})/(p^n - 1)$).

Consider the following sequence of cylinders K_j, $j = 1, 2, ...$, $\bigcup_j K_j = \mathbb{A}$,

$$K_j = [-j, j] \times \prod_{i=1}^{j} (p_i^{-j} \mathbb{Z}_{p_i}) \times \prod_{i=j+1}^{\infty} \mathbb{Z}_{p_i}, \text{ where } p_i \text{ is } i\text{-th prime.}$$

Image $\pi_n(K_j)$ is also a cylinder, $\pi_n(K_j) = T_{j\infty} \times T_{j2} \times ... \times T_{jp} \times ...$, where $T_{jp} \subset \mathbb{Q}_p$. Its measure is equal to product of corresponding measures in $\mathbb{Q}_p$: $\lambda(\pi_n(K_j)) = \lambda_\infty(T_{j\infty}) \cdot \prod_p \lambda_p(T_{jp})$. This product contains infinitely many multipliers not greater than $1/n \leq 1/2$. Therefore

$$\lambda(\pi_n(\mathbb{A})) = \lambda(\pi_n(\bigcup_{j=1}^{\infty} K_j)) = \lambda(\bigcup_{j=1}^{\infty} \pi_n(K_j)) = 0. \quad \square$$

References

1. W. Shikhof, *Ultrametric Calculus*. Cambridge University Press, Cambridge, 1984.
2. Vladimirov V.S., Volovich I.V., and Zelenov I.I., *P-adic Analysis and Mathematical Phisycs*. World Scientific Publ., Singapore, 1993.
3. Gelfand I.M., Graev M.I., Pjatetski-Shapiro I.I., *Representation Theory and Automorphic Functions*. Nauka, Moscow, 1966. (Russian).
4. Khrennikov A., *Non-Archimedean Analisys: Quantum Paradoxes, Dynamical Systems and Biological Models*. Kluwer Academic Publisher, Dordrecht, 1997.

Functional Analysis and its Applications
V. Kadets and W. Żelazko

Invariant Subspaces of Dissipative Operators in Krein Space and Sobolev Problem on a Rotating Top ⋆

A. A. Shkalikov [a]

[a]Moscow Lomonosov State University, Department of Mechanics and Mathematics, 119992, Moscow, Russia

Abstract

Let $\mathcal{H}$ be separable Hilbert space with fundamental symmetry $J = P_+ - P_-$ where $P_\pm$ are mutually orthogonal projectors such that J^2 is the identity operator. A short proof of the following result is presented in the paper. If A is a maximal dissipative operator in the Krein space $\mathcal{K} = \{\mathcal{H}, J\}$, the domain of A contains $P_+(\mathcal{H})$ and the operator P_+AP_- is compact then there is an A-invariant maximal nonnegative subspace $\mathcal{L}$ such that the spectrum of the restriction $A|_\mathcal{L}$ lies in the closed upper-half plane. This theorem contains the well-known results on this subject of Pontrjagin, Langer, Krein and Azizov.

As an application of this theorem the criterion for the stability of a rotaing top with a cavity filled with an ideal liquid is given.

Key words: Pontrjagin and Krein spaces, dissipative operators, invariant subspaces, fluid mechanics
2000 MSC: 47B37, 47B44

Studying in 1940 a problem on a rotating top with a cavity filled with an ideal incompressible fluid, S. L. Sobolev met an interesting class of operators which became self-adjoint after multiplication by a self-adjoint operator with one negative eigenvalue. Sobolev paid attention of L. S. Pontrjagin to such a class of operators (the

⋆ The work is supported by RFFI, grants No 01-01-00691 and No 01-01-00958.
Email address: ashkalikov@yahoo.com (A. A. Shkalikov).

investigations of Sobolev were published only 20 years later in [19]). In 1944 Pontrjagin published a fundamental paper [15] on properties of self-adjoint operators in spaces endowed by indefinite metric with finite number of negative squares.

One says that a Hilbert space $\mathcal{H}$ is endowed by a fundamental symmetry or an indefinite metric, if a quadratic form $[x,x] = (Gx,x)$ is introduced in $\mathcal{H}$, where $(\cdot,\cdot)$ is the usual scalar product and G is symmetric operator in $\mathcal{H}$. This metric is called *regular* if G is a bounded and boundedly invertible operator. Without loss of generality one can assume in this case that the metric is generated by the operator $J = P_+ - P_-$, where $P_\pm$ are orthogonal projectors in $\mathcal{H}$ such that $P_+ + P_- = I$. The space $\mathcal{H}$ equipped with indefinite scalar product $[\cdot,\cdot]$ is called the Pontrjagin space (we denote it by $\Pi_\varkappa = \{\mathcal{H}, J\}$) if $\operatorname{rank} P_+$ or $\operatorname{rank} P_-$ equals $\varkappa < \infty$. It is called the Krein space (we denote it by $\mathcal{K} = \{\mathcal{H}, J\}$) if the both operators P_+ and P_- are infinite dimensional. A subspace $\mathcal{L}$ in Pontrjagin or Krein space is called nonnegative (nonpositive) if $[x,x] \geqslant 0$ $(\leqslant 0)$ for all $x \in \mathcal{L}$. A subspace $\mathcal{L}$ is said to be maximal nonnegative if there is no nontrivial nonnegative extensions of $\mathcal{L}$.

Pontrjagin [15] proved in 1944 the following fundamental result.

Pontrjagin theorem. *Let A be a self-adjoint operator in the space $\Pi_\varkappa$, $\varkappa = \operatorname{rank} P_+ < \infty$. Then there exists an A-invariant maximal nonnegative subspace $\mathcal{L}$ of dimension $\varkappa$ such that the spectrum of the restriction $A|_{\mathcal{L}}$ lies in the closed upper half-plane.*

Starting from the paper [15] the problem on the existence of invariant subspaces stays in the center of the operator theory in the spaces with indefinite metric. Important generalizations of Pontrjagin theorem were obtained by Langer [10,11] and Krein [7]. The problem on the existence of maximal semi-definite invariant subspaces was also positively solved for the class of the so-called definitizable operators in Krein space (see [12,13]). Moreover, Krein and Langer [8] and Azizov [1] proved that Pontrjagin theorem remains valid for maximal dissipative operators in Pontrjagin space.

Here we shall prove the theorem on existence of maximal nonnegative subspaces for an m-dissipative operator A in Krein space, provided that the domain of the operator $D(A)$ contains $\mathcal{H}_+ = P_+(\mathcal{H})$ and the operator P_+AP_- is compact (moreover, we relax the last condition by showing that the theorem remains valid if the operator P_+AP_- is only relatively compact wih respect to P_-AP_-). From one hand this theorem generalizes the result of Langer [11] to the case of a maximal dissipative operator (instead of a self-sdjoint one), on the other hand it generalizes the result of Krein, Langer and Azizov [8,1] for the case of Krein space (instead of $\Pi_\varkappa$). Note that there is an analogue of this result for nonexpansive operators in Krein spaces obtained by Azizov and Khoroshavin (see [2, Ch. 3,§3]). The connection of this result with our version can be realized by the Cayley transform, although it is not immediate (this was communicated to the author by Azizov). However, the main advantage of this presentation is that we give a direct and short proof. An important remark is that we borrow the basic ideas from the original paper of Pontrjagin [15], although we essentially simplify those from the technical point of view.

In the second part of the paper we demonstrate an application of the theorem on invariant subspaces to Sobolev problem on a rotating top with a cavity filled with a liquid. The first part of this paper is based on the author's paper [17] (which has not been translated into English). In the second part we use the results of the paper [5].

1. Theorem on invariant nonnegative subspaces for dissipative operators in Krein space

Let us represent the space $\mathcal{H}$ in the form $\mathcal{H} = \mathcal{H}_+ \oplus \mathcal{H}_-$ where $\mathcal{H}_\pm = P_\pm(\mathcal{H})$. Consider a closed operator A in $\mathcal{H}$ and assume that its domain $\mathcal{D}(A)$ is such that $\mathcal{D}(A) \supset \mathcal{H}_+$. Then the operator A admits the following matrix representation with respect to the decomposition $\mathcal{H} = \mathcal{H}_+ \oplus \mathcal{H}_-$:

$$A = \begin{pmatrix} A_{11} & A_{12} \\ A_{21} & A_{22} \end{pmatrix} = \begin{pmatrix} P_+AP_+ & P_+AP_- \\ P_-AP_+ & P_-AP_- \end{pmatrix}.$$

Since the operator AP_+ is closed and well defined on the whole space $\mathcal{H}$, it is bounded. Consequently, the operators A_{11} and A_{21} are bounded.

An operator A acting in Krein space $\mathcal{K} = \{\mathcal{H}, J\}$ is said to be *dissipative* if

$$\operatorname{Im}[Ax, x] \geqslant 0 \qquad \forall x \in \mathcal{D}(A).$$

It is called *uniformly dissipative* if there is $\varepsilon > 0$ such that

$$\operatorname{Im}[Ax, x] \geqslant \varepsilon(x, x) \qquad \forall x \in \mathcal{D}(A).$$

A dissipative operator A in $\mathcal{K}$ is said to be *maximal dissipative* (or *m-dissipative*) if the operator JA with the domain $\mathcal{D}(A)$ is m-dissipative in Hilbert space $\mathcal{H}$, i. e. if the open lower half plane belongs to the resolvent set of JA (see [4, Ch. V] for other equivalent definitions). Our nearest goal is to prove the following result.

Theorem 1. *Let A be a maximal dissipative operator in Krein space $\mathcal{K} = \{\mathcal{H}, J\}$ such that $\mathcal{D}(A) \supset \mathcal{H}_+$ and let the operator $A_{12} = P_+AP_-$ be compact in the separable space $\mathcal{H}$. Then there exists an A-invariant maximal nonnegative subspace $\mathcal{L}$ such that the spectrum of the restriction $A|_{\mathcal{L}}$ lies in the closed upper half-plane.*

First we shall prove several lemmas.

Lemma 1. *Let A be a maximal dissipative (uniformly m-dissipative) operator in Krein space $\mathcal{K}$ and $\mathcal{D}(A) \supset \mathcal{H}_+$. Then A_{22} is m-dissipative (uniformly m-dissipative) operator in $\mathcal{H}_-$ and for all λ from the open half-plane $\mathbb{C}^+$ ($\forall\lambda$ from the half-plane $\operatorname{Im}\lambda \geqslant -\varepsilon$ with some $\varepsilon > 0$) the operator*

$$G(\lambda) := A_{12}(A_{22} - \lambda)^{-1}$$

is bounded. Moreover, given any $\delta > 0$ the operator function $G(\lambda)$ is uniformly bounded in the half-plane $\operatorname{Im}\lambda > \delta$ ($\operatorname{Im}\lambda > -\varepsilon + \delta$).

Proof. The first statement that A_{22} is m-dissipative in $\mathcal{H}_-$ provided that A is m-dissipative in $\mathcal{K}$ one can find in [2, Th. II.2.9]. We will get this fact independently. Notice that if $x \in \mathcal{D}(A_{22})$ then $x \in \mathcal{D}(A)$ (since $\mathcal{D}(A) \supset \mathcal{H}_+$). Therefore, for all $x \in \mathcal{D}(A_{22})$ we have $\operatorname{Im}(A_{22}x, x) = \operatorname{Im}[Ax, x] \geqslant 0$, hence A_{22} is dissipative. Further, fix a number $a > 2\|AP_+\|$ (under the assumptions of the lemma the operator AP_+ is bounded) and consider the operator function

$$S(\lambda, a) = \begin{pmatrix} ia & A_{12} \\ 0 & -A_{22} + \lambda \end{pmatrix} = (JA + ia) + (\lambda - ia)P_- - AP_+.$$

The operator JA is m-dissipative in $\mathcal{H}$, therefore, $JA + ia$ is invertible and $\|(JA + ia)^{-1}\| \leqslant a^{-1}$. Since $a > 2\|AP_+\|$, the operator $S(ia, a)$ is invertible. Obviously,

$$\operatorname{Im}(S(\lambda, a)x, x) \geqslant (a/2)\|x\|^2 \qquad \forall x \in \mathcal{D}(A)$$

and for all $\lambda \in \mathbb{C}_a^+ = \{\lambda \mid \operatorname{Im}\lambda \geqslant a\}$. This implies that $S(\lambda, a)$ is invertible for all $\lambda \in \mathbb{C}_a^+$ and $\|S^{-1}(\lambda, a)\| \leqslant 2a^{-1}$. Hence, $A_{22} - \lambda$ is invertible for $\lambda \in \mathbb{C}_a^+$. Since A_{22} is dissipative, $A_{22} - \lambda$ is invertible for $\lambda \in \mathbb{C}^+$ and so is $S(\lambda, a)$. Now, it follows from the identity

$$G(\lambda) = A_{12}(A_{22} - \lambda)^{-1} = aP_+S^{-1}(\lambda, a)P_-$$

that the operator $G(\lambda)$ is bounded for all $\lambda \in \mathbb{C}^+$ and $\|G(\lambda)\| \leqslant 2$ for $\lambda \in \mathbb{C}_a^+$. From the Hilbert identity we have

$$G(\lambda) = G(\lambda + ia) - iaG(\lambda + ia)\,(A_{22} - \lambda)^{-1}.$$

Hence, given any $\delta > 0$ the estimate

$$\|G(\lambda)\| \leqslant 2 + 2a\delta^{-1}$$

holds for all $\lambda \in \mathbb{C}_\delta^+$. The same arguments can be applied in the case of the uniformly m-dissipative operator A. □

Lemma 2. *A subspace $\mathcal{L}$ is maximal nonnegative in the space $\mathcal{K} = \{\mathcal{H}, J\}$ if and only if one of the following conditions holds:*

(i) there is an operator $K : \mathcal{H}_+ \to \mathcal{H}_-$, $\|K\| \leqslant 1$, such that

$$\mathcal{L} = \{x \mid x = x_+ + Kx_+,\ x_+ \in \mathcal{H}_+\}; \tag{1}$$

(ii) $\mathcal{L}$ is nonnegative and there is no nonzero element $y_+ \in \mathcal{H}_+$ such that $[x, y^+] = (x, y_+) = 0$ for all $x \in \mathcal{L}$.

Proof. (See [15]). Let $\mathcal{L}$ be a nonnegative subspace in $\mathcal{K}$. For all $x = x_+ + x_- \in \mathcal{L}$ we have $\|x_+\| \geqslant \|x_-\|$, hence $P_+ : \mathcal{L} \to P_+(\mathcal{L})$ is a bijection. Now we get the representation

$$\mathcal{L} = \{x \mid x = x_+ + Kx_+,\ x_+ \in P_+(\mathcal{L}),\ \|K\| \leqslant 1\},$$

where $K = P_+^{-1} - I$. Obviously, $\mathcal{L}$ has no proper nonnegative extensions if and only if $P_+(\mathcal{L}) = \mathcal{H}_+$. The equalities $(x, y_+) = 0$ for all $x \in \mathcal{L}$ imply $y_+ \perp P_+(\mathcal{L})$. Thus conditions (i) and (ii) are equivalent. □

The operator K participating in representation (1) is called the *angle operator* of the subspace $\mathcal{L}$.

Lemma 3. *Let A be a maximal uniformly dissipative operator in $\mathcal{K}$ and $\mathcal{D}(A) \supset \mathcal{H}_+$. Then the real axis $\mathbb{R}$ belongs to the resolvent set of A and the spectrum of A in $\mathbb{C}^+$ is bounded. If a contour Γ_+ surrounds the spectrum of A in $\mathbb{C}^+$ and*

$$Q_+ = \frac{1}{2\pi i} \int_{\Gamma_+} (\lambda - A)^{-1} d\lambda \tag{2}$$

is the corresponding Riesz projector, then $\mathcal{L}_+ = Q_+(\mathcal{H})$ is the A-invariant maximal nonnegative subspace.

Proof. The invariance of $\mathcal{L}_+$ under A and the inclusion $\mathcal{L}_+ \subset \mathcal{D}(A)$ follow from the properties of the Riesz projectors [16, Ch. 7]. By assumption JA (and hence $JA - \mu J$ for all $\mu \in \mathbb{R}$) is maximal uniformly dissipative operator in Hilbert space $\mathcal{H}$. Then $JA - \mu J$ is boundedly invertible for $\mu \in \mathbb{R}$, hence, so is $A - \mu$ for the same values of μ.

Denote

$$G(\lambda) = A_{12}(A_{22} - \lambda)^{-1}, \quad F(\lambda) = (A_{22} - \lambda)^{-1} A_{21}.$$

By Lemma 1 these operators are bounded for all λ from the half-plane $\operatorname{Im} \lambda > -\varepsilon$, and we can write the Frobenius–Shur factorization

$$-(\lambda - A) = \begin{pmatrix} I & G(\lambda) \\ 0 & I \end{pmatrix} \begin{pmatrix} M(\lambda) & 0 \\ 0 & A_{22} - \lambda \end{pmatrix} \begin{pmatrix} I & 0 \\ F(\lambda) & I \end{pmatrix}, \tag{3}$$

where

$$M(\lambda) = A_{11} - \lambda - A_{12}(A_{22} - \lambda)^{-1} A_{21}$$

is the transfer function of the matrix A in $\mathcal{H}_+ \oplus \mathcal{H}_-$. Consequently, the operator $\lambda - A$ is invertible in $\overline{\mathbb{C}^+}$ if and only if the same is true for $M(\lambda)$. For $\lambda \in \overline{\mathbb{C}^+}$ and $|\lambda| \to \infty$ we have

$$M^{-1}(\lambda) = (A_{11} - \lambda)^{-1} \left(I - G(\lambda) A_{21} (A_{11} - \lambda)^{-1}\right)^{-1} = -\lambda^{-1} + O(\lambda^{-2}),$$

since by Lemma 1 $G(\lambda)$ is uniformly bounded in $\overline{\mathbb{C}^+}$. Thus the spectrum of A in the upper half-plane is bounded.

Take a contour Γ_+ in (2) consisting of a segment $[-R, R]$ and the half of the circle $C_R = \{\lambda \,\big|\, |\lambda| = R, \ \operatorname{Im} \lambda \geqslant 0\}$. The operator $A_+ = A|_{\mathcal{L}^+}$ is bounded in the subspace $\mathcal{L}_+ = Q_+(\mathcal{H})$. Therefore, we have as $R \to \infty$

$$\frac{1}{2\pi i} \int_{C_R} (\lambda - A_+)^{-1} d\lambda = \frac{1}{2\pi i} \int_{C_R} \left(\lambda^{-1} - \lambda^{-2} A_+ (I - \lambda^{-1} A_+)^{-1}\right) d\lambda = \frac{1}{2} I + O(R^{-1}),$$

where I is the identity operator. Let $x = Q_+ x \in \mathcal{L}_+$ and $y = (\lambda - A_+)^{-1} x$. Then we have

$$[x,x] = \mathrm{Re}\,[Q_+x,x] = \frac{1}{2}[x,x] + \frac{1}{2\pi}\int_{-R}^{R} \mathrm{Im}\,[y,(\lambda - A)y]d\lambda + O(R^{-1}). \quad (4)$$

For $\lambda \in \mathbb{R}$

$$\mathrm{Im}\,[y,(\lambda - A)y)] = \mathrm{Im}\,[Ay,y] \geqslant \varepsilon(y,y).$$

Passing to the limit as $R \to \infty$ and taking into account the inequality $\|x\| \leqslant \|\lambda - A_+\|\,\|y\|$, we obtain

$$\frac{1}{2}[x,x] = \frac{1}{2\pi}\int_{-\infty}^{\infty} \mathrm{Im}\,[Ay,y]\,d\lambda \geqslant \varepsilon\|x\|^2 \int_{-\infty}^{\infty} \|\lambda - A_+\|^{-2}\,d\lambda.$$

It is left to prove that the subspace $\mathcal{L}_+$ is maximal positive. It follows from (3) that

$$(\lambda - A)^{-1} = \begin{pmatrix} -M^{-1}(\lambda) & M^{-1}(\lambda)G(\lambda) \\ F(\lambda)M^{-1}(\lambda) & F(\lambda)M^{-1}(\lambda)G(\lambda) - (A_{22}-\lambda)^{-1} \end{pmatrix}.$$

Hence, for all $z \in \mathcal{H}_+$

$$((\lambda - A)^{-1}\,z,z) = -(M^{-1}(\lambda)z,z).$$

Integrating this equality along the contour Γ_+, using relation (4) and tending $R \to \infty$, we get

$$(Q_+z,z) = [Q_+z,z] = \frac{1}{2}(z,z) + \frac{1}{2\pi}\int_{-\infty}^{\infty} \mathrm{Im}\,[Ay,y]\,d\lambda,$$

where $y = (\lambda - A)^{-1}z$. Consequently, $2(Q_+z,z) \geqslant (z,z)$ for all $z \in \mathcal{H}_+$. Hence there is no nonzero $z \in \mathcal{H}_+$ such that $z \perp \mathcal{L}_+ = Q_+(\mathcal{H})$. By Lemma 2 the subspace $\mathcal{L}_+$ is maximal positive. □

Lemma 4. *Let the operators A_{11} and A_{21} in the canonical representation of the operator A be bounded and $\mathcal{D}(A_{12}) \supset \mathcal{D}(A_{22})$. Then a maximal nonnegative subspace $\mathcal{L}$ is A-invariant if and only if its angle operator K possesses the properties*

(i) the image of K lies in $\mathcal{D}(A_{22})$;

(ii) the operator K satisfies the operator Riccati equation

$$A_{21} + A_{22}K - KA_{11} - KA_{12}K = 0. \quad (5)$$

Proof. (See [15,6]). Let $\mathcal{L}$ have representation (1) and be A-invariant. Then the image of K lies in $\mathcal{D}(A_{22})$ and for any $x_+ \in \mathcal{H}_+$ there is an element $y_+ \in \mathcal{H}_+$ such that

$$\begin{pmatrix} A_{11} & A_{12} \\ A_{21} & A_{22} \end{pmatrix}\begin{pmatrix} x_+ \\ Kx_+ \end{pmatrix} = \begin{pmatrix} y_+ \\ Ky_+ \end{pmatrix}. \quad (6)$$

Substituting $y_+ = A_{11}x_+ + A_{12}Kx_+$ in the second equation we find

$$(A_{12} + A_{22}K)x_+ = K(A_{11} + A_{12}K)x_+ \qquad \forall x_+ \in \mathcal{H}_+,$$

and (5) follows. Conversely, if the image of K lies in $\mathcal{D}(A_{22})$ and (5) holds, then equation (6) is valid with $y_+ = A_{11}x_+ + A_{12}Kx_+$. This implies A-invariance of $\mathcal{L}$. □

Lemma 5. *Let T be an operator in $\mathcal{H}$ and let a domain Ω with possible exception of isolated points belong to the resolvent set of T. Let T_n be a sequence of operators such that $\|T_n - T\| \to 0$ as $n \to \infty$ and Ω lies in the resolvent set of all operators T_n. Then the whole domain Ω belongs to the resolvent set of T.*

Proof. Obviously, if a point μ belongs to $\Omega \cap \rho(T)$ then there is a neighbourhood $\mathcal{U}_\delta(\mu)$ such that $\mathcal{U}_\delta(\mu) \in \rho(T_n)$ for sufficiently large n and

$$\|(T_n - \lambda)^{-1} - (T - \lambda)^{-1}\| \to 0 \qquad \text{as } n \to \infty$$

uniformly for $\lambda \in \mathcal{U}_\delta(\mu)$. Take a contour γ in Ω which does not intersect the isolated spectrum points of T. From the last relation we get that the Riesz projector corresponding to γ equals to zero (since the corresponding Riesz projectors for T_n equal to zero). This implies that the spectrum of T in Ω is empty. □

The proof of Theorem 1. Consider the family of operators

$$A(\varepsilon) = A + i\varepsilon J = \begin{pmatrix} A_{11} + i\varepsilon & A_{12} \\ A_{21} & A_{22} - i\varepsilon \end{pmatrix}, \qquad \varepsilon > 0.$$

Since A is m-dissipative in $\mathcal{K}$ and

$$\operatorname{Im}[A(\varepsilon)x, x] = \operatorname{Im}[Ax, x] + \varepsilon[Jx, x] \geqslant \varepsilon(x, x),$$

the operators $A(\varepsilon)$ are uniformly m-dissipative in $\mathcal{K}$. By Lemmas 2, 3 there is an $A(\varepsilon)$-invariant uniformly positive subspace $\mathcal{L}(\varepsilon)$ and the corresponding angle operator $K(\varepsilon)$ satisfies the Riccati equation

$$A_{21} + (A_{22} - i\varepsilon)K(\varepsilon) - K(\varepsilon)(A_{11} + i\varepsilon) - K(\varepsilon)A_{12}K(\varepsilon) = 0. \tag{7}$$

As $\|K(\varepsilon)\| < 1$, there is a sequence of operators $K_n = K(\varepsilon_n)$ such that $K_n \rightharpoonup K$ (converges weakly to K). This follows from the fact that the unit sphere of the separable space $\mathcal{H}$ is compact in the weak topology. Certainly, $\|K\| \leqslant 1$.

Observe that $K_n \rightharpoonup K$ implies $K_nB \rightharpoonup KB$ and $BK_n \rightharpoonup BK$ for any bounded operator B. One can not guarantee the convergence $K_nBK_n \rightharpoonup KBK$, however, such convergence does hold if B is a compact operator. It suffices to check this only for one-dimensional operator B, since a compact operator can be approximated in the uniform operator topology by operators of finite rank. Indeed, if $B = (\cdot, v)u$ then for any f, $g \in \mathcal{H}_+$ we have

$$(K_nBK_nf, g) = (K_nf, v)(K_nu, g) \to (Kf, v)(Ku, g) = (KBKf, g).$$

By the assumption of Theorem 1 the operator A_{12} is compact. Then all the summands in (7) have the weak limits with a possible exception of the sequence $A_{22}K_n$. However, it follows from the Riccati equation that

$$A_{22}K_n \rightharpoonup KA_{11} + KA_{12}K - A_{21} =: S.$$

In particular, for all $f \in \mathcal{H}_+$, $g \in \mathcal{D}(A_{22}^*)$ we have

$$(A_{22}K_n f, g) = (K_n f, A_{22}^* g) \to (Kf, A_{22}^* g) = (Sf, g).$$

From the last relation we conclude that $Kf \in \mathcal{D}(A_{22})$ and $A_{22}K = S$. Hence the limit operator K satisfies the Riccati equation (5) and by Lemma 2 there is an A-invariant nonnegative subspace $\mathcal{L}$ with the angle operator K.

It is left to show that the spectrum of the restriction $A|_{\mathcal{L}}$ lies in the closed upper half-plane. Obviously, the operator $F : \mathcal{L} \to \mathcal{H}_+$ defined by $Fx = x_+$ for $x = x_+ + Kx_+ \in \mathcal{L}$ is invertible and

$$A|_{\mathcal{L}} = F^{-1}(A_{11} + A_{12}K)F.$$

It is left to show that the spectrum of $A_{11} + A_{12}K$ lies in $\overline{\mathbb{C}}^+$. Since A is m-dissipative in $\mathcal{K}$, the operator $-A_{11}$ is m-dissipative in $\mathcal{H}_+$. Then the spectrum of A_{11} lies in $\overline{\mathbb{C}^+}$ and by theorem on holomorphie operator function the spectrum of $A_{11}+A_{12}K$ is discrete in $\mathbb{C}^-$ (as the operator $A_{12}K$ is compact). Applying Lemma 5 we complete the proof. □

Remark 1. *One can weaken the assumption that A_{12} is compact (cf. [2, Ch. 3, § 3, Problem 9]). It suffices to assume that A_{12} is A_{22}-relatively compact, i. e. that the operator $T = A_{12}(A_{22} - \lambda)^{-1}$ is compact for some (and then for all) $\lambda \in \rho(A_{22})$. Indeed, assume that T is compact for some $\lambda \in \mathbb{C}^-$. We can choose λ such that $\|T\| < 1$. Denote $S_n = (A_{22} - \lambda)K_n$. The compactness condition of A_{12} was not used in Lemmas; it was only used while passing to the limit in the Riccati equation. Let us rewrite equation (8) in the form*

$$(I - K_nT)S_n = (i\varepsilon_n - \lambda)K_n + K_0(A_{11} + i\varepsilon_n) - A_{21}.$$

Since $\|(I - K_nT)^{-1}\| \leqslant (1 - \|T\|)^{-1}$, it follows from the last equation that the norms of S_n are uniformly bounded. Hence, there is a sequence S_{n_j} which converges weakly to an operator S. Since T is compact, we have $K_{n_j}TS_{n_j} \rightharpoonup KTS$. Obviously, $S = (A_{22} - \lambda)K$, hence, K satisfies Riccati equation (6), and the corresponding nonnegative subspace is A-invariant. Further, $A_{12}K = TS$ is compact operator. Therefore, the spectrum of $A|_{\mathcal{L}}$ coincides with the spectrum of $A_{11} + A_{12}K$ and lies in $\overline{\mathbb{C}^+}$.

Theorem 1 does not contain information about the spectrum of the restriction $A|_{\mathcal{L}}$ on the real axis. In applications such an information is often quite important (see [18]).

Below we formulate a simple version of a general theorem, which can be found in [18]. Here we omit the proof, although for such a simple formulation it can be easily carried out independently.

Theorem 2. *Let the assumption of Theorem 1 hold and μ be a real isolated eigenvalue of A. Let $\mathcal{L}$ be an A-invariant nonnegative subspace and $A^+ = A|_{\mathcal{L}}$. Then μ belongs to the spectrum of A^+ if and only if there is a nonzero element $x \in \ker(A-\mu)$ such that $[x,x] \geqslant 0$. In particular, if $\ker(A-\mu)$ is a negative subspace in $\mathcal{K}$, then μ belongs to the resolvent set of A^+.*

2. Sobolev problem on a rotating top with a cavity filled with a viscous liquid

2.1. *Notations and the equation of motion*

We consider a system "rigid body + fluid". A rigid body occupies a bounded region Ω_1 in $\mathbb{R}^3$ and contains a cavity Ω inside, i. e. $\Omega \subset \Omega_1$. The cavity Ω is entirely filled with a viscous incompressible liquid. The body is rotating around a fixed origin of coordinates. Consider the orthonormal frame $Oe_0e_1e_2$ rigidly attached to the body in such a way that the unit vectors of this frame coincide with the principal axes of inertia of the body with frozen liquid. Consider also a fixed orthonormal frame $Of_0f_1f_2$. In the nonperturbed state the axis e_0 coincides with f_0. The motion of the system is described by the variables $(z, w, v(x))$, where z is the coordinates of f_0, w is the angular velocity of the rigid body and $v(x)$ is the velocity field of the fluid. Here $x = (x_0, x_1, x_2) \in \Omega$ are the coordinates of a point. All variables z, w, $v(x)$ and x are vectors in the frame $Oe_0e_1e_2$.

We have to recall some basic notations and operators of fluid mechanics (see [9,14], for example). The finitness of kinetic energy implies that the values of the velocity field $v(x)$ belongs to the subspace $J_0(\Omega) \subset L_2^3(\Omega)$. The subspace $J_0(\Omega)$ is the closure of smooth compactly supported in Ω solenoidal vectors, i. e. of vectors $v(x)$ such that $\operatorname{div} v(x) = 0$. The vector

$$B^*v = \int\limits_{\Omega} [x, v(x)]\, dx$$

is called the *gyrostatic moment* of the fluid (here $[\cdot,\cdot]$ denotes the vector product if $\mathbb{R}^3$). One can consider B^* as a bounded operator $B^* : J_0(\Omega) \to \mathbb{R}^3$. Then the adjoint operator B is defined by $Bw = P_0[w, x]$, where P_0 is the orthogonal projection of $L_2^3(\Omega)$ onto $J_0(\Omega)$. It is easy to see that $\operatorname{rot} Bw \equiv 2w$.

Let Δ be the Laplace operator acting in $J_0(\Omega)$ on the domain consisting of smooth compactly supported in Ω functions. The operator $R = P_0\Delta$ is nonnegative and admits the Friedrichs extension, which is also denoted by R. It is called the *Stokes operator*. It is known [9] that the Stokes operator is uniformly positive and has discrete spectrum.

Let $\mathbb{C}^3$ be the complexification of $\mathbb{R}^3$. Define in $\mathbb{C}^3$ the operators by the quadratic forms

$$(A_0 w, w) = \int_{\Omega_1 \setminus \Omega} \|[w, x]\|^2 \, d\rho(x),$$

$$(A w, w) = \int_{\Omega} \|[w, x]\|^2 \, d\rho(x),$$

where $\rho(x)$ is the distribution of mass of the rigid body in $\Omega_1 \setminus \Omega$ and $d\rho(x) = \rho$ is the density of the fluid in Ω. Without loss of generality it is assumed further that the density of the fluid equals 1. The matrices corresponding to these operators are called inertia tensors of the body and the body with the frozen fluid, respectively. According to the definition of the frame $Oe_0e_1e_2$ we have

$$A = \operatorname{diag}(a_0, a_1, a_2)$$

where a_j are the moments of inertia of the body with frozen fluid with respect to the principal axis of inertia Oe_j. Further we assume $a_0 \geqslant a_1 \geqslant a_2$. Define the operator H in $\mathbb{C}^3$ by

$$Hx = -i[x, e_0] = \begin{pmatrix} 0 & 0 & 0 \\ 0 & 0 & -i \\ 0 & i & 0 \end{pmatrix}.$$

The operator

$$G = -2P_0 H$$

acting in the space $J_0(\Omega)$ is called the *gyroscopic operator*. It is easy to see that G is bounded and self-adjoint.

By ν is denoted the viscosity of the fluid. Denote also $k = glm$, where m is the mass of the system, g is the gravitational acceleration and l is the distance from the fixed point O to the center of gravity of the system.

The evolution of the system is described by the equations (see [14], for example)

$$\begin{aligned} \dot z &= [z, w], \\ A\dot w + B^*\dot v &= [Aw + B^* v, w] + k[z, e_0], \\ \nabla p + \nu \Delta v &= \dot v + (v \cdot \nabla) v + 2[w, v] + [w, [w, x]] + [\dot w, x], \qquad \operatorname{div} v(x) = 0. \end{aligned}$$

The first equation is known as the *kinematic relation*, the second equation describes the evolution of the kinetic moment of the system with respect to the point O and the third one is the Navier–Stokes equation for the fluid in the coordinate system attached to the rigid body. Using the first two equations the last one can be rewritten in the form (see [5])

$$B\dot w + \dot v = P_0([v, (2w + \operatorname{rot} v)]) + \nu P_0 \Delta v.$$

Obviously,

$$(z, w, v(x)) \equiv (e_0, \omega e_0, 0)$$

is the stationary solution of the obtained system of equations, provided ω is constant (angular velocity of the nonperturbed rotation). Linearing the system of nonlinear equations at this stationary solution, we obtain the system

$$\dot{z} = i\omega(Hz - \omega^{-1}Hw),$$
$$A\dot{w} + B^*\dot{v} = i\omega(k\omega^{-1}Hz - H(a_0I - A)w + HB^*v,$$
$$B\dot{w} + \dot{v} = i\omega(-2P_0H + i\nu\omega^{-1}P_0\Delta)v,$$

where I is the identity operator.

2.2. *The operator form of the linearized equations*

Consider the Hilbert space

$$\mathcal{H} = \mathbb{C}^3 \times \mathbb{C}^3 \times J_0(\Omega),$$

whose elements are columns $u = (z, w, v)^t$, where t denotes the transposition. Then the last system of the linearized equations is recast in the operator form as

$$\mathbf{W}\dot{u} = i\omega\mathbf{M}\mathbf{u}, \qquad \mathbf{u} \in \mathcal{H}, \tag{8}$$

where

$$\mathbf{W} = \begin{pmatrix} I & 0 & 0 \\ 0 & A & B^* \\ 0 & B & I \end{pmatrix}, \qquad \mathbf{M} = \begin{pmatrix} H & -\omega^{-1}H & 0 \\ -k\omega^{-1}H & H(A - a_0I) & HB^* \\ 0 & 0 & D \end{pmatrix}$$

and the operator D acting in $J_0(\Omega)$ is defined by

$$D = G + i\frac{\nu}{\omega}R.$$

The domain of $\mathbf{M}$ is assumed to be

$$\mathcal{D}(\mathbf{M}) = \mathbb{C}^3 \times \mathbb{C}^3 \times \mathcal{D}(R).$$

Equation (8) is not convenient for the study, since the operator $\mathbf{M}$ is neither symmetric nor dissipative. There is a trick [5] which reduces this equation to a nicer form. Consider the operator

$$\mathbf{S} = \begin{pmatrix} \omega^2(A+N) & -\omega I & 0 \\ -\omega A & I & 0 \\ -\omega B & 0 & I \end{pmatrix}, \qquad \text{where } N = \begin{pmatrix} 1 & 0 & 0 \\ 0 & n_1 & 0 \\ 0 & 0 & n_2 \end{pmatrix}$$

with

$$n_1 = a_0 - a_1 - k/\omega^2, \qquad n_2 = a_0 - a_2 - k/\omega^2.$$

A direct calculation shows that

$$\mathbf{J} := \mathbf{SW} = \begin{pmatrix} \omega^2(A+N) & -\omega A & -\omega B^* \\ -\omega A & A & B^* \\ -\omega B & B & I \end{pmatrix},$$

$$\mathbf{L} := \mathbf{SM} = \begin{pmatrix} (\omega^2 a_0 - 2k)H & -\omega HA + k\omega^{-1}H & -\omega HB^* \\ -\omega AH + k\omega^{-1}H & AH + HA - a_0 H & HB^* \\ -\omega BH & BH & D \end{pmatrix}.$$

Obviously, the operator $\mathbf{S}$ is bounded.

It follows from the relation

$$\begin{pmatrix} \omega^2(A+N) & -\omega I \\ -\omega A & I \end{pmatrix} = \begin{pmatrix} I & -\omega I \\ 0 & I \end{pmatrix} \begin{pmatrix} \omega^2 N & 0 \\ 0 & I \end{pmatrix} \begin{pmatrix} I & 0 \\ -\omega A & I \end{pmatrix}$$

that $\mathbf{S}$ is boundedly invertible if and only if both numbers n_1 and n_2 are nonzero. In this case the operator $\mathbf{J}$ is also boundedly invertible. Hence, if n_1, $n_2 \neq 0$ the linearized system (8) can be rewritten in the equivalent form

$$\dot{\mathbf{u}} = i\omega \mathbf{T}\mathbf{u}, \qquad \mathbf{T} = \mathbf{J}^{-1}\mathbf{L}. \tag{9}$$

We can represent the operator $\mathbf{L}$ in the form

$$\mathbf{L} = \mathbf{V} + \mathbf{D}, \qquad \mathbf{D} := \mathrm{diag}\{I, I, D\}, \qquad D = G + i\nu\omega^{-1}R,$$

where $\mathbf{V}$ is finite-dimensional symmetric operator, $G = G^*$ is gyroscopic operator and $R = R^*$ is the Stokes operator. Now it is easily seen that the operator $\mathbf{L}$ is maximal uniformly dissipative operator (it is assumed that $\omega > 0$ and $\nu > 0$).

Obviously, $\mathbf{J}$ is a symmetric bounded operator having finitely many nonpositive eigenvalues. We shall prove that the number of its negative eigenvalues coincides with those of the matrix N.

First we notice that $\mathbf{W}$ is positive. Indeed,

$$(\mathbf{Wu}, \mathbf{u}) = (z, z) + (A_0 w, w) + \int_\Omega ([u, x] + v(x), [u, x] + v(x))\, dx \geqslant 0.$$

If $\mathbf{Wu} = 0$ then $z = 0$ and $w = 0$ (we recall that $A_0 > 0$), therefore, $v(x) \equiv 0$. Hence, $\mathbf{W} > 0$.

Consider the operators

$$\mathbf{J}_0 = \begin{pmatrix} \omega^2 N & 0 & 0 \\ 0 & A - B^*B & 0 \\ 0 & 0 & I \end{pmatrix}, \qquad \mathbf{S}_0 = \begin{pmatrix} I & 0 & 0 \\ -\omega I & I & 0 \\ -\omega B & B & I \end{pmatrix}.$$

As the operator $\mathbf{W}$ is positive, so is the operator

$$\begin{pmatrix} A & B^* \\ B & I \end{pmatrix} = \begin{pmatrix} I & 0 \\ B & I \end{pmatrix} \begin{pmatrix} A - B^*B & 0 \\ 0 & I \end{pmatrix} \begin{pmatrix} I & B^* \\ 0 & I \end{pmatrix}.$$

Therefore, $A - B^*B > 0$ and the number $\pi_-(\mathbf{J}_0)$ of negative eigenvalues of $\mathbf{J}_0$ coincides with the corresponding number $\pi_-(N)$ of the matrix N. Now, using the equality $\mathbf{J} = \mathbf{S}_0^*\mathbf{J}_0\mathbf{S}_0$, we obtain an important result:

$$\pi_-(\mathbf{J}) = \pi_-(N)$$

and $\mathbf{J}$ *is invertible provided that* $n_1, n_2 \neq 0$.

Hence, we reduced the equations of motion to form (9), where $\mathbf{T}$ is m-dissipative operator in Pontrjagin space $\Pi_\varkappa = \{\mathcal{H}, J\}$ with $\varkappa = \pi_-(N)$.

Now we shall use the fact which is left without proof in this paper (the reader can find it in [5]). *The operator* $\mathbf{L}$ *has no nonzero real eigenvalues. Zero is an eigenvalue of* $\mathbf{L}$ *of multiplicity* 2 *and the corresponding null subspace is the linear span of the vectors* $\mathbf{x}_0 = (e_0, 0, 0)^t$ *and* $x_1 = (0, e_0, 0)^t$. *This subspace is* $\mathbf{J}$*-positive, i. e.* $(\mathbf{Jx}, \mathbf{x}) > 0$ *for all* $\mathbf{x} = \alpha\mathbf{x}_0 + \beta\mathbf{x}_1 \neq 0$. This allows to obtain the following important result.

Theorem 3. *Assume that* $n_1, n_2 \neq 0$. *Then the operator* $\mathbf{T}$ *has exactly* $\pi_-(N)$ *eigenvalues in the upper half-plane. In particular, the problem in question is stable if*

$$\omega > (a_0 - a_1)/k$$

and is not stable if

$$\omega < (a_0 - a_1)/k.$$

Proof. The first assertion follows directly from Theorem 2 of Section 1. If $\omega > (a_0 - a_1)/k$, then both numbers n_1 and n_2 are positive (we recall that by assumption $a_1 > a_2$, therefore $n_2 > n_1$). In this case $\mathbf{J} > 0$ and $\mathbf{T} = \mathbf{J}^{-1/2}\mathbf{L}_1\mathbf{J}^{1/2}$ is similar to the m-dissipative operator $\mathbf{L}_1 = \mathbf{J}^{-1/2}\mathbf{L}\mathbf{J}^{-1/2}$. If $\omega < (a_0 - a_1)/k$, then $\mathbf{T}$ has at least one eigenvalue from the lower half-plane, and this implies the instability of the problem.

This theorem has quite clear physical sense: the shape of the body and the angular velocity of nonperturbed motion are responsible for the stability. One can guarantee the stability if the body has an oblong shape, i. e. if $a_0 \gg a_1 \geqslant a_2$. Otherwise the stability holds only with large nonperturbed angular velocity.

Historical remarks about the above problem and other results can be found in the book [14] and the paper [5].

References

1. T. Ja. Azizov, Dissipative operators in Hilbert space with indefinite metric, *Izv. Acad. Nauk SSSR, Ser. Mat.* **37** no 3 (1973) (in Russian). English transl. in *Math. USSR Izv.* **7** (1973).
2. T. Ja. Azizov, I. S. Iohvidov, *Linear operators in space with indefinite metric.* John Wiley, Chichester, 1989.

3. I. Gohberg and M. G. Krein, *Introduction to the theory of linear nonselfadjoint operators in Hilbert space.* AMS, Providence, Rhode Island, 1969.
4. T. Kato, *Perturbation theory for linear operators* (2-nd edition). Springer-Verlag, New-York, 1976.
5. A. G. Kostyuchenko, A. A. Shkalikov, M. Yu. Yurkin, On the stability of a top with a cavity filled with a viscous Fluid, *Funct. Anal. and its Appl.* **32** (2) (1998) 100–113.
6. M. G. Krein, On an application of the fixed point principle in the theory of operators in space with indefinite metric, *Uspekhi Matem. Nauk* **50** (2) (1950)180–190 (in Russian).
7. M. G. Krein, On a new application of fixed point principle in the theory of operators in space with indefinite metric, *Dokl. Acad. Nauk SSSR* **154** (5) 1023–1026 (in Russian).
8. M. G. Krein, H. Langer, On definite subspaces and generalized resolvents of Hermitian Operatorss in spaces $\Pi_{\varkappa}$, *Funkz. Anal. i Prilozh.* **5** (2) (1971), 59–71, **5** (3) 54–69 (in Russian). English transl. in *Funct. Anal. and Appl.* **5** (1971).
9. O. A. Ladyzhenskaya, *The mathematical theory of incompressible flow.* Gordon and Breach, New York, 1964.
10. H. Langer, On J–hermitian operators, *Dokl. Acad. Nauk SSSR* **134** (2) (1962) 263–266 (in Russian).
11. H. Langer, Eine Veralgemeinerung eines Satzes von L. S. Pontrjagin, *Math. Ann.* **152** (5) (1963) 434–436.
12. H. Langer, Invariant subspaces of a class of operators in spaces with indefinite metric, *J. Funct. Anal.* **19** (3) (1975) 232–241.
13. H. Langer, Spectral functions of definitizable operators in Krein spaces, *Lecture Notes in Mathematics* **948** (1982) 1–46.
14. N. N. Moiseev, V. V. Rumyantsev, *The dynamics of a body with fluid-filled cavities*. Nauka, Moscow, 1965 (in Russian).
15. L. S. Pontrjagin, Hermitian operators in spaces with indefinite metric, *Izv. Acad. Nauk SSSR, Ser. Mat.* **8** (1944) 243–280 (in Russian).
16. F. Riesz, B. Sz–Nagy, *Leçons d'analyse fonctionelle.* Akademai KIADO, Budapest, 1972.
17. A. A. Shkalikov, On the existence of invariant subspaces of dissipative operators in spaces with indefinite metric, *Fund. i prikl. matem.* (2) (1999) (in Russian).
18. A. A. Shkalikov, Operator pencils arising in elasticity and hydrodynamics: the instability index formula, *Operator Theory: Advances and Applications* **87**, Birkhäuser, (1996) 258–285.
19. S. L. Sobolev, Motion of a symmetric top with a cavity filled with a fluid. *Zh. Prikl. Mekh. Tekhn. Fiz.* (3) (1960) 20–55 (in Russian).

V. Kadets and W. Żelazko

Property of Maximality for Spectral Components

Alexey Tikhonov

Mathematical analysis division, Taurida National University, 4, Yaltinskaia str., Simferopol, Crimea, 95007, Ukraine

Abstract

The aim of this paper is to show that operators closed to normal ones with spectrum on a curve possess certain maximal invariant subspaces, which are related to the interior and the exterior of the curve. We generalize the corresponding results (for the real axis and the unit circle) established by Naboko and Makarov.

Key words: invariant subspace, functional model, operator, Hilbert space
2000 MSC: 47A15, 47A10, 47A25

1. Introduction

In the paper [1] Naboko introduced the spectral components $N_{\pm}$ for operators closed to self-adjoint ones. Several years later considering the parallel case of the almost unitary operators Makarov [2] found the following characterization for these invariant subspaces: the subspace N_+ is the maximal among those rationally invariant subspaces H' of an operator S for which the restriction $S|H'$ is a deformation of a weak completely non-unitary (c.n.u.) contraction. Recall that an operator S is called a deformation of an operator T (we shall write $S \prec T$) if there exists a quasiaffinity X such that $SX = XT$ (a bounded operator X is called a quasiaffinity if $\operatorname{Ker} X = \{0\}$, $\operatorname{Ker} X^* = \{0\}$). Operators S and T are called quasisimilar if $S \prec T$ and $T \prec S$. The aim of this paper is to generalize the above characterization to the case of trace class perturbations of normal operators with spectrum on a curve.

Email address: tikhonov@club.cris.net (Alexey Tikhonov).

We pass to precise statements. Let C be a simple closed $C^{4+\varepsilon}$-smooth curve, $G_+ = \operatorname{Int} C$ ($G_- = \operatorname{Ext} C$) be the interior (exterior) of the curve C, $\varphi_\pm$ be conformal maps of the unit disk $\mathbb{D}$ onto $G_\pm$. Let $U_0 \in [H]$ be a normal (that is, $U_0^* U_0 = U_0 U_0^*$) operator acting in a Hilbert space H, and $\sigma(U_0) \subset C$. We shall consider trace class perturbations $S = U_0 + V$, $V \in \mathfrak{S}_1$ such that $\rho(S) \cap G_+ \neq \emptyset$. For this class of operators, we define the spectral components:

$$N_\pm(S) = \operatorname{clos}\{\, f \in H : \forall g \in H \quad ((S-z)^{-1} f, g) \in E^2(G_\pm)\,\},$$

where $E^2(G_\pm)$ are Hardy-Smirnov spaces [3]. This (weak) description is purely in terms of the operator S and does not use any addition information like a representation of S as a perturbation of some auxiliary operator. If S is almost unitary, these subspaces coincide with the corresponding subspaces defined in [1,2]. For readers familiar with Sz.-Nagy-Foiaş's theory [4], we note that $N_+(T_0) = H_{\cdot 1}$, $N_-(T_0) = H$ whenever T_0 is a c.n.u. contraction. Moreover, $N_-(\varphi_+(T_0)) = H$.

To formulate the main result of this paper we need to introduce the following notation $\mathcal{G}_\pm = \{\, T : T = \varphi_\pm(T_0),\ T_0 \text{ is a c.n.u. contraction}\,\}$ for the classes of operators that are "conformal maps" of c.n.u. contractions onto domains $G_\pm$ and $\operatorname{Lat}_\pm S = \{\, H' \in \operatorname{Lat} S : \exists T \in \mathcal{G}_\pm \ \ S|H' \prec T\,\}$ for invariant subspaces of the operator S such that the restrictions of S to them are deformations of operators from $\mathcal{G}_\pm$.

Theorem . 1) $N_\pm(S) = \bigvee_{H' \in \operatorname{Lat}_\mp S} H'$; 2) $N_\pm(S) \in \operatorname{Lat}_\mp S$;
3) S is quasisimilar to an operator from $\mathcal{G}_\pm$ iff $N_\mp(S) = N_\mp(S^*) = H$.

Thus the subspaces $N_\pm(S)$ are the *minimal* invariant subspaces that contain all invariant subspaces on which the operator S induces operators that are deformations of operators from $\mathcal{G}_\mp$. On the other hand, the subspaces $N_\pm(S)$ are the *maximal* among those invariant subspaces on which the operator S induces operators that are deformations of operators from $\mathcal{G}_\mp$.

2. Functional model

Our basic tool to obtain above mentioned characterizations of $N_\pm(S)$ is the functional model for trace class perturbations of normal operators with spectrum on a curve [5]. Here we briefly survey the construction of this model, which is a generalization of the well-known functional model of Sz.-Nagy-Foiaş-Naboko [4,6,1,7].

Let $\mathfrak{N}$ and $\mathcal{H}$ be separable Hilbert spaces. We shall consider pairs $\Pi = (\pi_+, \pi_-)$ of operators $\pi_\pm \in [L^2(C, \mathfrak{N}), \mathcal{H}]$ satisfying the following conditions:

$(i)_1$ $\forall \psi \in L^\infty(C,[\mathfrak{N}])$ $(\pi_\pm^* \pi_\pm)\psi = \psi(\pi_\pm^*\pi_\pm)$; $(ii)_1$ $(\pi_-^\dagger \pi_+)z = z(\pi_-^\dagger \pi_+)$;

$(i)_2$ $(\pi_\pm^*\pi_\pm)^{-1} \in [L^2(C,\mathfrak{N})]$; $(ii)_2$ $P_-(\pi_-^\dagger \pi_+)P_+ = 0$;

(iii) $\operatorname{Ran}\pi_+ \vee \operatorname{Ran}\pi_- = \mathcal{H}$,

where $\pi_-^\dagger$ is the Moore-Penrose inverse for π_-, $P_\pm$ are the (non-orthogonal) projections onto the Hardy-Smirnov spaces $E^2(G_\pm,\mathfrak{N})$ such that $\operatorname{Ran}P_\pm = E^2(G_\pm,\mathfrak{N})$, $\operatorname{Ker}P_\pm = E^2(G_\mp,\mathfrak{N})$. The conditions $U\pi_\pm = \pi_\pm z$ uniquely determine the normal operator U with absolutely continuous spectrum on the curve C. For this operator, we have $U\tau_\pm = \tau_\pm z$, $\pi_\pm^\dagger U = z\pi_\pm^\dagger$, $\tau_\pm^\dagger U = z\tau_\pm^\dagger$, where $\tau_\pm^\dagger = (\Delta^\pm)^{-1}(\pi_\mp^\dagger - \Theta^\pm \pi_\pm^\dagger)$, $\Delta^\pm = (I - \Theta^\pm\Theta^\mp)^{1/2}$, $\Theta^\pm = \pi_\mp^\dagger \pi_\pm$. We put

$$\mathcal{K}_\Theta = \{\, f \in \mathcal{H} : \pi_+^\dagger f \in E^2(G_-,\mathfrak{N}),\ \pi_-^\dagger f \in E^2(G_+,\mathfrak{N}) \,\},$$

$$P_\Theta = (I - q_+)(I - q_-), \quad Q_+ = q_+(I - q_-), \quad Q_- = q_-,$$

where $q_\pm = \pi_\pm P_\pm \pi_\pm^\dagger$. For them, we have $\mathcal{K}_\Theta = \operatorname{Ran}P_\Theta$, $P_\Theta^2 = P_\Theta = (I - Q_+)(I - Q_-) = I - Q_+ - Q_-$, $Q_\pm^2 = Q_\pm$, $Q_\pm Q_\mp = 0$. It is easy to show that $\Theta^\pm \in L^\infty(C,[\mathfrak{N}])$. Moreover, $\Theta^+ \in B(G_+,[\mathfrak{N}])$, that is, Θ^+ admits a bounded analytic continuation to the domain G_+. We also define operator-valued functions $\Theta_\pm^-$ by formula $\Theta_\pm^-(z)n = (P_\pm\Theta^- n)(z), n \in \mathfrak{N}$, $z \in G_\pm$. Under certain addition assumptions, which are always satisfied in the case of functional model for trace class perturbations of normal operators, we have $\Theta_\pm^- \in B(G_\pm,[\mathfrak{N}])$.

Let $\varkappa \in [\mathfrak{N}]$. We put $\varkappa_-^l = \varkappa_-^r = \varkappa$, $\varkappa_+^l = -(I + \varkappa\Theta_+^-)$, $\varkappa_+^r = -(I + \Theta_+^-\varkappa)$; $\Theta_{\cdot\varkappa}^- = I - \Theta_-^-\varkappa$, $\Theta_{\cdot\varkappa}^+ = -\varkappa + \Theta^+ + \Theta^+\Theta_+^-\varkappa$; $\Theta_{\varkappa\cdot}^- = I - \varkappa\Theta_-^-$, $\Theta_{\varkappa\cdot}^+ = -\varkappa + \Theta^+ + \varkappa\Theta_+^-\Theta^+$. We have $\Theta_{\cdot\varkappa}^\pm \in (B:N)(G_\pm,[\mathfrak{N}])$. (We write $W \in (B:N)(G,[\mathfrak{N}])$ if $W \in B(G,[\mathfrak{N}])$ and $W^{-1} \in N(G,[\mathfrak{N}])$, that is, $W^{-1} = 1/\delta\,\Omega$, $\delta \in B(G)$, $\Omega \in B(G,[\mathfrak{N}])$. In the terminology of [4] the latter means that W admits a scalar multiple.) In the paper [5] the following theorem is established.

Theorem . Let C be a simple closed $C^{4+\varepsilon}$-smooth curve, U_0 be a normal operator such that $\sigma(U_0) \subset C$, and let $S - U_0 \in \mathfrak{S}_1$, $\rho(S) \cap \operatorname{Int}C \neq \emptyset$. Then there exist a pair $\Pi = (\pi_+,\pi_-)$ satisfying the conditions (i), (ii), (iii), an operator $\varkappa \in [\mathfrak{N}]$, and an invertible operator $Z \in [H,\mathcal{K}_\Theta]$ such that $\widehat{S}Z = ZS$, where $\widehat{S}f_0 = Uf_0 + (\pi_+^\dagger\varkappa_+^r + \pi_-^\dagger\varkappa_-^r)\Theta_{\cdot\varkappa}^-(\infty)^{-1}(\pi_+^\dagger U f_0)(\infty)$, $f_0 \in \mathcal{K}_\Theta$.

This theorem allow us to proceed with functional model objects only. Working with the functional model we have the advantage to produce explicit formulas. We consider the following subspaces of $\mathcal{H}$:

$$\begin{aligned} N_\pm(\Pi,\varkappa) &= \{\, f\in\mathcal{H}:\ \varkappa^l_\pm\pi^\dagger_\mp P_\Theta f = \Theta^\pm_{\varkappa\cdot}\pi^\dagger_\pm Q_\pm f\,\};\\ N(\Pi,\varkappa) &= \{\, f\in\mathcal{H}:\ \Theta^\pm_{\varkappa\cdot}\pi^\dagger_\pm f = \varkappa^l_\pm\Delta^\pm\tau^\dagger_\pm f\,\};\\ M(\Pi,\varkappa) &= \{\, f\in\mathcal{H}:\ \Theta^-_{\cdot\varkappa}(z)^{-1}(\pi^\dagger_+ f)(z) = \Theta^+_{\cdot\varkappa}(z)^{-1}(\pi^\dagger_- f)(z)\,\};\\ NM_\pm(\Pi,\varkappa) &= \{\, f\in N_\pm(\Pi,\varkappa):\ \tau^\dagger_\mp f = 0\,\}. \end{aligned}$$

Here and throughout of the paper we refer a reader to [5] for more details. These "lifted" spectral components are crucial for us in this paper (we define the operators $S_\pm$ (see Section 3) acting in $N_\pm(\Pi,\varkappa)$) as well as for studying the duality problem of spectral components in [5]. One of consequences of that duality is the fact of coinciding weak and model spectral components $X(\widehat{S}) = \operatorname{clos} P_\Theta X(\Pi,\varkappa)$, $X\in\{N_\pm, N, NM_\pm\}$ and $M(\widehat{S}) = M(\Pi,\varkappa)\cap\mathcal{K}_\Theta$. Note also that $N(\Pi,\varkappa) = N_+(\Pi,\varkappa)\cap N_-(\Pi,\varkappa)$ and $N_\pm(\Pi,\varkappa) = NM_\pm(\Pi,\varkappa)\vee N(\Pi,\varkappa)$.

3. Property of maximality

Proof (of the main theorem).

1) The inclusion $\bigvee_{H'\in\mathrm{Lat}_\mp S} H' \subset N_\pm(S)$ can be easily derived from the following properties of $N_\pm$: $T\in\mathcal{G}_\pm \Rightarrow N_\mp(T) = H$; $X\in[H]$, $XT = SX \Rightarrow XN_\pm(T)\subset N_\pm(S)$. The inverse inclusion follows from statement 2.

2) We introduce the following subspaces and operators:

$$\begin{aligned} &N'_\pm = N'_\pm(\Pi,\varkappa) = (I-Q_\mp)N_\pm(\Pi,\varkappa)\subset N_\pm(\Pi,\varkappa)\,,\\ &S_\pm f = Uf + \pi^\dagger_\mp\varkappa^r_\mp\Theta^-_{\cdot\varkappa}(\infty)^{-1}(\pi^\dagger_+ Uf_0)(\infty),\ f\in N'_\pm(\Pi,\varkappa)\,. \end{aligned}$$

To prove that $N_\pm(S)\in\mathrm{Lat}_\mp S$ we shall check that $S|N_\pm(S)$ are deformations of $S_\pm$ and, in turn, that $S_\pm$ are deformations of operators from $\mathcal{G}_\mp$. For this purpose we present the following list of properties:

a) $P_\Theta|N'_\pm = (I-Q_\pm)|N'_\pm$, $\operatorname{Ker}(P_\Theta|N'_\pm) = \{0\}$;

b) $S_\pm\in[N'_\pm]$; **c)** $\sigma(S_\pm)\subset\sigma(S)\setminus G_\pm$; **d)** $SP_\Theta|N'_\pm = P_\Theta S_\pm$;

e) $(S_\pm - a)^{-1} = (I-Q_\mp)(U-a)^{-1}|N'_\pm$, $a\in G_\pm$,

from which it follows that $S|N_\pm(S)$ are deformations of $S_\pm$. The proofs, which are mainly based on computations with the model, are omitted.

To prove that $S_\pm$ are deformations of operators from $\mathcal{G}_\mp$ we need to dwell on the spectral structure of operators $S_\pm$. To this end we consider the subspaces:

$$\begin{aligned} &N'_{e\pm} = \operatorname{clos}\widetilde{N}'_{e\pm},\quad \widetilde{N}'_{e\pm} = \widetilde{N}'_{e\pm}(\Pi,\varkappa) = (I-Q_\mp)N(\Pi,\varkappa);\\ &NM'_\pm = NM'_\pm(\Pi,\varkappa) = (I-Q_\mp)NM_\pm(\Pi,\varkappa) = \{\, f\in N'_\pm:\ \tau^\dagger_\mp f = 0\,\}, \end{aligned}$$

which are absolutely continuous and singular components, respectively, for the operators $S_\pm$ (see Remark 1 for the exact statement). For $N'_{e\pm}$, we have

$$\textbf{f)}\ \operatorname{Ker}(I-Q_\mp)|N(\Pi,\varkappa)=\{0\};\quad \textbf{g)}\ N'_{e\pm}\in \operatorname{Lat}S_\pm;$$

$$\textbf{h)}\ \sigma(S_\pm|N'_{e\pm})\subset C;\quad \textbf{i)}\ S_\pm(I-Q_\mp)f=(I-Q_\mp)Uf,\ f\in N(\Pi,\varkappa)\,.$$

Whence we can conclude that the restrictions $S_\pm|N'_{e\pm}$ are deformations of the normal operator $U|N(\Pi,\varkappa)$. It can easily be shown that there exist operators $T_{e\pm}\in[H_{e\pm}]\cap\mathcal{G}_\mp$ such that the operators $T_{e\pm}$ are similar to $U|N(\Pi,\varkappa)$. Denote by $V_{e\pm}$ an operator from $[H_{e\pm},N(\Pi,\varkappa)]$ that realized this similarity $UV_{e\pm}=V_{e\pm}T_{e\pm}$. So, $S_\pm|N'_{e\pm}$ are deformations of $T_{e\pm}\in\mathcal{G}_\mp$. Further, for $NM'_\pm$, we have

$$\textbf{j)}\ NM'_\pm\in\operatorname{Lat}S_\pm;\quad \textbf{k)}\ \sigma(S_\pm|NM'_\pm)\subset\sigma(S_\pm);\quad \textbf{l)}\ NM'_\pm=\pi_\mp\varkappa^r_\mp E_{\Theta^\mp_{\cdot\varkappa}}\,,$$

where $E_{\Theta^\mp_{\cdot\varkappa}}=\{\,v_\pm\in E^2(G_\pm,[\mathfrak{N}]):\ \Theta^\mp_{\cdot\varkappa}v_\pm\in E^2(G_\mp,[\mathfrak{N}])\,\}$. To justify the latter property, one can make use of the following assertion: $f\in NM'_\pm\ \Rightarrow\ \pi^\dagger_\mp f=\varkappa^r_\mp v_\pm$, where $-v_+=(\Theta^+_{\cdot\varkappa})^{-1}\pi^\dagger_- P_\Theta f=(\Theta^-_+\Theta^+-I)\pi^\dagger_+Q_+f+\Theta^-_+\pi^\dagger_-P_\Theta f\in E^2(G_-,[\mathfrak{N}])$, $-v_-=(\Theta^-_{\cdot\varkappa})^{-1}\pi^\dagger_+P_\Theta f=\Theta^-_-\pi^\dagger_-Q_-f+\pi^\dagger_+P_\Theta f\in E^2(G_-,[\mathfrak{N}])$. Next, we have

$$\textbf{m)}\ (\pi_\mp\varkappa^r_\mp|E_{\Theta^\mp_{\cdot\varkappa}})^{-1}=-(\Theta^\pm_{\cdot\varkappa})^{-1}\pi^\dagger_\mp P_\Theta\in[NM'_\pm,E_{\Theta^\mp_{\cdot\varkappa}}]\,.$$

Since we have $-\pi_\mp\varkappa^r_\mp(\Theta^\pm_{\cdot\varkappa})^{-1}\pi^\dagger_\mp P_\Theta f=f$, $f\in NM'_\pm$, it suffices to check that $\operatorname{Ker}(\pi_\mp\varkappa^r_\mp|E_{\Theta^\mp_{\cdot\varkappa}})=\{0\}$. Indeed, let $\pi_\mp\varkappa^r_\mp v_\pm=0$. Then $\varkappa^r_\mp v_\pm=0$. Consequently, 1) $\varkappa^r_-v_+=0\ \Rightarrow\ \varkappa v_+=0\ \Rightarrow\ v_+=(I-\Theta^-_-\varkappa)v_+=\Theta^-_{\cdot\varkappa}v_+\in E^2(G_-,[\mathfrak{N}])\cap E^2(G_+,[\mathfrak{N}])\ \Rightarrow\ v_+=0$; 2) $\varkappa^r_-v_+=0\ \Rightarrow\ \varkappa v_-=(\varkappa+\Theta^+\varkappa^r_+)v_-=-\Theta^+_{\cdot\varkappa}v_-\in E^2(G_+,[\mathfrak{N}])\cap E^2(G_-,[\mathfrak{N}])\ \Rightarrow\ \varkappa v_-=0\ \Rightarrow\ v_-=(I+\Theta^-_+\varkappa)v_-=-\varkappa^r_+v_-=0$. Besides, we have

$$\textbf{n)}\ (S_\pm-a)^{-1}\pi_\mp\varkappa^r_\mp v_\pm=\pi_\mp\varkappa^r_\mp\frac{v_\pm-v_\pm(a)}{z-a}\,,\quad a\in G_\pm\quad v_\pm\in E_{\Theta^\mp_{\cdot\varkappa}}\,,$$

that is, the restrictions $S_\pm|NM'_\pm$ are similar to operators $T_{i\pm}\in[H_{i\pm}]\cap\mathcal{G}_\mp$, where $H_{i\pm}=E_{\Theta^\mp_{\cdot\varkappa}}$ and $(T_{i\pm}-a)^{-1}=P_\pm\frac{1}{z-a}|E_{\Theta^\mp_{\cdot\varkappa}}$, $a\in G_\pm$.

We have found that the parts $S_\pm|N'_{e\pm}$ and $S_\pm|NM'_\pm$ are deformations of operators $T_{i\pm}$ and $T_{e\pm}$ from $\mathcal{G}_\mp$. To show that $S_\pm$ are deformations of operators from $\mathcal{G}_\mp$ it remains to check that the operators $X_\pm=\pi_\mp\varkappa^r_\mp\oplus(I-Q_\mp)V_{e\pm}$ are quasiaffinity. This follows from the properties:

$$\textbf{o)}\ N'_{e\pm}\cap NM'_\pm=\{0\}\,;\quad \textbf{p)}\ N'_{e\pm}\vee NM'_\pm=N'_\pm\,.$$

First, we check o). Let $f\in N'_{e\pm}\cap NM'_\pm$. Then $f=\lim\limits_{n\to\infty}f_n$, $f_n\in\widetilde{N}'_{e\pm}$. Since $\widetilde{N}'_{e\pm}\subset N_\pm(\Pi,\varkappa)$, we have $\pi^\dagger_\pm P_\Theta f_n\in\Theta^\mp_{\cdot\varkappa}E^2(G_\mp,[\mathfrak{N}])\subset\Theta^\mp_{\cdot\varkappa i}E^2(G_\mp,[\mathfrak{N}])\ \Rightarrow\ (\Theta^\mp_{\cdot\varkappa i})^{-1}\pi^\dagger_\pm P_\Theta f\in E^2(G_\mp,[\mathfrak{N}])$, where $\Theta^\mp_{\cdot\varkappa}=\Theta^\mp_{\cdot\varkappa i}\Theta^\mp_{\cdot\varkappa e}$ is the inner-outer factorization. Whence we have

$$(\Theta^\mp_{\cdot\varkappa e})^{-1}(\Theta^\mp_{\cdot\varkappa i})^{-1}\pi^\dagger_\pm P_\Theta f=(\Theta^\mp_{\cdot\varkappa})^{-1}\pi^\dagger_\pm P_\Theta f=(\Theta^\pm_{\cdot\varkappa})^{-1}\pi^\dagger_\mp P_\Theta f\in E^2(G_\pm,[\mathfrak{N}]).$$

Since $\Theta^{\mp}_{\cdot\varkappa\mathrm{e}}$ are two-sided outer functions, we have $(\Theta^{\mp}_{\cdot\varkappa\mathrm{i}})^{-1}\pi^{\dagger}_{\pm}P_{\Theta}f = 0$. Then $\pi^{\dagger}_{\pm}P_{\Theta}f = 0 \Rightarrow \pi^{\dagger}_{\mp}P_{\Theta}f = 0 \Rightarrow P_{\Theta}f = 0 \Rightarrow f = 0$. Then we have $N'_{\mathrm{e}\pm} \cap NM'_{\pm} = \{0\}$. To check the property p) we note that $N'_{\pm} = (I - Q_{\mp})N_{\pm}(\Pi, \varkappa) = (I - Q_{\mp})(NM_{\pm}(\Pi, \varkappa) \vee N(\Pi, \varkappa)) = NM'_{\pm} \vee N'_{\mathrm{e}\pm}$.

3) The necessity of the conditions $N_{\mp}(S) = N_{\mp}(S^*) = H$ is obvious. We shall prove the sufficiency. We have already shown $S_{\pm} \prec T_{\pm} = T_{\mathrm{i}\pm} \oplus T_{\mathrm{e}\pm}$ and we will show that $T_{\pm} \prec S_{\pm}$. To this end we take (regularization of the unbounded operators $X^{-1}_{\pm}$) $Y_{\pm} = X^{-1}_{\mp}\delta_{\mp}(S_{\pm})$, where $\delta_{\mp}(S_{\pm}) = (I - Q_{\mp})\delta_{\mp}(U)|N'_{\pm}$, $\delta_{\pm} \in B(G_{\pm})$ are scalar outer functions such that $(\Theta^{\pm}_{\varkappa\cdot})^{-1} = (1/\delta_{\pm})\,\Omega^{\pm}$ and operator valued functions $\Omega^{\pm} \in L^{\infty}(C, [\mathfrak{N}])$. It is easy to see that $Y_{\pm}S_{\pm} = T_{\pm}Y_{\pm}$ and $\operatorname{clos}\operatorname{Ran}Y_{\pm} = H_{\mathrm{i}\pm} \oplus H_{\mathrm{e}\pm}$. Since $\delta_{\mp}$ are outer functions, we get $\operatorname{Ker}\delta_{\mp}(U)|N'_{\pm} = \{0\}$. Hence $\operatorname{Ker}\delta_{\mp}(S_{\pm}) = \{0\}$ and $\operatorname{Ker}Y_{\pm} = \{0\}$. It remains to show that the operators $Y_{\pm}$ are bounded. Let $f \in N'_{\pm}$. Take $f_{\mathrm{e}\pm} = \pi_{\mp}\Omega^{\mp}\varkappa^{l}_{\mp}\Delta^{\mp}\tau^{\dagger}_{\mp}f + \tau_{\mp}\delta_{\mp}\tau^{\dagger}_{\mp}f$ and $f_{\mathrm{i}\pm} = \delta_{\mp}(U)f - f_{\mathrm{e}\pm} = \pi_{\pm}(\delta_{\mp}\pi^{\dagger}_{\pm} - \Omega^{\mp}\varkappa^{l}_{\mp}\Delta^{\mp}\tau^{\dagger}_{\mp})f$. For them, it is easy to check that $f_{\mathrm{e}\pm} \in N(\Pi, \varkappa)$ and $f_{\mathrm{i}\pm} \in NM_{\pm}(\Pi, \varkappa)$. Then we have $(I - Q_{\mp})f_{\mathrm{i}\pm} \in NM'_{\pm}$ and $\delta_{\mp}(S_{\pm})f = (I - Q_{\mp})\delta_{\mp}(U)f = (I - Q_{\mp})f_{\mathrm{i}\pm} + (I - Q_{\mp})f_{\mathrm{e}\pm} = -X_{\pm}(\Theta^{\pm}_{\cdot\varkappa})^{-1}\pi^{\dagger}_{\mp}P_{\Theta}(I - Q_{\mp})f_{\mathrm{i}\pm} + X_{\pm}V^{-1}_{\mathrm{e}\pm}f_{\mathrm{e}\pm}$. Whence the operators $Y_{\pm}$ are bounded. Thus $T_{\pm} \prec S_{\pm}$ and the operators $S_{\pm}$ and $T_{\pm}$ are quasisimilar.

In the same way, in the dual model for an adjoint operator we have the quasisimilarity of $S_{*\pm}$ and $T_{*\pm}$. We claim that the operators $(T_{\pm})^*$ and $T_{*\pm}$ are similar. This is the case for the parts $(T_{\mathrm{e}\pm})^*$ and $T_{*\mathrm{e}\pm}$. In order to make sure that $(T_{\mathrm{i}\pm})^*$ and $T_{*\mathrm{i}\pm}$ are similar, we observe that the operators $T_{\mathrm{i}\pm}$ and $(T_{*\mathrm{i}\pm})^*$ (or, to be more precise, their resolvents) are defined by the same formula. (In the case when $G_{+} = \mathbb{D}$, the operators $T_{\mathrm{i}\pm}$ are standard restricted backward shifts [4,6]). The difference is in the spaces where they act: $E_{\Theta^{\mp}_{\cdot\varkappa}}$ and $E_{\Theta^{\mp}_{\varkappa\cdot}}$. Recall definitions for operator valued functions $\Theta^{-}_{\cdot\varkappa} = I - \Theta^{-}_{-}\varkappa$ and $\Theta^{-}_{\varkappa\cdot} = I - \varkappa\Theta^{-}_{-}$. These analytical in the domain G_{-} functions differ only by the order of the multipliers. To cover both cases (for the domains $G_{\pm}$) we need the following lemma.

Lemma . Let $P^2 = P$, $Q = I - P$, $PDQ = PGQ = PFQ = PCQ = 0$, $CD = DC$, $CG = GC$, $CF = FC$. Then $(I + PFG)|H_1 \in [H_1, H_2]$, $((I + PFG)|H_1)^{-1} = (I + PFD)|H_2$, $PC|H_1 \in [H_1], PC|H_2 \in [H_2]$, $(I + PFG)PC|H_1 = PC(I + PFG)|H_1$, where $H_1 = \operatorname{Ker}Q \cap \operatorname{Ker}P(D + G + DFG)$, $H_2 = \operatorname{Ker}Q \cap \operatorname{Ker}P(D + G + GFD)$.

Note if we choose in the lemma $P = I$, $D = I + A$, $G = I + B$, $F = -I$, we obtain the well-known identities $\dim\operatorname{Ker}(I - AB) = \dim\operatorname{Ker}(I - BA)$ and $\sigma_p(AB) \setminus \{0\} = \sigma_p(BA) \setminus \{0\}$. In our cases the conditions like $PDQ = 0$, $CD = DC$ mean that D is an analytic operator-valued function.

If we choose in the lemma $P = P_{-}$, $D = \Theta^{+}$, $G = -\varkappa$, $F = -\Theta^{-}_{+}$, $C = \frac{1}{z-a}$, $a \in G_{-}$, then we obtain that the operators $P_{-}\frac{1}{z-a}|E_{\Theta^{+}_{\varkappa\cdot}}$ and $P_{-}\frac{1}{z-a}|E_{\Theta^{+}_{\cdot\varkappa}}$ are similar. If we choose $P = P_{+}$, $D = I + \Theta^{-}_{-}$, $G = I + \varkappa$, $F = -I$, $C = \frac{1}{z-a}$, $a \in G_{+}$, then we obtain the similarity for the operators $P_{+}\frac{1}{z-a}|E_{\Theta^{-}_{\varkappa\cdot}}$ and

$P_+ \frac{1}{z-a} | E_{\Theta_\varkappa^-}$.

Thus we have established the property:

q) *$S_{*\pm}$ and $S_\pm^*$ are quasisimilar .*

Therefore we have $S \prec S_\pm$ and $S^* \prec S_\pm^*$. This completes the proof of the statement. □

Remark 1. Note also that **r**) $M(S_\pm) = NM'_\pm(\Pi, \varkappa)$; **s**) $N(S_\pm) = N'_{e\pm}(\Pi, \varkappa)$.

Remark 2. One can specify statement 2 of the theorem : we have $S|N_\pm(S) \prec \varphi_\mp(T_\pm)$, where the operators $T_\pm$ are c.n.u. contractions such that their characteristic function admit a scalar multiple.

Remark 3. There are analogous theorems for the spectral components $N(S)$ and $NM_\pm(S)$. The corresponding assertions can be obtained if we replace $\mathcal{G}_\pm$ by $\mathcal{G}_e = \{T : T = \varphi(U_0),\ U_0^*U_0 = U_0U_0^* = I,\ M(U_0) = \{0\}\}$ or by $\mathcal{G}_{i\pm} = \{T : T = \varphi_\pm(T_0),\ T_0$ *is c.n.u. contraction of class* $C_0\}$, respectively.

References

1. S. N. Naboko, On spectral analysis of non-selfadjoint operators, *Dokl. Akad. Nauk SSSR*, **232** (1977) 36-39; English transl. in: *Sov. Math. Dokl.* **18** (1977) 32-36.
2. N. G. Makarov, Canonical subspaces of almost unitary operators, in: *A. Haar Memorial Conference. Vol. 1, 2 (Budapest, 1985), Colloq. Math. Soc. János Bolyai, vol. 49*. North-Holland, Amsterdam–New York, 1985, 611–621.
3. P. L. Duren, *Theory of H^p spaces*. Academic Press, New York–London, 1970.
4. B. Sz.-Nagy and C. Foiaş, *Harmonic analysis of operators on Hilbert space*. North-Holland, Amsterdam–London, 1970.
5. A. S. Tikhonov, Functional model and duality of spectral components for operators with continuous spectrum on a curve, *Algebra i Analiz*, **14** (4) (2002) 158–195; English transl. in: *St.Petersburg Math.J.*, **14** (4) (2003).
6. N. K. Nikol'skiĭ and V. I. Vasyunin, Elements of spectral theory in terms of the free functional model I. Basic constructions, in: Sh. Axler, J. McCarthy, D. Sarason, eds., *Holomorphic Spaces (Berkeley, CA, 1995)*. Cambridge Univ. Press, Cambridge, 1998, 211–302.
7. N. G. Makarov and V. I. Vasyunin, A model for noncontraction and stability of the continuous spectrum, in: *Lecture Notes in Math., vol. 864*. Springer, Berlin–New York, 1981, 365–412.

Functional Analysis and its Applications
V. Kadets and W. Żelazko

A General View on Approximation Ideals

Nicolae Tiţa

Transilvania University of Braşov, Faculty of Sciences, Department of Mathematics, 2200 Brasov – Romania

Abstract

By means of the symmetric norming functions and the abstract concept of approximation numbers of an element in a normed algebra, we present a general view of approximation ideals. Some properties and applications are also presented.

Key words: Approximation ideals, symmetric norming functions
2000 MSC: 46A32, 47B06

Introduction. The purpose of this paper is to present a new study of the approximation ideals in a Banach algebra X. This approach generalizes the approximation ideals for operators and sequence spaces (see the examples below). To this end the generalized approximation numbers of an element from the Banach algebra X and the symmetric norming functions of R. Schatten are considered. If the algebra X is $L(H)$, the Banach algebra of all linear and bounded operators $T : H \to H$, where H is a separable Hilbert space, these ideals are the Schatten classes of operators, [2], [5], [7], [10], [11]. If $B : X \times X \to X$ is a bounded bilinear operator, it is proved that, for some approximations ideals A, the induced operator $B : A \times A \to A$ is also bounded, but not for all approximation ideals.

1. Approximation numbers

Let $(X, \|\cdot\|)$ be a normed (Banach) algebra and let $\| x \|^*$: $X \to N \cup \{+\infty\}$ be a group norm

$$(\|\cdot\|^* > 0 \text{ if } x \in X, x \neq 0, \ \| x+y \|^* \leq \| x \|^* + \| y \|^*, \| x \|^* = \| -x \|^*)$$

Email address: n.tita@info.unitbv.ro (Nicolae Tiţa).

such that $\| xy \|^* \leq \min\{\| x \|^*, \| y \|^*\}$. For all $x \in X$ we define a sequence $(a_n(x))$ as follows:

$$a_n(x) = \inf\{\| x - \overline{x} \|, \ \overline{x} \in X, \ \| \overline{x} \|^* < n\}, \ n = 1, 2, \ldots$$

It is obvious that: $\| x \| = a_1(x) \geq a_2(x) \geq \ldots \geq 0$.

Proposition 1.1. *The next relations are true:*

(1). $\sum_{n=1}^{k} a_n (x_1 + x_2) \leq 2 \sum_{n=1}^{k} (a_n(x_1) + a_n(x_2)), \ k = 1, 2, \ldots$

(2). $\sum_{n=1}^{k} a_n (x_1 x_2) \leq 2 \sum_{n=1}^{k} (a_n(x_1) a_n(x_2)), \ k = 1, 2, \ldots$

Proof. By the definition of the numbers $a_n(x)$ for all $\epsilon > 0$ there are $\overline{x}_1, \overline{x}_2 \in X$ such that $\| \overline{x}_i \|^* < n$ and $\| x_i - \overline{x}_i \| \leq a_n(x_i) + \frac{\epsilon}{2}, \ i = 1, 2$. Because $\| \overline{x}_1 + \overline{x}_2 \|^* < 2n - 1$ we obtain:

$$a_{2n-1} (x_1 + x_2) \leq \| (x_1 + x_2) - (\overline{x}_1 + \overline{x}_2) \| \leq \| x_1 - \overline{x}_1 \| + \| x_2 - \overline{x}_2 \| \leq$$
$$\leq a_n(x_1) + a_n(x_2) + \epsilon.$$

Since ϵ is arbitrary it implies :

$$a_{2n-1} (x_1 + x_2) \leq a_n(x_1) + a_n(x_2).$$

Now we can write:

$$\sum_{n=1}^{k} a_n (x_1 + x_2) \leq \sum_{n=1}^{2k} a_n (x_1 + x_2) \leq 2 \sum_{n=1}^{k} a_{2n-1} (x_1 + x_2) \leq$$
$$\leq 2 \sum_{n=1}^{k} (a_n(x_1) + a_n(x_2)).$$

This proves (1).

To prove the relation (2) we remark at first that $a_{2n-1} (x_1 x_2) \leq a_n(x_1) a_n(x_2)$, because, if $\overline{x}_1, \overline{x}_2$ are as above, we have: $\| \overline{x}_1 (x_2 - \overline{x}_2) + x_1 \overline{x}_2 \|^* < 2n - 1$ and hence

$$a_{2n-1} (x_1 x_2) \leq \| x_1 x_2 - [\overline{x}_1 (x_2 - \overline{x}_2) + x_1 \overline{x}_2] \| = \| (x_1 - \overline{x}_1)(x_2 - \overline{x}_2) \| \leq$$
$$\leq \left(a_n(x_1) + \frac{\epsilon}{2}\right) \left(a_n(x_2) + \frac{\epsilon}{2}\right).$$

Since ϵ is arbitrary we obtain the purpose relation.

Now, (2) results in the same way as (1), namely:

$$\sum_{n=1}^{k} a_n(x_1 x_2) \leq 2 \sum_{n=1}^{k} a_{2n-1}(x_1 x_2) \leq 2 \sum_{n=1}^{k} a_n(x_1) a_n(x_2).$$

Examples. We present two examples of normed algebras on which we can take a norm $\| \cdot \|^*$.

(1). Let $(\ell_\infty, \|\cdot\|_\infty)$ be the normed algebra[1] of all bounded sequences (real or complex numbers), with $\| x \|_\infty = \sup_n |x_n|$. Here $\| x \|^* = card(x) = card\{i \in N : x_i \neq 0\}$. In this case $a_n(x)$ are the approximation numbers of the sequence x, $a_n(x) = \inf\{\| x - \overline{x} \|_\infty : card(\overline{x}) < n\}$.

(2). Let $(L(E), \|\cdot\|)$ be the normed algebra of all linear and bounded operators $T : E \to E$, where E is a normed (Banach) space and $\|\cdot\|$ is the uniform norm. The norm $\|\cdot\|^*$ is $\| T \|^* = rankT$ and hence $a_n(A), A \in L(E)$, is the $n-$ th approximation number of the operator A,

$$a_n(A) = \inf\{\| A - \overline{A} \| : \overline{A} \in L(E),\ rank\overline{A} < n\}.$$

Remark. If E is a separable Hilbert space and A is a compact operator, it is well known, [2], [11], that $a_n(A) = \lambda_n (AA^*)^{\frac{1}{2}}$, where $\{\lambda_n (AA^*)^{\frac{1}{2}}\}$ is the sequence of eigenvalues of $(AA^*)^{\frac{1}{2}}$ in decreasing order. In the sequel we denote by K the set of all real sequences x such that $card(x)$ is finite and $x_1 \geq x_2 \geq \ldots \geq 0$. The functions ϕ (symmetric norming functions) of R. Schatten, [5], [6], [7], [11] are defined as follows: $\phi : K \to R$ such that ϕ is a norm on the cone K, $\phi(1,0,0,\ldots) = 1$; and moreover if $x, y \in K$ and $\Sigma_1^k x_i \leq \Sigma_1^k y_i$, $k = 1,2,\ldots$ then $\phi(x) \leq \phi(y.)$ If $x \in \ell_\infty$ and $x_1 \geq x_2 \geq \ldots \geq 0$, we denote $\phi(x) = \lim_n \phi(x_1, \ldots, x_n, 0, 0, \ldots)$. We remark that in the sequel X will be an algebra with the unit element e.

2. Approximation ideals $A_\phi(X)$

For $x \in X$ and ϕ a symmetric norming function we define

$$A_\phi(X) = \{x \in X : \phi(\{a_n(x)\}) < \infty\}.$$

Proposition 2.1. *$A_\phi(x)$ is an ideal (bilateral) in X.*

Proof. At first we prove that, if $x_i \in A_\phi(X)$, $i = 1,2$, then $x_1 + x_2 \in A_\phi(X)$. This is a corollary of the proposition 1.1. and the properties of ϕ. Precisely, since $\Sigma_1^k a_n(x_1 + x_2) \leq 2\Sigma_1^k (a_n(x_1) + a_n(x_2))$, we have:

$$\phi(\{a_n(x_1 + x_2)\}) \leq 2\phi(\{a_n(x_1) + a_n(x_2)\}) \leq 2(\phi(a_n(x_1)) + \phi(a_n(x_2))).$$

Now, since $a_n(x_1 x_2 x_3) \leq \| x_1 \| a_n(x_2) \| x_3 \|$ (from the proof of proposition 1.1., (2)), we obtain:

$$\phi(\{a_n(x_1 x_2 x_3)\}) \leq \| x_1 \| \phi(\{a_n(x_2)\}) \| x_3 \|$$

and hence, if $x_2 \in A_\phi(X)$, $x_1, x_3 \in X$, we have $x_1 x_2 x_3 \in A_\phi(X)$. So, $A_\phi(X)$ is an ideal in X.

Remarks. 1). We remark that $\| x \|_\phi = \phi(\{a_n(x)\})$ is a quasinorm on $A_\phi(X)$ and this is a quasinormed vector space. The properties $\| x \|_\phi > 0$, $x \neq 0$ and

[1] For $x, y \in \ell_\infty, xy = (x_1 y_1, x_2 y_2, \ldots)$

$\| x+y \|_\phi \leq 2\,(\| x \|_\phi + \| y \|_\phi)$ are obvious (see above). If α is a scalar ($\alpha \neq 0$), from the above proof, we obtain:

$$\| \alpha x \|_\phi = \| \alpha e x \|_\phi \leq \| \alpha e \| \cdot \| x \|_\phi = |\alpha| \cdot \| x_\phi \|_\phi .$$

On the other hand

$$\| x \|_\phi = \| \frac{1}{\alpha} \cdot \alpha x \|_\phi \leq \frac{1}{|\alpha|} \| \alpha x \|_\phi ,$$

i.e., $|\alpha| \, \| x \|_\phi \leq \| \alpha x \|_\phi$. This means $\| \alpha x \|_\phi = |\alpha| \, \| x \|_\phi$ for all scalars α, since for $\alpha = 0$ the equality is also true.

2). $A_\phi(X)$ is quasinormed subalgebra of X, because

$$\| xy \|_\phi \leq \| x \|_\phi \cdot \| y \|_\phi$$

(from the definition of $\| x \|_\phi$ it results that $\| x \| \leq \| x \|_\phi$.)

Proposition 2.2. *If $x_1 \in A_\phi(X)$ and $x_2 \in A_{\phi^*_\psi}(X)$ then $x_1 x_2 \in A_\psi(X)$, with $\| x_1 x_2 \|_\psi \leq 2 \| x_1 \|_\phi \cdot \| x_2 \|_{\phi^*_\psi}$, where ϕ, ψ are symmetric norming functions and ϕ^*_ψ is the dual function of ϕ relative to ψ.* [2]

Proof. From the proposition 1.1. we obtain:

$$\| x_1 x_2 \|_\psi = \psi\left(\{a_n(x_1 x_2)\}\right) \leq 2\psi\left(\{a_n(x_1) a_n(x_2)\}\right) \leq$$
$$\leq 2\phi\left(\{a_n(x_1)\}\right) \cdot \phi^*_\psi\left(\{a_n(x_2)\}\right) = 2 \| x_1 \|_\phi \cdot \| x_2 \|_{\phi^*_\psi} .$$

For the special case when $X = L(E)$, the results are presented in [8], [11], and for the particular case when E is a Hilbert space one can see [2], [5], [7]. For some particular cases of functions ϕ we can obtain the special classes of Lorentz or Lorentz - Zygmund ideals [9], [11] and more particular if ϕ is $\phi_p : (a_n(x)) \to (\Sigma a_n^p(x))^{\frac{1}{p}}$, $1 \leq p < \infty$, we obtain the ideals $A_p(X)$, respectively $L_p(X)$ for operators. In this last case, the conclusion of the proposition 2.2 is

$$\| x_1 x_2 \|_p = \left(\Sigma a_n^p \left(x_1 x_2\right)\right)^{\frac{1}{p}} \leq 2 \| x_1 \|_{p_1} \cdot \| x_2 \|_{p_2} ,$$

where $\frac{1}{p} = \frac{1}{p_1} + \frac{1}{p_2}$, because $(\phi_{p_1})^*_{\phi_p} = \phi_{p_2}$, [11].

3. Bilinear operators on $A_\phi(X)$

In this section we suppose that $B : \ X \times X \to X$ is a bilinear and bounded operator such that $\| B(x_1, x_2) \|^* \leq n^2$ if $\| x_i \|^* \leq n, \ n = 1, 2, \ldots$ It is of great interest to know if the induced operator

$$B : A_\phi(X) \times A_\phi(X) \to A_\phi(X)$$

is also bounded. The answer will be affirmative only for some functions ϕ, namely for the functions $\overline{\phi} : \ (a_n(x)) \to \phi\Big(\{\frac{a_n(x)}{n}\}\Big)$, where ϕ is a symmetric norming function. To this end we prove the following Lemma.

---

[2] $\phi^*_\psi(x) = \sup_{(y_i) \in K} \frac{\psi(\{x_i y_i\})}{\phi(\{y_i\})}$, [5] [10], [11].

Lemma 3.1. *The sequence* $(a_n(B(x,y)))$ *satisfies inequality*

$$\Sigma_{n=1}^{k}\frac{a_n(B(x_1,x_2))}{n}\leq 6M\Sigma_{n=1}^{k}\frac{\| x_2 \| a_n(x_1)+\| x_1 \| a_n(x_2)}{n},\ k=1,2,\ldots,$$

where M *is constant.*

Proof. Let $M>0$ be a constant such that $\| B(x_1,x_2) \|\leq M \| x_1 \| \cdot \| x_2 \|$ and let $\overline{x}_i$ be such that $\| \overline{x}_i \|^* < n, i=1,2,$ and $\| x_i-\overline{x}_i \|\leq a_n(x_i)+\epsilon$. Since $\| B(\overline{x}_1,\overline{x}_2) \|^* < n^2$, we obtain:

$$\begin{aligned} a_{n^2}(B(x_1,x_2)) &\leq \| B(x_1,x_2)-B(\overline{x}_1,\overline{x}_2) \| = \| B(x_1-\overline{x}_1,x_2)+B(\overline{x}_1,x_2-\overline{x_2}) \| \\ &\leq M(\| x_1-\overline{x}_1 \| \cdot \| x_2 \| + \| \overline{x}_1 \| \cdot \| x_2-\overline{x}_2 \|) \leq \\ &\leq M[(a_n(x_1)+\epsilon) \| x_2 \| + (\| \overline{x}_1-x_1 \| + \| x_1 \|)(a_n(x_2)+\epsilon)]. \end{aligned}$$

Since ϵ is arbitrary it results

$$a_{n^2}(B(x_1,x_2))\leq 2M[\| x_2 \| a_n(x_1)+\| x_1 \| a_n(x_2)].$$

Then we can write

$$\begin{aligned} \sum_{n=1}^{k}\frac{a_n(B(x_1,x_2))}{n} &\leq \sum_{n=1}^{(k+1)^2-1}\frac{a_n(B(x_1,x_2))}{n} \\ &= \sum_{n=1}^{k}\sum_{i=n^2}^{(n+1)^2-1}\frac{a_i(B(x_1,x_2))}{i}\leq 3\sum_{n=1}^{k}\frac{a_{n^2}(B(x_1,x_2))}{n} \\ &\leq 6M\sum_{n=1}^{k}\frac{\| x_2 \| a_n(x_1)+\| x_1 \| a_n(x_2)}{n}. \end{aligned}$$

The proof is complete.

Theorem 3.2. *If* B *is an operator as above, then the induced operator* $B: A_{\overline{\phi}}\times A_{\overline{\phi}}\to A_{\overline{\phi}}$ *is bounded.*

Proof.

$$\begin{aligned} \| B(x_1,x_2) \|_{\overline{\phi}} &= \overline{\phi}(\{a_n(B(x_1,x_2))\})=\phi\left(\left\{\frac{a_n(B(x_1,x_2))}{n}\right\}\right)\leq \\ &\leq 6M\phi\left(\left\{\frac{\| x_2 \| a_n(x_1)+\| x_1 \| a_n(x_2)}{n}\right\}\right)\leq \\ & 6M\left(\| x_2 \| \cdot \| x_1 \|_{\overline{\phi}}+\| x_1 \| \cdot \| x_2 \|_{\overline{\phi}}\right). \end{aligned}$$

Since $\| x_i \|\leq\| x_i \|_{\overline{\phi}}$, we obtain:

$$\| B(x_1,x_2) \|_{\overline{\phi}} \leq 12M \| x_1 \|_{\overline{\phi}} \| x_2 \|_{\overline{\phi}}$$

and hence B is bounded.

Remark. The results are also true when $B: X\times Y\to Z$, where X,Y,Z are Banach spaces.

Applications. Let $T_i : E \to E, i = 1, 2$, be linear and bounded operators where E is a normed space, then operator $T_1 \bigotimes_\alpha T_2 : E \bigotimes_\alpha E \to E \bigotimes_\alpha E$ is bounded if α is a tensor norm ($\| T_1 \bigotimes T_2 \|_\alpha \leq \| T_1 \| \cdot \| T_2 \|$) [3] From the theorem 3.2. we obtain:

Corollary 3.3. *If* $T_i \in L_{\overline{\phi}}(E)$, $i = 1, 2$, *then* $T_1 \bigotimes T_2 \in L_{\overline{\phi}}(E \bigotimes_\alpha E)$.

This result is well-known [9], [10], [11].

Acknowledgement. The author thanks the referee for his helpful comments and advice.

References

1. J. Diestel, J. Uhl jr., Vector Measures, *Math. Surveys, A.M.S.* **15** (1977).
2. I. Gohberg, M. Krein, *Introduction to the theory of non-selfadjoint operators*. A.M.S. Providence, 1969.
3. G. Köthe, *Topological Vector spaces II*. Springer Verlag, 1979.
4. J. Peetre, G. Sparr, Interpolation of normed abelian groups, *Ann. Mat. pura et Appl.* **92** (1972) 217 -262.
5. N. Salinas, Symmetric norm ideals and relative conjugate ideals, *Trans. A. M.S.* **188** (1974) 213-240.
6. R. Schatten, *A theory of cross-spaces*. Princeton Univ. 1950.
7. R. Schatten, *Normideals of completely continuous operators*. Springer Verlag 1960.
8. N. Ţiţa, On a class of $\ell_{\phi,\varphi}$ operators, *Collect. Math.* **32** (1981) 275 -279.
9. N. Ţiţa, Approximation spaces and biliniar operators, *Studia Univ. Babes-Bolyai, Math.* **35** (1990) 89-92.
10. N. Ţiţa, On the approximation numbers of the tensor product operator, *Ann. Univ. "Al. I. Cuza" Iaşi* **40** (1994) 329 - 331.
11. N. Tita, *Ideale de operatori generate de s numere*. Ed. Univ. "Transilvania", Brasov, 1998.
12. K. Yosida, *Functional analysis*. Springer Verlag, 1965.

3 All standard norms on the tensor product of two normed spaces are tensor norms, [1], [3], [10].

Functional Analysis and its Applications
V. Kadets and W. Żelazko

Inversion of the local Pompeiu transform

Natalia P. Volchkova

Department of Mathematical Analysis and Function Theory, Donetsk National University, 24, Universitetskaya str., Donetsk, 83055, Ukraine

Abstract
The construction of inversion of the local Pompeiu transform for some classes of cylinders is obtained.

Key words: Fourier series, convolution, Bessel functions, spherical functions
2000 MSC: 42C10, 44A35, 33C10, 43A90

1. Introduction

Let $\mathbb{R}^n$ be a real Euclidean space of dimension $n \geq 2$ with the Euclidean norm $|\cdot|$, let $M(n)$ be the group of Euclidean motions in $\mathbb{R}^n$, and let $\mathcal{F} = \{\mu_i\}_{i=1}^k$ be a finite family of distributions with compact supports in $\mathbb{R}^n$. For fixed $g \in M(n)$ we consider the distribution $g\mu_i$ acting on $C^\infty(\mathbb{R}^n)$ by the rule

$$\langle g\mu_i, f\rangle = \langle \mu_i, f \circ g\rangle, \quad f \in C^\infty(\mathbb{R}^n).$$

The (global) Pompeiu transform $\mathcal{P}_\mathcal{F}$ is the map

$$\mathcal{P}_\mathcal{F}: C^\infty(\mathbb{R}^n) \to C^\infty(M(n))^k$$

given by

$$\mathcal{P}_\mathcal{F}(f)(g) = (\langle g\mu_1, f\rangle, \ldots, \langle g\mu_k, f\rangle), \quad g \in M(n). \tag{1}$$

Similarly, for an open set $U \subset \mathbb{R}^n$ the local Pompeiu transform maps $C^\infty(U)$ into the Cartesian product $C^\infty(\Lambda(U,\mu_1)) \times \ldots \times C^\infty(\Lambda(U,\mu_k))$ by the formula (1), where $\Lambda(U,\mu_i) = \{g \in M(n): \mathrm{supp} g\mu_i \subset U\}$.

For given $\mathcal{F}$ and U the following problems arise (see [4]).

Problem 1. Find out whether $\mathcal{P}_{\mathcal{F}}$ is an injective map. If it is not, describe the kernel of $\mathcal{P}_{\mathcal{F}}$.

Problem 2. In the case $\mathcal{P}_{\mathcal{F}}$ is injective, find the converse map.

Many authors have studied the injectivity of the Pompeiu transform and related problems for special $\mathcal{F}$ and U (see the survey papers [4,13], which contain an extensive bibliography, and also [2,3,6,8–12]). The most interesting case is the case when $U = B_R(y) = \{x \in \mathbb{R}^n : |x - y| < R\}$, and $\mathcal{F} = \{\chi_E\}$ is the characteristic function (indicator) of a compact set $E \subset B_R(y)$ of positive measure. For this family $\mathcal{F}$ and the set $E \subset \overline{B_r(x_0)}$ having a hyperbolic point $x_1 \in E \cap \partial B_r(x_0)$ and the global Pompeiu property, the Pompeiu transform $\mathcal{P}_{\mathcal{F}}$ is injective with respect to U if $R > 2r$ (see [2,3], and also [6,9–11], where for some E the minimal value of R for which $\mathcal{P}_{\chi_E}$ is injective is found). In [3] for the sets E with the above properties the construction of the inversion of the transform $\mathcal{P}_{\chi_E}$ in the ball $B_R(y)$, $R > 3r$ is obtained. Besides, for the case where E is a square a function $f \in C^\infty(B_R(y))$ is recovered in [3] from its Pompeiu transform $\mathcal{P}_{\chi_E}(f)$ also for $R > 2r$. In connection with this, the inversion of the transform $\mathcal{P}_{\chi_E}$ in the ball $B_R(y)$ of radius $R > 2r$ for other compacts E is of great interest. In the present paper the solution of this problem for some class of cylinders is obtained.

2. Statement of the main result

Let $\mathbb{S}^{n-1} = \{x \in \mathbb{R}^n : |x| = 1\}$, let ρ and σ be the polar coordinates in $\mathbb{R}^n$ (for each $x \in \mathbb{R}^n$ we set $\rho = |x|$ and if $x \neq 0$, then we set $\sigma = \frac{x}{\rho} \in \mathbb{S}^{n-1}$). Let $\left\{Y_s^{(k)}(\sigma)\right\}$, $1 \leq s \leq d_k$, be a fixed orthonormal basis in the spase $\mathcal{H}_k$ of spherical harmonics of degree k, regarded as a subspace of $L^2\left(\mathbb{S}^{n-1}\right)$ (see [7, Chapter 4, section 2]). To every function $f \in L_{loc}(B_R)$ we assign its Fourier series

$$f(x) \sim \sum_{k=0}^{\infty} \sum_{s=1}^{d_k} f_{ks}(\rho) Y_s^{(k)}(\sigma), \quad 0 < \rho < R,$$

where

$$f_{ks}(\rho) = \int_{\mathbb{S}^{n-1}} f(\rho\sigma) \overline{Y_s^{(k)}(\sigma)} \, d\sigma.$$

To reconstruct f it is sufficient to find the coefficients f_{ks} of its Fourier series.

Further, as usual, $\mathcal{D}(\mathbb{R}^n)$ is the space of infinitely differentiable functions on $\mathbb{R}^n$, with compact supports, and $\mathcal{D}'(\mathbb{R}^n)$ is the space of distributions on $\mathbb{R}^n$. Let $\mu_1 * \mu_2$ be the convolution of two distributions on $\mathbb{R}^n$ one of which is compactly supported. We need a concept of circular symmetrization of a distribution defined as follows: for any $\mu \in \mathcal{D}'(\mathbb{R}^n)$, we define $\mathcal{R}\mu$ setting

$$\langle \mathcal{R}\mu, \varphi \rangle = \langle \mu(x), \int\limits_{\mathbf{SO}(n)} \varphi(kx)\,\mathrm{d}k \rangle, \quad \varphi \in \mathcal{D}(\mathbb{R}^n), \tag{2}$$

where $\mathbf{SO}(n)$ is the group of rotations of $\mathbb{R}^n$, and $\mathrm{d}k$ is the normalized Haar measure on $\mathbf{SO}(n)$.

Let $\alpha \in (0, 2\pi)$, $\beta_\alpha = \begin{cases} \alpha/2 \ , \ \alpha \in (0,\pi], \\ \pi \ , \ \alpha \in (\pi, 2\pi), \end{cases}$ $Sg = \{z \in \mathbb{C}: |z| \le 1, |\arg z| \le \beta_\alpha, \operatorname{Re} z \ge \cos\frac{\alpha}{2}\}$ be the circular segment of angular measure α, $H = (Sg - h_\alpha) \times [-b_3, b_3] \times \ldots \times [-b_n, b_n]$, where $h_\alpha = \begin{cases} \cos\dfrac{\alpha}{2} \ , \ \alpha \in (0,\pi], \\ 0 \ , \ \alpha \in (\pi, 2\pi), \end{cases}$ $b_k > 0$, $k = 3, \ldots, n$.

Further we need the following differential operators: $\Delta = \frac{\partial^2}{\partial x_1^2} + \cdots + \frac{\partial^2}{\partial x_n^2}$; $D^{\varkappa} = \frac{\partial^{|\varkappa|}}{\partial x_1{}^{\varkappa_1} \ldots \partial x_n{}^{\varkappa_n}}$ $(\varkappa = (\varkappa_1, \ldots, \varkappa_n) \in \mathbb{Z}_+^n, |\varkappa| = \varkappa_1 + \cdots + \varkappa_n)$, $D_1 = \frac{\partial}{\partial x_1} + \mathrm{i}\frac{\partial}{\partial x_2}$, $D_2 = \frac{\partial^{n-2}}{\partial x_3 \ldots \partial x_n}$, $D_{ij}(a) = (x_i + a)\frac{\partial}{\partial x_j} - x_j\frac{\partial}{\partial x_i}$, $(a \in \mathbb{R}^1)$. We denote $\mathcal{R}_k = \mathcal{R}(D_2 D_1^k \mu)$, where $\mu = \frac{\partial^2}{\partial x_2^2} D_{12}(h_\alpha) D_2 \chi_H$.

Let r_0 be the radius of the least closed ball containing Sg, $r = \sqrt{r_0^2 + b_3^2 + \cdots + b_n^2}$, $R > 2r$, $f \in C^\infty(B_R)$, where $B_R = B_R(0)$. For $x \in B_{R-r}$ we set $\check{f}(x) = f(-x)$, $f_1(x) = (f * \mathcal{R}\chi_H)(x)$, $f_i(x) = \left(\check{f} * \nu_i\right)(x)$, $i = 2, 3$, where $\nu_2 = \mathcal{R}_1$,

$$\nu_3 = \begin{cases} \mathcal{R}_5 + \dfrac{4}{9n^2 r^2}(3n-4)(3n+2)\Delta\mathcal{R}_1 - \dfrac{4}{9n^2}\Delta^2\mathcal{R}_1 \ , \ 3n\sin^2\dfrac{\alpha}{2} = 2r^2, \\ \mathcal{R}_3 + \dfrac{2}{3n}\Delta\mathcal{R}_1 \ , \ 3n\sin^2\dfrac{\alpha}{2} \ne 2r^2, \end{cases}$$

for $\alpha \in (0, \pi]$, and $\nu_3 = \mathcal{R}_2$ for $\alpha \in (\pi, 2\pi)$.

The main result of this paper is the following

Theorem 1. Let $R > 2r$. Then for any $k \in \mathbb{Z}_+$, $1 \le s \le d_k$, $\rho \in (0, R)$ there exist distributions $\mathcal{U}_{l,i}$, $(l \in \mathbb{N}, i = 1, 2, 3, 4)$ with the following properties:
(1) supp$\mathcal{U}_{l,i} \subset B_{R-r}$ $(l \in \mathbb{N}, i = 1, 2, 3)$, supp$\mathcal{U}_{l,4} \subset B_R$ $(l \in \mathbb{N})$;
(2) for any $f \in C^\infty(B_R)$

$$\left(\Delta^n \check{f}\right)_{ks}(\rho) = \lim_{l\to\infty} \left(\langle \mathcal{U}_{l,1}, f_2\rangle + \langle \mathcal{U}_{l,2}, f_3\rangle\right), \tag{3}$$

$$\check{f}_{ks}(\rho) = \lim_{l\to\infty} \left(\langle \mathcal{U}_{l,3}, f_1\rangle + \langle \mathcal{U}_{l,4}, \Delta^n \check{f}\rangle\right). \tag{4}$$

Let us make some remarks. Lemmas 1 - 3 (see section 3 below) show how to reconstruct the functions f_i, $i = 1, 2, 3$ in terms of $\mathcal{P}_{\chi_H}(f)$. Therefore, applying the equalities (3), (4) we can compute the Fourier coefficients of $\check{f}$ in terms of the Pompeiu transform of f. Thus, in Theorem 1 the procedure for the local inversion of the Pompeiu transform $\mathcal{P}_{\chi_H}$ is obtained. This method enabled us to obtain similar results for other compact sets E. In particular, analogues of Theorem 1 can be proved for cylinders of the form $K \times [-b_3, b_3] \times \ldots \times [-b_n, b_n]$, where K belongs to some class of polygons or circular domains.

3. Auxiliary assertions

Let $1 \leq i < j \leq n$, $h, \theta \in \mathbb{R}^1$, let $\tau_{i,h}$ be a shift on h along x_i axis, and let $k_{i,j,\theta}$ be the rotation in the plane (x_i, x_j) by an angle θ.

Lemma 1. Let $E \subset \overline{B}_{r_1}$, $R > r_1$, $f \in C^\infty(B_R)$. Then

$$\mathcal{P}_{\chi_E}\left(\frac{\partial f}{\partial x_i}\right)(g) = \frac{\mathrm{d}}{\mathrm{d}h}\left(\mathcal{P}_{\chi_E}(f)(\tau_{i,h} \circ g)\right)\Big|_{h=0}, \tag{5}$$

$$\mathcal{P}_{\chi_E}\left(D_{ij}(0)f\right)(g) = \frac{\mathrm{d}}{\mathrm{d}\theta}\left(\mathcal{P}_{\chi_E}(f)(k_{i,j,\theta} \circ g)\right)\Big|_{\theta=0}, \tag{6}$$

where $gE \subset B_R$.

Proof. For small h we have

$$\mathcal{P}_{\chi_E}(f)(\tau_{i,h} \circ g) = \mathcal{P}_{\chi_E}(f \circ \tau_{i,h})(g). \tag{7}$$

Differentiating (7) with respect to h and setting $h = 0$ we obtain (5). Equality (6) is proved analogously. □

Lemma 2. Let $R > r_1$, $E \subset \overline{B}_{r_1}$, $\nu = D^{\varkappa} D_{ij}(a)\chi_E$. Then for any $f \in C^\infty(B_R)$ and $x \in B_{R-r_1}$ we have

$$\left(\check{f} * \mathcal{R}\nu\right)(x) = (-1)^{|\varkappa|+1} \int\limits_{\mathbf{SO}(n)} \left(\mathcal{P}_{\chi_E}\left(D_{ij}(a)D^{\varkappa}\right)(y)\left(f(ky - x)\right)\right)(\mathbf{e})\,\mathrm{d}k, \tag{8}$$

where $\mathbf{e}$ is the unit element of $\mathbf{SO}(n)$.

Proof. Using (2), we have

$$\left(\check{f} * \mathcal{R}\nu\right)(x) = \langle \nu(y), \int\limits_{\mathbf{SO}(n)} f(ky - x)\,\mathrm{d}k \rangle. \tag{9}$$

By the relation (9) and the definition of $D_{ij}(a)$, we obtain

$$\left(\check{f} * \mathcal{R}\nu\right)(x) = (-1)^{|\varkappa|+1} \int\limits_{E} \int\limits_{\mathbf{SO}(n)} \left(D_{ij}(a)D^{\varkappa}\right)(y)\left(f(ky - x)\right)\mathrm{d}k\,\mathrm{d}y.$$

Hence, from (1) we obtain the required assertion. □

Remark. Let $x \in B_{R-r_1}$, $k \in \mathbf{SO}(n)$ be fixed, and let g_1 be the element of $\mathbf{M}(n)$ acting by the formula $g_1 y = ky - x$. Then $\mathcal{P}_{\chi_E}(f(ky-x))(g) = \mathcal{P}_{\chi_E}(f)(g_1 g)$, where $gE \subset B_{R-|x|}$. Due to this relation and Lemma 1 the values $\check{f} * \mathcal{R}\nu$ can be recovered from Pompeiu transform $\mathcal{P}_{\chi_E}(f)$ via formula (8).

Let δ be the Dirac delta measure at the origin of $\mathbb{R}^n$.

Lemma 3 ([3]). Let $R > 2r_1$, $E \subset \overline{B}_{r_1}$, $\mu(\varkappa) = \mathcal{R}(D^{\varkappa}\chi_E)$. Then for any $f \in C^{\infty}(B_R)$, and every $x \in B_{R-r_1}$ we have

$$(f * \mu(\varkappa))(x) = \int\limits_{\mathbf{SO}(n)} \langle D^{\varkappa}\delta(y), (\mathcal{P}_{\chi_E} f)\left(\begin{Vmatrix} -k^{-1} & x - ky \\ 0 & 1 \end{Vmatrix}\right)\rangle \, \mathrm{d}k,$$

where $\mathbf{M}(n)$ is considered as the group of $(n+1)\times(n+1)$ matrices of the form

$$\begin{Vmatrix} k & x \\ 0 & 1 \end{Vmatrix}, \quad k \in \mathbf{SO}(n), \quad x \in \mathbb{R}^n,$$

and $\mathbb{R}^n$ is identified with the affine subspace $\{x_{n+1} = 1\}$ of $\mathbb{R}^{n+1}$.

For each $m \in \{1, \dots, n\}$ let η_m be a map $\mathbb{R}^n \to \mathbb{R}^n$ acting as follows: if $x = (x_1, \dots, x_n) \in \mathbb{R}^n$, then $\eta_m x = ((\eta_m x)_1, \dots, (\eta_m x)_n)$, where $(\eta_m x)_k = x_k$ for $k \neq m$ and $(\eta_m x)_m = -x_m$. Let G_+^n (respectively, G_-^n) be the set of maps $\mathbb{R}^n \to \mathbb{R}^n$ representable as a composition of an even (respectively, odd) number of maps η_m, $1 \le m \le n$.

Lemma 4. For any $f \in C^{n+1}(H)$ we have

$$\int\limits_H D_2 D_{12}(h_\alpha) \frac{\partial^2}{\partial x_2^2} f(x_1, \dots, x_n) \, \mathrm{d}x_1 \dots \mathrm{d}x_n =$$

$$= \left(\sum_{\eta \in G_+^{n-2}} - \sum_{\eta \in G_-^{n-2}}\right) \Big[f(z_1, \eta b) - f(z_2, \eta b) + \sin\frac{\alpha}{2}\frac{\partial f}{\partial x_2}(z_1, \eta b) +$$

$$+ \sin\frac{\alpha}{2}\frac{\partial f}{\partial x_2}(z_2, \eta b)\Big],$$

where

$$z_1 = \begin{cases} -\mathrm{i}\sin\dfrac{\alpha}{2}, & \alpha \in (0, \pi], \\ \mathrm{e}^{-\mathrm{i}\frac{\alpha}{2}}, & \alpha \in (\pi, 2\pi), \end{cases} \quad z_2 = \overline{z}_1, \quad b = (b_3, \dots, b_n).$$

Proof. Let $S_\alpha = Sg - h_\alpha$, $u \in C^3(S_\alpha)$. Then

$$\int_{S_\alpha} \left(D_{12}(h_\alpha) \frac{\partial^2}{\partial x_2^2} u \right) (x_1, x_2)\, \mathrm{d}x_1\, \mathrm{d}x_2 \tag{10}$$

$$= \int_{\cos \frac{\alpha}{2} - h_\alpha}^{1-h_\alpha} \left[(\mathcal{A}u) \left(x_1, \sqrt{1-(x_1+h_\alpha)^2} \right) \right.$$

$$\left. - (\mathcal{A}u) \left(x_1, -\sqrt{1-(x_1+h_\alpha)^2} \right) \right] \mathrm{d}x_1,$$

where $\mathcal{A} = \frac{\partial}{\partial x_1} - x_2 \frac{\partial^2}{\partial x_1 \partial x_2} + (x_1 + h_\alpha) \frac{\partial^2}{\partial x_2^2}$. Bearing in mind that

$$(\mathcal{A}u) \left(x_1, \sqrt{1-(x_1+h_\alpha)^2} \right)$$

$$= \frac{\partial}{\partial x_1} \left[u \left(x_1, \sqrt{1-(x_1+h_\alpha)^2} \right) \right.$$

$$\left. - \sqrt{1-(x_1+h_\alpha)^2} \frac{\partial u}{\partial x_2} \left(x_1, \sqrt{1-(x_1+h_\alpha)^2} \right) \right],$$

$$(\mathcal{A}u) \left(x_1, -\sqrt{1-(x_1+h_\alpha)^2} \right)$$

$$= \frac{\partial}{\partial x_1} \left[u \left(x_1, -\sqrt{1-(x_1+h_\alpha)^2} \right) \right.$$

$$\left. + \sqrt{1-(x_1+h_\alpha)^2} \frac{\partial u}{\partial x_2} \left(x_1, -\sqrt{1-(x_1+h_\alpha)^2} \right) \right]$$

from (10) we find

$$\int_{S_\alpha} \left(D_{12}(h_\alpha) \frac{\partial^2 u}{\partial x_2^2} \right) (x_1, x_2)\, \mathrm{d}x_1\, \mathrm{d}x_2 \tag{11}$$

$$= u(z_1) - u(z_2) + \sin \frac{\alpha}{2} \frac{\partial u}{\partial x_2}(z_1) + \sin \frac{\alpha}{2} \frac{\partial u}{\partial x_2}(z_2).$$

Since for every $v \in C^{n-1}\left([-b_3, b_3] \times \ldots \times [-b_n, b_n] \right)$

$$\int_{-b_3}^{b_3} \cdots \int_{-b_n}^{b_n} (D_2 v)(x_3, \ldots, x_n)\, \mathrm{d}x_3 \ldots \mathrm{d}x_n = \sum_{\eta \in G_+^{n-2}} v(\eta b) - \sum_{\eta \in G_-^{n-2}} v(\eta b),$$

from (11) we obtain the assertion of Lemma 4. □

Let J_q be the Bessel function of the first kind of order $q \geq 0$, and let $j_q(z) = J_q(z)/\, z^q$. The spherical transform of the radial distribution μ with the compact support in $\mathbb{R}^n$ is defined by

$$\widetilde{\mu}(\lambda) = \langle \mu(x), j_{\frac{n-2}{2}}(\lambda |x|) \rangle, \quad \lambda \in \mathbb{C}. \tag{12}$$

Lemma 5. Let $k \in \mathbb{Z}_+$. Then

$$\widetilde{\mathcal{R}}_k(\lambda) = (-1)^{n-1}2^{n-2}b_3 \ldots b_n \lambda^{2(k+n-2)} \left\{ C_{1k} j_{\frac{3n+2k-6}{2}}(\lambda r) + \lambda^2 C_{2k} j_{\frac{3n+2k-4}{2}}(\lambda r) \right\},$$

where $C_{1k} = z_1^k - z_2^k + \mathrm{i}k \sin\frac{\alpha}{2}\left(z_1^{k-1} + z_2^{k-1}\right), C_{2k} = \sin^2\frac{\alpha}{2}\left(z_1^k - z_2^k\right)$.

Proof. Since $j_q'(t) = -tj_{q+1}(t)$ (see [1, Chapter 7, § 7.2, formula (51)]), we have

$$\frac{\partial}{\partial x_1}\left((x_1 + \mathrm{i}x_2)^k j_q(\lambda|x|)\right) \tag{13}$$
$$= k(x_1 + \mathrm{i}x_2)^{k-1} j_q(\lambda|x|) - \lambda^2 x_1 (x_1 + \mathrm{i}x_2)^k j_{q+1}(\lambda|x|),$$

$$\frac{\partial}{\partial x_2}\left((x_1 + \mathrm{i}x_2)^k j_q(\lambda|x|)\right) \tag{14}$$
$$= \mathrm{i}k(x_1 + \mathrm{i}x_2)^{k-1} j_q(\lambda|x|) - \lambda^2 x_2 (x_1 + \mathrm{i}x_2)^k j_{q+1}(\lambda|x|).$$

Using induction on k from (13), (14) we find

$$(-1)^k D_1^k\left(j_q(\lambda|x|)\right) = \lambda^{2k}(x_1 + \mathrm{i}x_2)^k j_{q+k}(\lambda|x|).$$

Hence (see (12))

$$\widetilde{\mathcal{R}}_k(\lambda) = \lambda^{2(k+n-2)} \langle \mu, x_3 \ldots x_n (x_1 + \mathrm{i}x_2)^k j_{\frac{3n+2k-6}{2}}(\lambda|x|) \rangle,$$

where $\mu = \frac{\partial^2}{\partial x_2^2} D_{12}(h_\alpha) D_2 \chi_H$. Using Lemma 4, we obtain the required assertion. □

By the Paley – Wiener theorem [5, Theorem 7.3.1] there exist radial distributions μ_1 and μ_2 with supports in B_r, for which $\widetilde{\mu}_1(\lambda) = (-1)^n \widetilde{\nu}_2(\lambda)/\lambda^{2n}$, $\widetilde{\mu}_2(\lambda) = (-1)^n \widetilde{\nu}_3(\lambda)/\lambda^{2n}$. ¿From Lemma 5 and properties of Bessel functions (see [1, Chapter 7, § 7.9]) it follows that $\widetilde{\mu}_1(\lambda)$ and $\widetilde{\mu}_2(\lambda)$ have no common zeros. Further, we shall need a lower estimate for the function $\widetilde{\mu}_1(\lambda)\widetilde{\mu}_2(\lambda) j_{\frac{n}{2}+k-1}(\varepsilon\lambda)$, where $\varepsilon > 0$.

Lemma 6. Let $a_1, a_2, a_3 > 0$, $k \in \mathbb{Z}_+$,

$$\theta(\lambda) = j_{\frac{3n-2}{2}}(a_1\lambda) j_{\frac{3n}{2}}(a_2\lambda) j_{\frac{n}{2}+k-1}(a_3\lambda).$$

Then there are constants L_{1k}, L_{2k}, such that for any integer $l \geq L_{1k}$ there exists $\rho_l \in (l, l+1)$ such that if either $|z| = \rho_l$ or $|\operatorname{Im}\lambda| \geq 1$ and $|\lambda| \geq L_{1k}$, then

$$|\theta(\lambda)| \geq \frac{L_{2k}}{|\lambda|^{\frac{7n+2k-1}{2}}} \mathrm{e}^{(a_1+a_2+a_3)|\operatorname{Im}\lambda|}.$$

Proof. The function $\theta(\lambda)$ is an even entire function, therefore to prove the assertion of Lemma 6 we can assume that $\operatorname{Re}\lambda \geq 0$. From the asymptotic development of the Bessel function (see [1, Chapter 7, § 7.13, formula (3)]) we find

$$\theta(\lambda) = \frac{2}{\pi}\sqrt{\frac{2}{\pi}} \frac{(a_1)^{-\frac{3n-1}{2}}(a_2)^{-\frac{3n+1}{2}}(a_3)^{-\frac{n+2k-1}{2}}}{\lambda^{\frac{7n+2k-1}{2}}} \cos\left(a_1\lambda - (3n-1)\frac{\pi}{4}\right) \times$$

$$\times \cos\left(a_2\lambda - (3n+1)\frac{\pi}{4}\right)\cos\left(a_3\lambda - (n+2k-1)\frac{\pi}{4}\right) + \mathrm{O}\left(\frac{\mathrm{e}^{(a_1+a_2+a_3)|\operatorname{Im}\lambda|}}{|\lambda|^{\frac{7n+2k+1}{2}}}\right).$$

By the Łojasiewicz inequality (see [3]) we have

$$|\cos z| \geq \frac{1}{\pi\mathrm{e}} d(z,V)\mathrm{e}^{|\operatorname{Im} z|}, \tag{15}$$

where $V = \{(2l+1)\pi/2, l \in \mathbb{Z}\}$, $d(z,V) = \min(1, dist(z,V))$. Using (15) and repeating arguments from the proof of Lemma 7 in [3], we obtain the assertion of Lemma 6. □

In what follows let us assume that $R > 2r$. Choose a strictly increasing sequence of positive numbers $\{\varepsilon_m\}_{m=1}^{\infty}$ with limit $\frac{R}{2r} - 1$, and a corresponding increasing sequence of radii $R_m = 2r(1+\varepsilon_m)$, $m \geq 1$, $R_0 = 0$.

Lemma 7. Let $R > 2r$.Then for any $k \in \mathbb{Z}_+$, $m \in \mathbb{N}$, $t \in [R_{m-1}, R_m)$ there are two sequences of radial distributions satisfying the following conditions:

(1) $\operatorname{supp}\mu_{l,i} \subset B_{R_m - r}$, $i = 1,2$, $l \in \mathbb{N}$,

(2) there exist constants $L = L(k,R,r,\varepsilon_1,n), C = C(R,r,\varepsilon_1,n) > 0$ such that for $l \geq L$ the equality

$$\left|j_{\frac{n}{2}+k-1}(t\lambda) - (\widetilde{\mu}_1(\lambda)\widetilde{\mu}_{l,1}(\lambda) + \widetilde{\mu}_2(\lambda)\widetilde{\mu}_{l,2}(\lambda))\right| \leq$$

$$\frac{C(R,r,\varepsilon_1,n)}{l}\frac{\|\lambda\|^{-\frac{n}{2}-k+\frac{13}{2}}}{t^{\frac{n}{2}+k-1}}\mathrm{e}^{R_m|\operatorname{Im}\lambda|}, \quad \|\lambda\| = \max(1,|\lambda|),$$

holds.

To prove Lemma 7 it is sufficient to use Lemma 6 and repeat the arguments from the proof of Proposition 8 in [3].

4. The proof of main result

By Lemma 7 it follows (see [3, proof of Theorem 9]) that for any $m \in \mathbb{N}$, $\rho \in [R_{m-1}, R_m)$ there are distributions $\mathcal{U}_{l,i}$ $(l \in \mathbb{N}, i = 1,2)$ with supports in R_{R-r} such that for $l \geq L$ and any $f \in C^\infty(B_R)$ the following estimate is valid

$$|f_{ks}(\rho) - \langle \mathcal{U}_{l,1}, f * \mu_1\rangle - \langle \mathcal{U}_{l,2}, f * \mu_2\rangle| \tag{16}$$

$$\leq \frac{c_3}{l}\frac{\rho^{-\frac{n}{2}+1}}{(R-R_m)^M}\sup_{\substack{x\in B_{R'_m}\\ |\varkappa|\leq M}}\left|\frac{\partial^{|\varkappa|}}{\partial x^{\varkappa}}f(x)\right|,$$

where $R'_m = \frac{2}{3}R + \frac{1}{3}R_m$, $M = \left[\frac{n+13}{2}\right] + 1$, and the constant c_3 does not depend on R, r, ε_1, n. Substituting in (16) f for $\Delta^n \check{f}$ and having in mind that $\Delta^n \mu_i = \nu_{i+1}$,

$i = 1, 2$, we obtain the equality (3). Let now $T_1 = \mathcal{R}\chi_H$, $T_2 = \Delta^n \delta$. Then $\widetilde{T_1}(0) \neq 0$, $\widetilde{T_2}(\lambda) = (-1)^n \lambda^{2n}$. Thus, $\widetilde{T_1}$ and $\widetilde{T_2}$ have no common zeros. Moreover, $\widetilde{T_1}$ has an asymptotic behavior of the same type as that of the Bessel function (see [2,3]). Hence we can apply the same procedure as above to the two radial distributions T_1 and T_2. By the same reasons there exist distributions $\mathcal{U}_{l,i}$ ($l \in \mathbb{N}, i = 3, 4$) such that (4) holds. Theorem is proved.

References

1. H. Bateman and A. Erdélyi, *Higher transcendental functions.* **2** (Nauka, Moscow, 1974).
2. C.A. Berenstein and R. Gay, Le probléme de Pompeiu locale, *J. Anal. Math.* **52** (1989) 133–166.
3. C.A. Berenstein, R. Gay, A. Yger, Inversion of the local Pompeiu transform, *J. Anal. Math.* **54** (1990) 259–287.
4. C.A. Berenstein and D. Struppa, Complex analysis and convolution equations, Current Problems in Mathematics, Fundamental Directions, **54** (VINITI, Moscow, 1989) 5–111. (In Russian).
5. L. Hörmander, *Linear partial differential operators.* Springer–Verlag, Berlin 1963.
6. P.A. Masharov, A new Morera–type theorem on the disk, *Visnik Kharkiv Univ. Ser. Math. Prikl. Math. Mekh.* **475** (49) (2000) 126–132.
7. I. Stein and G. Weiss, *Introduction to Fourier analysis on Euclidean spaces.* Princeton University Press, Princeton, NJ, 1971.
8. V.V. Volchkov, A definitive version of the local two-radii theorem, *Mat. Sb.* **186** (6) (1995) 15–34.
9. V.V. Volchkov, An extremum problem related to Morera's theorem, *Mat. Zametki.* **60** (6) (1996) 804–809.
10. V.V. Volchkov, Extremal cases of the Pompeiu problem, *Mat. Zametki.* **59** (5) (1996) 671–680.
11. V.V. Volchkov, Extremal problems on Pompeiu sets II, *Mat. Sb.* **191** (5) (2000) 4–16.
12. V.V. Volchkov, Injectivity sets of the Pompeiu transform, *Mat. Sb.* **190** (11) (1999) 51–66.
13. L. Zalcman, A bibliographic survey of the Pompeiu problem, in: B.Fuglede et al. (eds), *Approximation by Solutions of Partial Differential Equations* (Kluwer Academic Publishers, Dordrecht, 1992) 185–194.

Functional Analysis and its Applications
V. Kadets and W. Żelazko

Space of Probability Measures and Absolute Extensors in the Asymptotic Category

Mykhajlo M. Zarichnyi

Department of Mechanics and Mathematics, Lviv National University, Universytetska 1, 79000 Lviv, Ukraine

Abstract

The main result asserts that there exists a proper metric space of asymptotic dimension zero such that its space of probability measures with respect to the Kantorovich metric is not an absolute extensor for the asymptotic category. This provides a negative answer to a question formulated by Dranishnikov.

Key words: Asymptotic dimension, probability measure, extensors, metric spaces
2000 MSC: 46E27, 46E30, 54C55, 54E35

1. Introduction

It is well-known that in the category of compact Hausdorff spaces the inclusion $P(\mathrm{AE}(0)) \subset \mathrm{AE}$ is valid, where P stands for the probability measure functor and AE (respectively AE(0)) denotes the class of absolute extensors (in dimension 0) for the class of paracompact spaces; see, e.g. the survey paper [9].

Every compact metrizable space is an AE(0), therefore, the space $P(X)$ is an absolute extensor for arbitrary compact metric space X. The question whether a counterpart of this fact is true in for the category of proper metric spaces and

⋆ The paper was finished during the author's visit to the University of Saskatchewan. He thanks the Department of Mathematics and Statistics for hospitality.
Email address: topos@franko.lviv.ua, mzar@litech.lviv.ua (Mykhajlo M. Zarichnyi).

the asymptotically Lipschitz maps (see the definitions below) was formulated by Dranishnikov [2, Problem 12] in connection with existence of the homotopy extension theorem in this category. An example providing the negative answer to this question was found in [10]. This example is of infinite asymptotic dimension in the sense of Gromov [3] and the aim of this note is to demonstrate that there is also a zero-dimensional counterexample. Moreover, we show that the space of probability measures of the mentioned space is not an absolute extensor for the class of proper metric spaces of the asymptotic dimension 0. Note also that the existence of an asymptotically zero-dimensional space whose space of probability measures is not an absolute extensor for the asymptotic category contradicts to the remark following Problem 12 in [2] .

This example also works for another constructions, in particular, for the spaces of measurable maps from a unit segment into a proper metric space and for its subspace consisting of step functions (the Hartman-Mycielski construction). That these spaces are absolute extensors in the category of metric spaces and continuous maps is a background for some theorems on extension of partial (pseudo)metrics, see [1], [11], [7]. Therefore, the corresponding problems on extension of partial (pseudo)metrics in the asymptotic category require alternative approach.

As in [10], the construction of the example is based on results of U. Lang [6] on extendability of asymptotically Lipschitz maps into Euclidean spaces.

2. Preliminaries

2.1. *Asymptotic category*

Together with Roe's category of proper metric spaces and coarse maps [8], the asymptotic category $\mathcal{A}$ introduced by A. Dranishnikov [2] turned out to be an important universe for developing asymptotic topology.

A typical metric will be denoted by d. A map $f\colon X \to Y$ between metric spaces is called (λ, ε)-*Lipschitz* for $\lambda > 0$, $\varepsilon \geq 0$ if $d(f(x), f(x')) \leq \lambda d(x, x') + \varepsilon$ for every $x, x' \in X$. A map is called *asymptotically Lipschitz* if it is (λ, ε)-Lipschitz for some $\lambda, \varepsilon > 0$.

The $(1, 0)$-Lipschitz maps are also called *Lipschitz* or *short*.

A metric space is *proper* if every closed ball in it is compact. A map of metric spaces is (metrically) *proper* if the preimages of the bounded sets are bounded. The objects of the category $\mathcal{A}$ are the proper metric spaces and the morphisms are the proper asymptotically Lipschitz maps.

A metric space Y (not necessarily an object of $\mathcal{A}$) is an *absolute extensor* (AE) for the category $\mathcal{A}$ if for every proper asymptotically Lipschitz map $f\colon A \to Y$ defined on a closed subset of a proper metric space X there exists a proper asymptotically Lipschitz extension $\bar{f}\colon X \to Y$ of f.

2.2. *Asymptotic dimension*

The notion of asymptotic dimension is introduced by M. Gromov [3].

Let X be a metric space. A family $\mathcal{C}$ of subsets of X is said to be *uniformly bounded* if there exists $M > 0$ such that $\operatorname{diam} A \leq M$ for every $A \in \mathcal{C}$. Given $D > 0$, we say that a family $\mathcal{C}$ of subsets of X is *D-disjoint* if $\inf\{d(a, a') \mid a \in A,\ a' \in A'\} > D$ for every $A, A' \in \mathcal{C}$, $A \neq A'$.

We say that the *asymptotic dimension* of X is $\leq n$ (written $\operatorname{asdim} X \leq n$) if for every $D > 0$ there exists a cover $\mathcal{U}$ of X such that $\mathcal{U} = \mathcal{U}^0 \cup \cdots \cup \mathcal{U}^n$, where every family $\mathcal{U}^i$ is uniformly bounded and D-disjoint.

If, in the definition of the absolute extensor we require that $\operatorname{asdim} X \leq n$, then the definition of the *absolute extensor in asymptotic dimension* n (briefly $\mathrm{AE}(n)$) is obtained.

It is easy to see that, for a proper metric space X, the inequality $\operatorname{asdim} X \leq 0$ is equivalent to the condition that for every $C > 0$ the diameters of the C-chains in X (i.e. the sequences $x_1, x_2, \ldots, x_k$ with $d(x_i, x_{i+1}) \leq C$ for every $i = 1, 2, \ldots, k-1$) are bounded from above.

2.3. *Spaces of probability measures*

For a metric space X let $P(X)$ denote the space of probability measures with compact supports on X. We naturally identify the probability measures with the corresponding functionals on the set $C(X)$ of continuous real-valued functions on X. Every point $x \in X$ is identified with the Dirac measure δ_x concentrated in X. The Kantorovich metric on $P(X)$ is defined by the formula:

$$d(\mu, \nu) = \sup\{|\mu(\varphi) - \nu(\varphi)| \mid \varphi \in C(X) \text{ is short}\},\ \ \mu, \nu \in P(X)$$

(see, e.g., [5]).

It is noted in [10] that the spaces $P(X)$ are not, in general, proper. Besides, it is remarked there that for every short map $f\colon X \to Y$ of proper metric spaces the naturally defined map $P(f)\colon P(X) \to P(Y)$ is short and that the barycenter map

$$b\colon P(\mathbb{R}^n) \to \mathbb{R}^n,\ b(\mu) = \int_{\mathbb{R}^n} x\, d\mu,$$

is short.

3. Example

The example below is a modification of the author's example of a proper metric space whose space of probability measures is not an AE in the asymptotic category [10]. The modification allows to achieve asymptotic zero-dimensionality of the example. For the sake of completeness we provide the details of the construction.

For every n, the Euclidean space $\mathbb{R}^n$ is naturally identified with the subspace $\{(x_i) \mid x_i = 0 \text{ for all } j > n\}$ of the space ℓ^2.

We endow the subspace $X' = \bigcup_{n\in\mathbb{N}}\{n^2\} \times \mathbb{R}^n \subset \mathbb{R} \times \ell^2$ with the metric

$$d((m,(x_i)),(n,(y_i))) = (|m-n|^2 + \|(x_i)-(y_i)\|^2)^{1/2}.$$

Obviously, X is a proper metric space. For every n we denote by $p_n\colon X' \to \mathbb{R}^n$ a map defined by the formula $p_n(m,(x_i)) = (x_1,\dots,x_n)$. Clearly, p_n is a short map.

It is shown in [6] (see Theorem 1.5 therein) that for any $n \geq 2$ there exists a metric space X_n that contains the Euclidean space $\mathbb{R}^n$ as a metric subspace and such that there is no (λ,ε)-Lipschitz retraction from X_n onto $\mathbb{R}^n$ with $\lambda < n^{1/4}$. In the sequel we will need an explicit construction of these spaces. Following [6], for every natural k and natural $n \geq 2$ we define graphs $G_{n,k}$ as follows: the set of vertices $V(G_{n,k})$ is the union of $I(G_{n,k})$ and $T(G_{n,k})$, where

$$\begin{aligned} I(G_{n,k}) &= \{x = (x_1,\dots,x_n) \in \mathbb{R}^n\colon |x_i| = k \text{ for all } i\},\\ T(G_{n,k}) &= \{x = (x_1,\dots,x_n) \in \mathbb{R}^n\colon |x_i| = 2k \text{ for all } i\}; \end{aligned}$$

the set of edges $E(G_{n,k})$ is defined by the condition: $\{x,y\} \in E(G_{n,k})$ if and only if $x,y \in V(G_{n,k})$ and either $\|x-y\| = 2k$ or $y = 2x$ (we suppose that the spaces $\mathbb{R}^n$ are endowed with the Euclidean metric).

The set $V(G_{n,k})$ is equipped with the metric $d = d_{n,k}$,

$$d(x,y) = \inf\left\{\sum_{i=1}^{l} \|x_{i-1}-x_i\|' \mid (x = x_0,x_1,\dots,x_l = y) \text{ is a path in } G_{n,k}\right\}$$

(here $\|x\|'$ denotes the max-norm of $x \in \mathbb{R}^n$.)

Define spaces X and Y as follows:

$$X = \bigcup_{n=2}^{\infty}\bigcup_{k=n}^{\infty}\{n^2\} \times T(G_{n^2,k^2}),\ Y = \bigcup_{n=2}^{\infty}\bigcup_{k=n}^{\infty}\{n^2\} \times V(G_{n^2,k^2})$$

where the metric on X is inherited from X' and the metric on Y is the maximal metric that agrees with the already defined metric on X and the metric $d_{n,k}$ on every $V(G_{n,k})$. It easily follows from the construction that X and Y are proper metric spaces, i.e. objects of the category $\mathcal{A}$.

We are going to show that $\operatorname{asdim}Y = 0$ (and consequently $\operatorname{asdim}X = 0$). Let $C > 0$ and suppose that $y_1,\dots,y_m$ is a C-chain in Y. Denote by k the minimal natural number such that $C < (k+1)^2$. If

$$\{y_1,\dots,y_n\} \subset \bigcup_{j=2}^{k}\bigcup_{l=j}^{k}\{j^2\} \times V(G_{j^2,l^2})$$

and $\operatorname{diam}\{y_1,\dots,y_n\} \leq \sqrt{(k^2)+(3k^2)^2} \leq \sqrt{10C}$. Otherwise

$$\{y_1,\dots,y_n\} \cap Y \setminus \bigcup_{j=2}^{k}\bigcup_{l=j}^{k}\{j^2\} \times V(G_{j^2,l^2}) \neq \emptyset$$

and $\{y_1,\dots,y_n\}$ is a singleton.

It is proved in [6] that the following holds for the spaces $X_{n^2} = \mathbb{R}^{n^2} \cup \bigcup_{k=n}^{\infty}\{n^2\} \times V(G_{n^2,k^2})$ endowed with the maximal metric that agrees with the initial metric on

$\mathbb{R}^{n^2}$ and the inherited from Y metric on $\bigcup_{k=n}^{\infty}\{n^2\}\times V(G_{n^2,k^2})$ (note that these two metrics coincide on the intersection of their domains): there is no (λ, e)-retraction of X_{n^2} to $\mathbb{R}^{n^2}$ with $\lambda < \sqrt{n}$.

Now, let $f\colon X \to P(X)$ be the map that sends $x \in X$ to $\delta_x \in P(X)$. The map f is an isometric embedding and we are going to show that there is no asymptotically Lipschitz extension of f onto Y. Assume the contrary and let $\bar{f}\colon Y \to P(X)$ be such an extension. We regard $\bar{f}$ as a map into $P(X') \supset P(X)$. Then there exist $\lambda > 0$ and $\varepsilon > 0$ such that

$$d(\bar{f}(x), \bar{f}(x')) \leq \lambda d(x, x') + \varepsilon$$

for all $x, x' \in Y$.

Let $n > \lambda^2$. Since the maps $P(p_{n^2})$ and $b\colon P(\mathbb{R}^{n^2}) \to \mathbb{R}^{n^2}$ are short, we conclude that the map $b \circ P(p_{n^2}) \circ \bar{f}|X_{n^2}$ is a (λ, ε)-Lipschitz retraction from X_{n^2} onto $\mathbb{R}^{n^2}$, which contradicts to the choice of λ. This demonstrates that X is not an AE for the category $\mathcal{A}$.

4. Space of measurable maps

For a Banach space B, by $L_1([0,1], B)$ we denote the set of Bochner integrable measurable maps from $[0,1]$ to B. Every metric space X can be isometrically embedded into a Banach space and this allows us to define $L_1([0,1], X)$ as a subspace in $L_1([0,1], B)$.

It is well-known that for any short map $f\colon X \to Y$ the induced map

$$f^*\colon L_1([0,1], X) \to L_1([0,1], Y),\ f^*(\varphi) = f\varphi,$$

is short as well. Also, the integration map

$$b\colon L_1([0,1], \mathbb{R}^n) \to \mathbb{R}^n,\ b(\varphi) = \int_0^1 \varphi(t)dt,$$

is short.

Let now X be an example from Section 3. Arguing similarly as in this example and using the above mentioned properties of the maps f^* and b, we conclude that the space $L_1([0,1], X)$ is not an AE for the category $\mathcal{A}$.

Recall that the Hartman-Mycielski construction is defined as follows (see [4]). Given a metric space X, denote by $HM(X)$ the set of all maps $\alpha\colon [0,1) \to X$ for which there exists a finite sequence $0 = t_0 < t_1 < \cdots < t_k = 1$ such that α is constant on every interval $[t_i, t_{i+1})$, $i = 0, \ldots, k-1$. The space $HM(X)$ is endowed with the metric $\hat{d}(\alpha, \beta) = \int_0^1 d(\alpha(t), \beta(t))dt$, which allows to consider $HM(X)$ as a subspace of $L_1([0,1], X)$. Again, one can similarly show that the space $HM(X)$, where X is an example from Section 3, is not an AE for the category $\mathcal{A}$.

5. Question

Let X be a proper metric space of finite asymptotic dimension and such that X is an absolute extensor for the category $\mathcal{A}$ (it is proved in [2] that $\mathbb{R}^n_+ = \{(x_1, \ldots, x_n) \in \mathbb{R}^n \mid x_n \geq 0\}$ is such a space). It is not known whether the space of probability measures $P(X)$ is an AE for the category $\mathcal{A}$. We leave this as an open question. Similar questions can be formulated for the spaces of measurable maps and the Hartman-Mycielski construction.

References

1. T. Banakh, C. Bessaga, On linear operators extending [pseudo]metrics, *Bull. Polish Acad. Sci. Math.* **48** (2000) 35-49.
2. A. Dranishnikov, Asymptotic topology, *Russian Math. Surveys.* **55** (6) (2000) 71-116.
3. M. Gromov, *Asymptotic invariants of infinite groups*. Geometric group theory, Vol. 2 (Sussex, 1991), 1–295, London Math. Soc. Lecture Note Ser., 182, Cambridge Univ. Press, Cambridge, 1993.
4. S. Hartman, J. Mycielski, On the imbedding of topological groups into connected topological groups, *Colloq. Math.* **5** (1958) 167–169.
5. J.E. Hutchinson, Fractals and self-similarity, *Indiana Univ. Math. J.* **30** (1981) 713-747.
6. U. Lang, Extendability of large-scale Lipschitz maps, *Trans. Amer. Math. Soc.* **351** (1999) 3975-3988.
7. E.D. Tymchatyn, M. Zarichnyi, On simultaneous linear extensions of partial (pseudo)metrics, Preprint (2002).
8. J. Roe, *Coarse cohomology and index theory for complete Riemannian manifolds*. Memoirs Amer. Math. Soc. no 497, 1993.
9. E.V. Shchepin, Functors and uncountable powers of compacta, (Russian) *Uspekhi Mat. Nauk* **36** no. 3(219), (1981) 3-62.
10. M. Zarichnyi, Asymptotic category and spaces of probability measures, *Visn. L'viv National University* (in press).
11. M. Zarichnyi, Regular linear operators extending metrics: a short proof, *Bull. Polish Acad. Sci. Math.* **44** (3) (1996) 267–269.

Functional Analysis and its Applications
V. Kadets and W. Żelazko

F-Algebras: Some Results and Open Problems

Wiesław Żelazko

Institute of Mthematics Polish Academy of Sciences, Śniadeckich 8, PO Box 21, 00-956 Warszawa 10, Poland

Key words: F-algebras, submultiplicative F-norm, B_o-algebras, Q-algebra, maximal ideals
2000 MSC: 46H05, 46H40

The first class of topological vector spaces (t.v.s.) considered in the famous Banach's book ([7]-[9]) is the class of the spaces of type F (the letter was chosen to honor the name of Maurice Fréchet), i.e. the class of completely metrizable t.v.s. The topology (metric) of such a space X can be given by means of an F-*norm* $||x||$, i.e. a functional $||\cdot||$ which satisfies
(i) $||x|| \geq 0$ and $||x|| = 0$ iff $x = 0$,
(ii) $||x+y|| \leq ||x|| + ||y||$,
(iii) $||\alpha x|| \leq ||\beta x||$,
(iv) $\lim \alpha_n = 0$ implies $||\alpha_n x|| = 0$,
(v) $\lim ||x_n|| = 0$ implies $\lim ||\alpha x_n|| = 0$,
(vi) $\lim ||x_n|| = 0$ and $\lim |\alpha_n| = 0$ implies $\lim ||\alpha_n x_n|| = 0$,
for all elements $x, y \in X$ and all scalars α, β (cf. [7]-[9] or [22]). The condition (vi) is a consequence of previous conditions. The distance (metric) in X is given by $||x-y||$.

A real or complex algebra A is said to be a topological (resp. semitopological) algebra, if it is a t.v.s. and the map $(x,y) \to xy$ is jointly (resp. separately) continuous. A commutative semitopological algebra is topological iff the map $x \to x^2$ is continuous (this fact fails to be true in a non commutative case). An F-algebra is defined as a semitopological algebra which is an F-space. For general informations on topological algebras the reader is referred to [20] and [30] (see also [13]).

Email address: zelazko@euro.impan.gov.pl (Wiesław Żelazko).

Proposition 1. (Arens [3]) *An F-algebra is a topological algebra.*

This Proposition follows immediately from the fact that a separately continuous bilinear map between F-spaces is automatically jointly continuous.

For incomplete metrizable semitopological algebras the Proposition 1 is false. One can have even a semitopological normed algebra which is not a topological algebra. Clearly, the multiplication in such an algebra does not extend to a separately continuous multiplication on its completion.

An example of F-algebra is the algebra S (already considered by Banach as an example of space of type F, cf [8], example 1 in the Introduction, §7) of all (classes of) Lebesgue measurable functions on the unit interval, provided with the topology of convergence in measure and with the pointwise algebra operations. Its F-norm can be given by the formula

$$||x|| = \int_0^1 \frac{|x(t)|}{1+|x(t)|} dt$$

The joint continuity of multiplication in an F-algebra does not lead to any reasonable inequality involving its F-norm. If A is a unital F-algebra and its F-norm is submultiplicative,

$$||xy|| \leq ||x||||y||,$$

or, more generally, its topology is given by a metric ρ satisfying

$$\rho(xy,0) \leq C\rho(x,0)\rho(y,0)$$

for all $x, y \in A$, $C > 0$, then the algebra in question is locally bounded, i.e. it has a bounded neighbourhood of the origin. In this case its topology can be given by means of a p-homogeneous, $0 < p \leq 1$, submultiplicative F-norm (i.e. $||\alpha x|| = |\alpha|^p||x||$ for each element x and each scalar α), satisfying moreover $||e|| = 1$, where e denotes the unity of A (cf [22] or [30]). The theory of locally bounded algebras is very similar to the (more narrow) theory of Banach algebras (cf [22], [27], [29], [30],). An example of such an algebra is the space l_p, $0 < p \leq 1$, of sequences $x = (\xi_i), 0 \leq i \leq \infty$ or $-\infty < i < \infty$, with convolution multiplication and the F-norm given by formula

$$||x|| = \sum_i |\xi_i|^p.$$

Other important classes of F-algebras are algebras of type B_o (according to the Banach School terminology, cf [18], [19], [29],[30]), i.e. locally convex F-algebras. The topology of such an algebra A can be given by means of an increasing sequence $|x|_1 \leq |x|_2 \leq \ldots$ of (homogeneous) seminorms satisfying

$$|xy|_i \leq |x|_{i+1}|y|_{i+1} \tag{1}$$

for all $i = 1, 2, \ldots$ and all elements $x, y \in A$. In the case when A is unital with the unity e, then we can require additionally $|e|_i = 1$ for all i. Example of such an algebra is constructed by Arens ([2]) algebra $L^\omega[0,1] = \bigcap_{p\geq 1} L_p[0,1]$ with pointwise algebra operations and the convergence in all norms $|x|_p = \left(\int_0^1 |x(t)|^p dt\right)^{1/p}, p \geq 1$

The same topology is given by any increasing sequence $p_i \to \infty$, and the sequence can be chosen so that the relations (1) are satisfied (see e.g. [30]).

An important subclass of the above one is the class of multiplicatively convex (m-convex) B_o-algebras, called also Arens-Michael B_o-algebras. These algebras were introduced independently in [4] and in [21]. Here we replace (1) by

$$|xy|_i \leq |x|_i|y|_i, \; i = 1, 2. \ldots.$$

Important examples of such algebras are algebras of analytic functions on a domain Ω (in C, C^n, or in a suitable manifold) provided with the pointwise algebra operations and the compact-open topology (topology of uniform convergence on compact subsets of Ω).

If instead of homogeneous seminorms we consider p-homogeneous seminorms, $0 < p \leq 1$ (p depending upon the seminorm in question), we obtain the class of *locally pseudoconvex* F-algebras. In this case the formula (1) has to be replaced by

$$(2) \qquad |xy|_i^{p_{i+1}} \leq |x|_{i+1}^{p_i}|y|_{i+1}^{p_i},$$

where p_i is the homogeneity exponent of the seminorm $|\cdot|_i$ (note that if $|\cdot|$ is a p-homogeneous seminorm, $0 < p \leq 1$, and $0 < q \leq 1$, then $|\cdot|^q$ is pq-homogeneous, i.e. it is subadditive too). We shall not go into details concerning these algebras and refer the reader to [25] and [30] (see also [17]). In this paper we shall be dealing only with questions concerning the general F-algebras. After discussing some long time open questions we shall concentrate mostly on the ideal theory of these algebras.

One of important still open questions (posed directly in [26]) is the problem whether the *Gelfand-Mazur* theorem is valid for the algebras of type F, i.e. whether every division algebra in this class is trivial (topologically isomorphic to C in case of a complex algebra and to either R, C, H in the real case; here H denotes the division algebra of real quaternions) The counterexample proposed in [25] does not work. Perhaps the most general positive result in this direction is due to Turpin and Waelbroeck ([24], see also [25]) and reads as follows

Theorem 2. *Let A be a complex (resp. real) topological division algebra with continuous inverse and possessing a continuous p-homogeneous seminorm, $0 < p \leq 1$. Then A is topologically isomorphic to C (resp. to either of R, C, H).*

The requirement that the operation $x \to x^{-1}$ is necessary for validity of the above result, but in the case of an F-algebra it is satisfied automatically. It follows from the following result proved by Arens in [3] and by Banach in [10] in the separable case, and in [29] in the general case.

Theorem 3. *Let A be a real or complex unital algebra of type F. Then the operation $x \to x^{-1}$ is continuous on the set $G(A)$ of all invertible elements in A iff the set $G(A)$ is a G_δ set.*

Note that the Gelfand-Mazur theorem fails for incomplete metrizable topological algebras even in the locally convex case. A suitable example was constructed by Williamson in [26]. Thus there is an infinite dimensional algebra of type B_o containing a dense division algebra. In this example the operation of taking an inverse is discontinuous (as follows from Theorem 2). However the mentioned above F-algebra S contains a dense field (the field of all rational functions) and it has a continuous inverse. Another example of algebra with discontinuous inverse is the mentioned above algebra $L^{\omega}[0,1]$ The multiplicatively-pseudoconvex algebras always have a continuous inverse, so that the Gelfand-Mazur theorem holds true for this class. It is also true for semitopological normed or p-normed algebras ([27]). However, even a finite dimensional Banach algebra can have a dense division ring: one can construct for an arbitrary natural n a division ring which is an algebra over rational numbers and which is algebraically isomorphic to a subring of $\mathbf{C}$, while topologically it is a dense subset of $\mathbf{C}^n$ (see [34]).

Another important problem is the famous Michael Problem ([21], Question 1; it was also independently posed by Mazur during his seminars; H. G. Dales writes in [13] that the problem was posed by Mazur around 1937). It is the question whether all multiplicative-linear functionals on an m-convex B_o-algebra are automatically continuous. The strongest up-to-now result in this direction is due to Arens [5]. We say that a topological algebra A is finitely generated if there is a finite subset of it such that A coincide with its smallest closed subalgebra containing this subset.

Theorem 4. *Let A be a commutative complex finitely generated m-convex B_o-algebra. Then all its multiplicative-linear functionals are continuous.*

The Michael-Mazur problem is also open for B_o-algebras or F-algebras even in the finitely generated or singly generated case. Using the Theorem 4 it is settled in positive in case of a commutative B_o-algebra for which the maximal ideal space is at most countable (cf. [13], Corollary 4.10.12, and [32]). For more information about this problems and also for some equivalent formulations see [14] and [16].

While for unital commutative Banach algebras or locally bounded F-algebras there is a 1-1 correspondence between maximal ideals and multiplicative-linear functionals (the maximal ideals coincide with kernels of these functionals), such a correspondence does not hold true for more general algebras. For m-convex B_o-algebras we have the following result ([31]). Recall that a unital semitopological algebra is said a Q-algebra if the group $G(A)$ of all its invertible elements is open.

Theorem 5. *Let A be a commutative unital complex m-convex algebra of type B_o. Then the following are equivalent.*

(i) All maximal ideals in A are closed.

(ii) All maximal ideals in A are of codimension one, i.e. they are kernels of multiplicative linear functionals.

(iii) The spectra of all elements in A are bounded.

(iv) The spectra of all elements in A are compact.

(v) A is a Q-algebra, i.e. the set $G(A)$ is open.

(vi) The maximal ideal space $\mathfrak{M}(A)$ is compact.

In the above $\mathfrak{M}(A)$ denotes the "maximal ideal space" of A, i.e. the set of all continuous multiplicative linear functionals on A provided with the Gelfand topology. The (algebraic) spectrum of an element x in A is given by the formula $\sigma(x) =$

$\{\lambda \in \mathbf{C} : x - \lambda e \notin G(A)\}$, where e denotes the unity of A.

It follows, in particular, that if a commutative m-convex unital B_o-algebra is not a Q-algebra, then it must have a dense maximal ideal of infinite codimension. Such an algebra, however, must have at least one closed maximal ideal which is a kernel of a multiplicative linear functional. In the non m-convex case it can happen that all maximal ideals are dense and of infinite codimension, it is so in the mentioned above algebra $L^\omega[0, 1]$ or in the Williamson's algebra, as well as in the algebra S. In the case of a commutative complex unital algebra of type B_o every closed maximal ideal is of codimension one and so it is a kernel of a multiplicative linear functional. It follows immediately from the Gelfand-Mazur theorem for these algebras (both facts are equivalent). For a general F-algebra both facts form open questions, as mentioned above. The equivalence of (i) and (v) in the above theorem is also true for general commutative F-algebras, as proved by Akkar and Nacir in [1]:

Theorem 6. *Let A be a real or complex commutative unital F-algebra. Then A has all maximal ideals closed iff it is a Q-algebra.*

We do not know whether the above result can be extended to non-commutative algebras and the one-sided ideals. An obvious conjecture (posed already in [38]) is the following one.

Conjecture 7. *Let A be a unital F-algebra. Then A has all maximal left ideals closed iff it is a Q_l-algebra (a left Q-algebra, i.e. an algebra for which the set $G_l(A)$ of all left-invertible elements is open).*

The difficulty lies in constructing a dense left ideal in the case when $G_l(A)$ is not open. The construction of such an ideal would not be difficult if we knew that A has a maximal commutative subalgebra which is not a Q-algebra. However, it is a still open problem whether a topological algebra whose all maximal commutative subalgebras are Q-algebras is itself such an algebra. We even do not know whether Q_l-algebras can really exist, i.e. whether one can have a Q_l-algebra which is not a Q-algebra.

We either do not know the answer to the following

Conjecture 8. *Let A be a real or complex F-algebra which is not a Q-algebra. Then A has either a (proper) dense left ideal or such a right ideal.*

We have only a two-sided version of the Akkar and Nacir Theorem (see [39], the proof is not yet published). To formulate it we need the following notation. Let x be an element of a unital algebra A, and denote by $id_2(x)$ the smallest two-sided ideal in A containing this element. Put $G_2(A) = \{x \in A : id_2(x) = A\}$. Call a semitopological algebra A a Q_2-algebra of the set $G_2(A)$ is a neighbourhood of the identity element in A (we do not know whether $G_2(A)$ is open when it has a non-empty interior, we know only that it is true for $G_l(A)$ or for $G_r(A)$). Clearly, a Q_l- or a Q_r-algebra is a Q_2-algebra.

Theorem 9. *Let A be a real or complex unital F-algebra. Then A has all maximal* *two-sided ideals closed iff it is a Q_2-algebra.*

As before, we do not know whether there exists a Q_2-algebra which is not a Q-algebra, but we suspect that such an algebra can exist.

For an infinite dimensional commutative F-algebra it can happen that not only maximal ideals are closed, but also all ideals are closed (in the Banach algebra case such a situation cannot exist, as follows from the mentioned below result of Grauert and Remmert [15]). For instance, consider the algebra A of all formal power series $x = \sum_0^\infty \xi_i t^i$ with usual multiplication (convolution multiplication of coefficients) and the seminorms

$$|x|_n = \sum_0^n |\xi_i|.$$

As a t.v.s it coincides with the space (s) considered by Banach in [8] (example 2 in §7 of the Introduction). Clearly these seminorms satisfy (2), so that s is a commutative m-convex B_o-algebra. It is not hard to see that every proper ideal of A is of the form $I_n = t^n A$, $n = 1, 2, \ldots$ (it follows from the fact that every element x in A, for which $\xi_0 \neq 0$, is invertible in A, in particular A is a Q-algebra), and so it is closed. The characterization of commutative F-algebras with all ideals closed was obtained independently by Choukri and El Kinani [12], and the author [37].

Theorem 10. *Let A be a real or complex unital F-algebra, then A has all ideals closed iff it is a Noetherian algebra.*

Recall that a commutative algebra A is said *Noetherian* if every its ideal is finitely generated, i.e. is of the form $I = x_1 A + x_2 A + \ldots + x_n A$ for some natural n, where $x_1, \ldots, x_n \in A$. Similarly we define a left or right Noetherian algebra in case when A is not commutative.

Thus every commutative Noetherian F-algebra must be a Q-algebra. The above mentioned example is even a principal ideal domain, i.e. every of its ideals is principal (i.e. singly generated). In a sense it is a trivial example because it is has only countably many ideals and is a radical algebra (it has only one maximal ideal). A much more involved example of an m-convex B_o algebra which is a principal ideal domain was constructed by Carboni and Larotonda [11]. It is a semisimple algebra with uncountably many maximal ideals (it consists of holomorphic functions in the unit disc in C, continuous on its boundary, and the unit disc is its maximal ideal space). The fact that the algebras in both examples are m-convex is forced by the fact that a commutative B_o-algebra, which is a Q-algebra, must be m-convex (see [28] for the complex case and [38] for the real case; however, in the non-commutative case we can have a non-m-convex Q-algebra of type B_o, cf [35]). For a Banach algebra case the theorem of Grauert and Remmert ([15], Appendix to §5) states that a commutative Noetherian Banach algebra must be finite dimensional. This result was extended to the non-commutative case by Sinclair and Tullo [23].

Again we do not know whether the non-commutative version of the Theorem 10 is true. An obvious conjecture is that an F-algebra has all left ideals closed iff it is left Noetherian (this condition is necessary, see [37]). The difficult part is in proving that a left Noetherian F-algebra must have all left ideals closed. We do not know either whether a left Noetherian F-algebra is automatically right Noetherian.

We pass now to the problem of existence of an F-algebra with all ideals dense (the other end of the scale). No such example is known. Aharon Atzmon [6] constructed a commutative unital complete locally convex algebra with all proper ideals dense, but it is non-metrizable (such an algebra cannot be multiplicatively convex). For non-unital algebras the problem is open even for Banach algebras (see e.g. [33]). Even it is not known whether a non-unital Banach algebra must have a proper closed subalgebra (unless it is of dimension one). If there is example of a (commutative) Banach algebra A without proper closed subalgebras, then its construction is not easy, since all operators of multiplication by elements of A have no proper closed invariant subspaces. Also no example of such an algebra is known in the class of all topological algebras. However, the author constructed in [36] an incomplete locally convex metrizable infinite dimensional semitopological algebra with all proper subalgebras dense (of course, such an algebra must be commutative and non-unital)

We close this paper with the discussion on the following Conjecture. The quoted below results are of a preliminary character and it is their first publication.

Conjecture 11. *A commutative unital F-algebra has a dense principal ideal iff the operation of taking an inverse is discontinuous.*

For our discussion we need the following concept. Call an element x of a unital commutative semitopological algebra A *topologically invertible* if there is a net (z_α) of elements of A such that $\lim_\alpha xz_\alpha = e$, where e is the unity of A. Equivalently, the principal ideal generated by x is dense. We could consider also such a concept in a non-commutative algebra and consider the left or light topological invertibility (it is unclear whether we can take the same net for an element which is both left and right topologically invertible), but here we limit ourselves to the commutative case and to the algebras of type F. Denote by $G^t(A)$ the set of all topologically invertible elements of such an algebra A. Clearly $G(A) \subset G^t(A)$ and an element $x \in A$ is in $G^t(A) \setminus G(A)$ iff x is a proper (i.e. non-invertible) topologically invertible element. We have $G^t(A)^2 \subset G^t(A)$, i.e. $G^t(A)$ is a semigroup under the multiplication. Also $xy \in G^t(A)$ implies that both x and y are in $G^t(A)$. The following result is true also in the non-commutative case (under a suitable definition of $G^t(A)$).

Proposition 12. *Let A be real or complex unital algebra of type F. Then the set $G^t(A)$ is a G_δ-set.*

The following Corollary gives half of the proof of the Conjecture 11.

Corollary 13. *Let A be as above and suppose that the operation $x \to x^{-1}$ is discontinuous. The A has a proper topologically invertible element and so a (proper) dense principal ideal.*

The proof follows immediately from Theorem 3 and Proposition 12. Under the assumptions of the Corollary 13, we can effectively construct a non-invertible element in $G^t(A)$. Namely, we can take a sequence of invertible elements $x_n \to e$ such that the sequence (x_n^{-1}) is divergent. Passing, if necessary, to a subsequence, we can assume that the product $x_1 x_2 \ldots$ is convergent to an element x. It can be proved that x is non-invertible and belongs to the set $G^t(A)$. In a similar way we can obtain another characterization of F-algebras with a continuous inverse.

Proposition 14. *Let A be a real or complex unital algebra of type F. Then the operation $x \to x^{-1}$ is continuous iff every convergent infinite product of invertible elements is invertible.*

It is not hard to see that the mentioned above algebra $L^\omega[0,1]$ has proper topologically invertible elements. To see it take, say, any continuous function $x(t)$ on the unit interval, $0 \leq x(t) \leq 1$, which decreases to zero on $[0, t_o]$ and increases to one on $[t_o.1]$, and such that the inverse $x^{-1}(t)$ is not in $L_p[0,1]$ for some $p \geq 1$. Thus x is in $L^\omega[0,1]$ and is non-invertible there. Put

$$z_n(t) = \begin{cases} x^{-1}(t), & \text{if } x^{-1}(t) \leq n; \\ n, & \text{otherwise.} \end{cases}$$

Clearly z_n is in $L^\omega[0,1]$ for all n. One can easily check that $xz_n \to e \equiv 1$ in the topology of $L^\omega[0,1]$.

On the other hand, it can be shown that all m-convex algebras, as well as the mentioned above algebra S fail to have such elements (these algebras have a continuous inverse). All these facts support the Conjecture 11.

The author does not know the answer to the following problem, interesting for the ideal theory of F-algebras.

Problem 15. *Let A be a commutative unital F-algebra and suppose that for some $x_1, \ldots, x_n \in A$ the ideal $x_1 A + \ldots + x_n A$ is proper and dense. Does it follow that A has proper topologically invertible elements ?*

One can define a *topological spectrum* σ^t in a complex unital F-alebra A, setting

$$\sigma_A^t(x) = \{\lambda \in \mathbf{C} : x - \lambda e \notin G^t(A).$$

It is smaller than the usual (algebraic) spectrum, but it has a certain advantage, as seen in the following

Proposition 16. *Let A be a unital complex metrizable topological algebra, and denote by $\overline{A}$ its completion (it is an F-algebra). Then for each x in A we have $\sigma_A^t(x) = \sigma_{\overline{A}}^t(x)$.*

The proof follows immediately from the formula $G^t(A) = G^t(\overline{A}) \cap A$. This concept of spectrum seems to be appropriate for incomplete metrizable topological algebras (even for normed algebras). It is not hard to see that it satisfies the spectral mapping theorem for (non-constant) polynomials, i.e. $\sigma^t(p(x)) = p(\sigma^t(x)$ for all x in the algebra in question. Here we assume $p(\emptyset) = \emptyset$, since for some elements this spectrum can be empty.

Clearly, the Proposition 16 fails for the algebraic spectrum, even if A is a normed algebra.

References

1. M. Akkar, C. Nacir, Continuité automatique dans les limites inductives localement convexes de Q-algébres de Fréchet, *Ann. Sci. Math. Québec* **19** (1995) 115-130.
2. R. Arens, The space L^ω and convex topological rings, *Bull. Amer. Math. Soc.* **52** (1946) 931-935.
3. R. Arens, Linear topological division algebras, ibidem 53 (1947) 623-630.
4. R. Arens, A generalization of normed rings, *Pacific J. Math.* **2** (1952) 455-471.
5. R. Arens, Dense inverse limit rings, *Michigan Math. J.* **5** (1958) 160-182.
6. A. Atzmon, An operator without invariant subspaces on a nuclear Fréchet space, *Ann. of Math.* **117** (2) (1983) 660-694.
7. S. Banach, *Teoria Operacyj, tom 1. Operacje liniowe.* (In Polish) Warszawa 1931.
8. S. Banach, *Théorie des opérations linéaires.* Warszawa, 1932.
9. S. Banach, *Kurs Funkcionalnogo Analizu (Liniini Operacii).* (In Ukrainian) Kiev 1948.
10. S. Banach, Remarques sur les corps et groupes metriques, *Studia Math.* **10** (1948), 178-181.
11. G. Carboni and A. Larotonda, An example of a Fréchet algebra which is a principal ideal domain, ibidem 138 (2000) 265-275.
12. R. Choukri and A. El Kinani, Topological algebras with ascending and descending chain conditions, *Arch. Math.* **72** (1999) 438-443.
13. H. G. Dales, Banach Algebras and Automatic Continuity, *London Math. Soc. Monographs* **24**, Oxford Univ. Press, 2000.
14. P. G. Dixon and J. Esterle, Michael's problem and the Poincaré-Fatou-Bieberbach phenomenon, *Bull. Amer. Math. Soc.* **15** (1986) 127-187.
15. H. Grauert, R. Remmert, *Analytische Stellenalgebren.* Springer 1971.
16. T. Husain, *Multiplicative Functionals on Topological Algebras.* Pitmann Books Ltd., London 1983.
17. A. Kokk and W. Żelazko, On vector spaces and algebras with maximal locally pseudoconvex topologies, *Studia Math.* **112** (1995) 195-201.
18. S. Mazur, W. Orlicz, Sur les espaces métriques linéaires I, *Studia Math.* **10** (1948) 184-205.
19. S. Mazur, W. Orlicz, Sur les espaces métriques linéaires II, ibidem 13 (1953), 137-179.
20. A. Mallios, *Topological Algebras. Selected Topics.* North-Holland, Amsterdam 1986.
21. E. Michael, Locally multiplicatively convex topological algebras, *Mem. Amer. Math. Soc.* **11** (1952).
22. S. Rolewicz, *Metric Linear Spaces.* PWN, Warszawa 1984.

23. A. M. Sinclair and A. W. Tullo, Noetherian Banach algebras are finite dimensional, *Math. Annalen* **211** (1974) 151-153.
24. P. Turpin, L Waelbroeck, Algèbres localement pseudo-convexex à inverse continu, *C. R. Acad. Sc. Paris* **267** (1968) 194-195.
25. L. Waelbroeck, Topological Vector Spaces and Algebras, *Lecture Notes in Math.* **230** Springer 1971.
26. J. H. Williamson, On topologizing the field $C(t)$, *Proc. Amer. Math. Soc.* **5** (1954) 729-734.
27. W. Żelazko, On the locally bounded and m-convex topological algebras, *Studia Math.* **19** (1960).
28. W. Żelazko, A theorem on B_o-division algebras, *Bull. Polon. Acad. Sc.* **8** (1960) 373-375.
29. W. Żelazko, *Metric generalizations of Banach algebras.* Rozprawy Mat. (Dissertationes Math.) **47** 1965.
30. W. Żelazko, Selected Topics in Topological Alghebras, *Aarhus Univ. Lecture Notes* **31** 1971.
31. W. Żelazko, On maximal ideals in commutative m-convex algebras, *Studia Math.* **58** (1976) 291-298.
32. W. Żelazko, Functional continuity of commutative m-convex B_o-algebras with countable maximal ideal spaces, *Coll. Math.* **51** (1987) 395 399.
33. W. Żelazko, On certain open problems in topological algebras, *Rend. Sem. Mat. Fis. Milano* **59** (1989) 49-58.
34. W. Żelazko, A subfield of a complex Banach algebra is not necessarily topologically isomorphic to a subfield of C, *Coll. Math.* **63** (1992) 135-137.
35. W. Żelazko, Concerning entire functions in B_o-algebras, *Studia Math.* **110** (1994) 283-290.
36. W. Żelazko, A semitopological algebra without proper closed subalgebras, *Coll. Math.* **74** (1997) 239-242.
37. W. Żelazko, A characterization of commutative Fréchet algebras with all ideals closed, *Studia Math.* **138** (2000) 293-300.
38. W. Żelazko, On m-convexity of commutative real Waelbroeck algebras, *Comm. Math.* **40** (2000) 235-240.
39. W. Żelazko, Some problems concerning Q-algebras, *Proc. ictaa 2000* (in print).